Sensory Systems of Animals

Perception in animals is a fascinating and challenging subject that calls to students from many disciplines. The aim of this book is to provide a knowledge base and unifying perspective on the field that will enable beginning researchers to chart their own course within it. The author describes, in a systematic but engaging way, the sensory systems of humans and other vertebrates, as well as arthropods and molluscs.

Why is it important to understand the senses of animals? One reason is that human activities are changing the perceptual world of animals in ways that expose them to danger, such as bright outdoor lighting that disorients migrating birds, and human sonar that drives whales to beach themselves. Reducing such dangers will require changes in human behavior and greater understanding of how animals react, physiologically and behaviorally, to anthropogenic changes in their environment.

The emphasis throughout the book is on research, in both the behavioral/ethological and neuroscientific traditions, that has led to important discoveries. The functional anatomy of each sensory system, from receptor cells to brain areas, is succinctly described, explaining how it underlies the animal's sensory abilities and behavior. Overall descriptions of a sense for a class of animals (e.g., hearing in spiders) are interspersed with expanded coverage of that sense in a particular animal, such as the ogre-faced web-casting spider that does a backflip to capture an insect buzzing overhead. Evolutionary themes are found throughout the book, for example, in describing the evolutionary development of the vertebrate ear, and the convergent evolution of the eyes of vertebrates and cephalopods.

With over 500 references and 80 illustrations, this textbook is primary reading intended for advanced undergraduates and beginning graduate students of veterinary science, lab animal science, and zoology. It would also be of great interest to professionals and academics working with animals such as veterinary professionals and zookeepers.

Sensory Systems of Animals

Biology and Behavior

Mark Hollins

CRC Press is an imprint of the
Taylor & Francis Group, an **informa** business

First edition published 2025
by CRC Press
2385 NW Executive Center Drive, Suite 320, Boca Raton FL 33431

and by CRC Press
4 Park Square, Milton Park, Abingdon, Oxon, OX14 4RN

CRC Press is an imprint of Taylor & Francis Group, LLC

Library of Congress Cataloging-in-Publication Data
Names: Hollins, Mark, author.
Title: Sensory systems of animals : biology and behavior / Mark Hollins.
Description: First edition. | Boca Raton, FL : CRC Press, 2025. | Includes bibliographical references and index. | Summary:"With over 500 references and 80 illustrations, this textbook is primary reading intended for advanced undergraduates and beginning graduate students of veterinary science, lab animal science, and zoology. It would also be of great interest to professionals and academics working with animals such as veterinary professionals and zookeepers"— Provided by publisher.
Identifiers: LCCN 2024024973 (print) | LCCN 2024024974 (ebook) | ISBN 9781032423319 (hbk) | ISBN 9781032423289 (pbk) | ISBN 9781003362319 (ebk)
Subjects: LCSH: Senses and sensation—Textbooks. | Perception in animals—Textbooks. | Animal behavior—Textbooks.
Classification: LCC QP434 .H65 2025 (print) | LCC QP434 (ebook) | DDC 591.5—dc23/eng/20240701
LC record available at https://lccn.loc.gov/2024024973
LC ebook record available at https://lccn.loc.gov/2024024974

ISBN: 978-1-032-42331-9 (hbk)
ISBN: 978-1-032-42328-9 (pbk)
ISBN: 978-1-003-36231-9 (ebk)

DOI: 10.1201/9781003362319

Typeset in Palatino Light
by Apex CoVantage, LLC

For my grandchildren

Hudson, Adrian, and Harper

Contents

Acknowledgments

I owe a debt to many for contributing to the interests in research and teaching, and then specifically in sensory science, that have prompted me to write this book. First, my parents. My father was an early environmentalist who eschewed plastic and pesticides, and my mother taught me the importance of backing up opinions with evidence.

Second, many teachers. My high school biology teacher, Dorothy Gregory, who first suggested that I pursue a career in science; Howard Baker, my honors advisor at Florida State University, who vividly explained the wonders of the vertebrate eye; Lorrin Riggs, my graduate advisor at Brown, who demonstrated in his own work the importance of physics in sensory research; and postdoctoral advisors Mathew Alpern at Michigan and Mitchell Glickstein at Brown. Mat emphasized the importance of combining experimentation with theory, and Mitch showed how human and animal research can complement each other.

Third, as a faculty member in the Department of Psychology and Neuroscience at the University of North Carolina at Chapel Hill, I have benefitted from the intellectually stimulating and supportive atmosphere created by many colleagues: faculty and graduate students who are devoted to research, teaching, and service. Their interest in discussing and fostering one another's work makes the department a true academic community.

Graduate and undergraduate courses in the human senses are a mainstay of the psychology curriculum, and I taught them regularly for many years. But in recent years, my students, with a broader environmental perspective than many of their predecessors, increasingly wanted to know how the senses of animals compare with our own. I realized that a separate course on this subject was needed. And when I taught it, I found that none of the available books on animal senses had the focus on basic principles that I wanted in a textbook. That is how I came to write this book, and I am grateful to the students whose enthusiasm for the subject prompted me to do so.

Three friends have helped me with valuable discussions related to this book. Horticulturist Eva Hoke answered many questions about bees and plants. Conservationist Mike Andrews increased my understanding of the impact of humans on the environment. And UNC colleague Sam Fillenbaum offered good advice on the writer's craft.

I am greatly indebted to Alice Oven, my primary editor at Taylor & Francis Group, who saw the promise of the book based on early chapters. Her steady helpfulness and sound judgment have been invaluable. Senior Editorial Assistant Shikha Garg provided expert guidance on a range of issues, especially with regard to figures and figure permissions. I am grateful for the care and precision with which she shepherded the manuscript through the editorial process. Thanks are also due to Jessica Ward of Whale Wise, and several anonymous reviewers, who provided helpful early advice.

Most of all, I am grateful to my beloved partner, Eleanor Leung-Hollins. In our more than 50 years of marriage, she has been a constant wellspring of support and wisdom. A gifted scientist and a creative problem-solver, she teaches me to think outside the box.

About the Author

Mark Hollins is Professor Emeritus of Psychology and Neuroscience at the University of North Carolina (UNC) at Chapel Hill. His lifetime interest in perception began when he received a View-Master stereoscope on his fifth birthday and wondered where the 3D space was coming from. At Brown University, he earned a PhD under the mentorship of pioneering vision researcher Lorrin Riggs. Mark's graduate school research dealt with color vision in the ground squirrel and brightness contrast in humans under nighttime conditions. After postdoctoral years at the University of Michigan and then back at Brown, he joined the faculty at UNC, where he continued his research on sensory topics. An interest in the perceptual abilities of people who are blind led to the book *Understanding Blindness: An Integrative Approach* (Erlbaum, 1989), which was reissued by Routledge (2022) as part of their Psychology Revivals series. He and graduate student Sliman Bensmaia collaborated on a series of studies showing that the perception of surfaces by touch relies on both rapidly and slowly adapting skin receptors; this now appears to be widely true among animals. For his work with student researchers, Mark received UNC's Distinguished Teaching Award for Post-Baccalaureate Instruction in 2007. Students' interest prompted him to create and teach a course on animal senses, but when he found that no existing book had the evolutionary perspective and research emphasis he wanted in a textbook, he decided to write this one. He hopes it will give students a unified framework for understanding the basic principles of sensory science and raise, for some, the possibility of a research career in this important and intriguing field.

Introduction

On a clear but moonless night in early spring, a bird flies north. He is guided by the stars. Somewhere in his brain is stored the knowledge that he must fly in the direction of a certain star, around which other stars rotate over the course of the night. The night is cool, and the sky is beautiful, motivating the bird to fly on despite fatigue. But as he flies over a city, problems arise. Light scattered upward from buildings and highways dims the stars, and pollution creates an additional veil. At the same time, bright points of light in the city, like fallen stars, add an element of confusion. By temporarily relying on other cues, the bird makes it past the city, but other cities and other challenges lie ahead.

Meanwhile, a young sea turtle is also making a journey. Hatched from an egg buried in sand on a Florida beach, she scrambles into the surf and reaches the relative safety of deep water. The newborn turtle soon enters a vast clockwise gyre that carries her around the North Atlantic, guided by the magnetic field of the Earth and an innate map in her brain. The map links the magnetic field at each point on the gyre with swimming instructions appropriate for that location. The turtle makes this journey repeatedly. When she reaches adulthood, she leaves the gyre and returns to her birthplace, guided by its unique magnetic signature. Here, she will mate in offshore waters and lay her eggs on the beach. But the incoming tide, which rises higher each year, now reaches almost to a seawall; her eggs may drown before they hatch.

Essential to the lives of these animals are their remarkable sensory abilities: the ability of the bird to perceive the patterns in the starry sky and of the turtle to perceive the Earth's magnetic field. In both cases, the brain links these perceptions—or neural representations of them—to courses of action. Without perception, there can be no meaningful action. This is true not just of animals that migrate over long distances but also of others who never stray far from home.

Consider the honeybee. She searches for nectar-bearing flowers, guided by their color and smell. Returning to the hive, she indicates the flowers' location with reference to the sun's location in the sky. The hive is a community ruled by a queen who controls the workers with a pheromone that she secretes. This rubs off on her attendants, who transfer it to others so that it eventually reaches every member of the hive. The queen lays eggs that can develop into either queens or workers. Queen pheromone prompts the workers to feed the developing larvae simple fare that will make them workers. But when the queen is dying, her secretion decreases, causing workers to feed some larvae royal jelly. They will develop into a new generation of potential queens. This process and many others in the hive depend on the chemical sensitivity—akin to our sense of smell—of the workers.

This book is about the senses of animals: the ways in which they acquire information about objects and events in their environment. For a field mouse, the objects could include blades of grass, edible bugs, and a hawk overhead; events might include the descent of the hawk, the squeaks of other mice, or the falling of rain. To be detected, these things must activate the animal's sensory systems. The grass pushes gently

DOI: 10.1201/9781003362319-1

against the mouse's skin, activating touch receptors; the squeaks of other mice reach its ears and stimulate auditory receptors; pheromone molecules released by potential mates stimulate olfactory receptors; and the small but expanding patch of sky blocked out by the hawk is registered by neurons in its visual system.

Like the mouse, all animals have sensory systems with receptor cells that respond to specific types of stimulation. Some types of environmental stimuli, such as light, sound, and pressure, are so informative that most animals, on different branches of the tree of life, possess sensory systems for detecting and processing them. Sensory systems for other forms of stimulation, such as magnetism, electricity, and infrared radiation, are widespread but not present in all animals.

THE EMERGENCE OF SENSORY SCIENCE

Sensory science is the study of these systems. It started in antiquity but then lay dormant until it was revived by Alhazen in the eleventh century, and by Isaac Newton and others during the Enlightenment. It has grown steadily since that time and, in the past half-century, has become a major field of research. Hundreds of researchers are currently engaged in studying sensory systems. Many are academics in departments of biology, neuroscience, psychology, and veterinary science. Many others work in conservation organizations, natural history museums, or wildlife sanctuaries, either as professionals or as volunteers. They are driven by scientific curiosity and, increasingly, by a desire to preserve animal species that are threatened by dwindling habitat, pollution, and climate change.

Sensory science includes the study of the human senses, since we are animals too. In fact, studying humans gives sensory researchers a special tool that other branches of systems biology, such as the study of the digestive system or the circulatory system, do not have. We have sensory experiences and can report them, so we are able to study the human senses "from the inside" as well as by means of objective approaches such as neuroimaging. This has helped researchers understand the functional significance of many anatomical and physiological findings.

A classic example of the usefulness of carefully examining our own sensory experiences is a discovery made by the Czech physiologist Jan Purkinje in 1819. On his strolls through a meadow, Purkinje often paused to admire wildflowers surrounded by greenery. In broad daylight, he noted that red flowers and green leaves looked about equally bright. But in the dim light of dawn, colors were pale or absent, and surprisingly, leaves were generally brighter (a lighter shade of gray) than flowers. Purkinje realized that this change in relative brightness, now called the *Purkinje shift*, indicates that the eye somehow works differently in bright and dim environments. This insight helped later scientists to grasp the significance of two types of receptor cells found in the retina of the eye, the rods and cones.

To be of maximum benefit, such observations need to be followed up with controlled experimentation. For example, to follow up Purkinje's observations, participants in a darkened laboratory can be shown a series of dim lights and asked whether each one is visible, and if so, what color it is. In this way, visual thresholds can be determined for lights of different wavelengths and at different levels of background illumination. Carefully controlled experiments along these lines have revealed properties of the pigments in rods and cones that absorb light and enable us to see.

The methods used in these experiments, including rigorous stimulus control and quantitative data analysis, are collectively known as *psychophysics*, a term introduced by German scientist and philosopher Gustav Fechner in 1860. Psychophysics became a major part of experimental psychology, a discipline founded by Wilhelm Wundt in 1879.

Like the human senses, the senses of other animals can be studied by both observation and experimentation. Researchers start by observing animals in their natural habitats to see how they use their senses. Once scientists have formed hypotheses about an animal's sensory abilities, they can test their ideas experimentally by manipulating stimuli in the animal's environment (either in the wild or in a laboratory setting) and recording its responses.

When sensory research with animals is carried out with carefully calibrated stimuli and limited response options for the animal, it is usually called *animal psychophysics*. For example, to study depth perception in falcons, researchers put red/green "3D" goggles on a bird and trained it to fly to geometrical patterns that contained stereoscopic depth (Fox et al., 1977).

On the other hand, when sensory research is carried out under more naturalistic conditions, with animals responding to complex stimuli in ways that are part of their normal behavioral repertoire, the work is considered part of *ethology*. This is especially true if the findings are viewed in an evolutionary context. For example, to study how some birds are able to tell the difference between their own eggs and those of parasitizing cuckoos, Marchetti (2000) surreptitiously placed Plasticine eggs of different sizes in the nests of Hume's warblers and observed the birds' reactions.

Regardless of the methods used, research on the senses of animals is deeply influenced by the insights of Charles Darwin (1809–1882). His concepts of adaptive radiation and natural selection explain how, and require that, an animal's sensory abilities have evolved to fit its needs. Hawks have extremely good visual acuity because the ability to see a mouse from high above has contributed to their survival. In contrast, moles are almost blind because they spend most of their time below ground; there is no evolutionary pressure for them to develop better vision than they need. Since different animals, even within a related group such as snails, have different habitats and lifestyles, it is to be expected that they will have different sensory abilities.

Conversely, learning about an animal's sensory abilities is an important part of understanding what its life is like. For example, if study of its eye shows that it does not have more than one visual pigment, we infer that it does not see the world in color. And less obviously but equally important, if the animal *does* have multiple visual pigments, then we know that it actually has—and benefits from—color vision. Otherwise, the resources used to produce and maintain those pigments would be a waste, and waste is not tolerated in the struggle for survival.

Biologist Jacob von Uexküll (1864–1944) influenced the field of animal perception by introducing the concept of an animal's *Umwelt*, its subjective world. Our Umwelt consists of all the people, places, and things that are part of our lives, from the distant horizon to an intimate conversation. It comprises not just the things we perceive but also our responses to them. An animal does not perceive everything in its environment. For example, von Uexküll observed that the jackdaw, a member of the crow family, readily seizes and eats caterpillars, but only if they are moving. A stationary caterpillar apparently goes unseen by the bird and is therefore not part of its Umwelt. Later research has confirmed that stimulus movement significantly enhances bird vision (e.g., Haller et al., 2014).

It is often assumed that we humans fully perceive our environment, but this is an illusion: We cannot, for example, perceive magnetic fields or the polarization of light, as many other animals do. The overall point is that each animal needs to be studied in the context of its own world. We don't perceive the world in exactly the same way as a dog, a bird, or an octopus.

Another pioneering student of animal senses was Karl von Frisch (1886–1982), an ethologist who shared the 1973 Nobel Prize in Physiology or Medicine with Konrad Lorenz and Niko Tinbergen. Von Frisch had a lifelong interest in the sensory abilities of animals and how they contribute to the animals' lives. His focus on the way honeybees use sensory cues to navigate their environment was the key to unlocking their complex behavior, including the waggle dance. More broadly, he showed how rigorous experimental methods can be applied to the study of animal senses not just in the laboratory but also in their natural environment. Von Frisch emphasized that a solid understanding of the senses of animals requires that knowledge of their behavior be combined with research on the anatomy and physiology of their sensory systems—what is now called *sensory neuroscience*.

Neuroscience as a distinct discipline emerged at the turn of the twentieth century, with the discovery by the Spanish anatomist Santiago

Ramón y Cajal (1852–1934) that the nervous system consists not of a continuous web of neural tissue, as had earlier been assumed, but of distinct cells that communicate only at special junctions called synapses. This *neuron doctrine* revolutionized understanding of the nervous system and became the fountainhead of a vast research field that continues to advance and expand.

Many neuroscientists have focused their attention on sensory systems. H. Keffer Hartline (1903–1983) was a physiologist who made the first recordings of the electrical activity of individual fibers in an animal's optic nerve; Georg von Békésy (1899–1972) was a biophysicist who showed how the inner ear transforms sound waves into neural activity; collaborators David Hubel (1926–2013) and Torsten Wiesel (born 1924) discovered how visual information is organized in the primary visual cortex of mammals; and George Wald (1906–1997) showed what happens at the molecular level when visual pigments absorb light. More recently, Linda Buck and Richard Axel discovered how hundreds of different olfactory receptor molecules help animals detect and identify odors; and David Julius and Ardem Patapoutian identified specialized molecules in the skin's sensory neurons that enable us to feel touch, temperature, and pain. These sensory researchers all won Nobel Prizes for their work, as did Cajal. Their key discoveries will be discussed later in the book.

ANIMALS DISCUSSED IN THIS BOOK

Scientists used to classify all living things as animals or plants. Animals were organisms that could move about under their own power; plants were sedentary but had the compensating advantage that they could make their own food. Even one-celled organisms could be divided up in this way. For example, amoebas were considered to be animals, and blue-green algae were plants.

As knowledge of the microscopic world increased, however, it became clear that there are many types of one-celled organisms that don't easily fit into what we think of as plant and animal categories. So one-celled living things are now said to belong to other kingdoms; the Animal Kingdom and the Plant Kingdom contain only multicellular organisms.

For reasons not fully understood, the development of a host of animals accelerated rather suddenly some 540 million years ago, early in a geological span of time called the Cambrian Period. During this "Cambrian explosion" of new life forms, several major animal groups, or *phyla*, developed. Some of these have no living representatives, but others thrived and diversified. In this book, we will limit ourselves to three of the most successful and familiar phyla: Chordata, Arthropoda, and Mollusca.

Chordates are animals that have a well-defined neural tube running down their back. Typically, this tube is enclosed in a spine consisting of bony segments called vertebrae, and animals that have these are called vertebrates. In this book, we will not be discussing the few primitive chordates that lack backbones, and so we will use the term vertebrates rather than chordates. Fish, amphibians, reptiles, birds, and mammals—including people—are *vertebrates*.

Arthropods are very different from vertebrates. Their body is supported, and its shape maintained, not by internal bones as in vertebrates but by a hard external skeleton, an *exoskeleton*. The word arthropod means "jointed limb." Our limbs are also jointed, but our joints are internal and therefore not as striking as those in an arthropod's legs. Arthropods include insects, spiders, scorpions, ticks, lobsters, crabs, and centipedes. With regard to both the number of species and the number of individual organisms, arthropods are the most numerous of all animals.

The third major group of animals we will consider is Phylum Mollusca, which includes snails, bivalves (oysters and clams), and cephalopods (squid, octopus, and cuttlefish). These *molluscs* have neither an internal nor an external skeleton. In fact, their name means "the soft ones." The difficulty of moving a soft body about on dry land probably explains why most molluscs are better adapted to the buoyancy of a water environment. Molluscs are characterized by a mantle, a large dorsal covering for the body cavity that houses the digestive system and other internal organs. Most molluscs also have a single large foot, although this has been greatly modified in some species. For example, the arms of squid and octopus are derived from the foot. Many molluscs have some type of shell that grows from the mantle, an obvious feature of bivalves and snails.

Vertebrates, arthropods, and molluscs all emerged around the start of the Cambrian period, and since that time, all three phyla have continued to diversify, despite intermittent setbacks. In the modern ("Anthropocene") Epoch, however, the number of species in all three phyla is shrinking, as a result of the degradation of the environment that humans are bringing about.

Limiting our discussion to these three groups of animals is not meant as a disparagement of other phyla. All prove by their continued existence that they have the qualities (including sensory abilities) that they need to pass the most crucial of all tests—the test of survival. Indeed, many have remarkable traits. For example, the starfish *Linckia laevigata* (a member of the phylum Coelenterata) can use the eye at the tip of each arm to return to its home reef from up to 2 m away (Garm & Nilsson, 2014). And the planarian *Dugesia japonica*, a member of the phylum Platyhelminthes, can learn not to fear light and can retain this lesson even if it loses its head and needs to grow a new one (Shomrat & Levin, 2013).

However, more elaborate sensory abilities require not only refined sense organs but also nervous systems with a degree of complexity found only in vertebrates, arthropods, and molluscs. Limiting the book to these three phyla will allow us to emphasize remarkable sensory behaviors—navigating by the stars, echolocating in the dark, detecting mates as well as predators and prey based on the chemicals they release—and to explain, to the extent that our present state of knowledge will allow, the receptoral and neural mechanisms on which these behaviors rest.

ORGANIZATION OF THE BOOK

This book is organized by sensory system. There are four chapters on vision, followed by one chapter each on touch, hearing, taste, and smell. Vision gets four chapters not because it is more important than other senses—a primacy which is true for some animals but not for a great many others—but because more is known about it. The book concludes with a chapter on senses that we humans don't have, followed by some final thoughts.

Most chapters begin with a description of the sensory system in humans, because this is what readers are most familiar with. This makes it possible for basic terms and concepts related to that sense to be easily explained. The different forms that the sensory modality takes in other animals are then presented, with other vertebrates discussed first, followed by arthropods and then molluscs.

There are millions of species of animals, but the sensory systems of only a small percentage have been studied. The approach taken in this book, for each sensory modality, is to explain general principles and then to focus on one or two species representative of each major group (fish, insects, etc.). These animals have fascinating stories to tell, based on clear and sometimes surprising research findings.

Throughout, the structure and function of sense organs are described, including their role in creating the animal's *Umwelt*. The central processing of sensory signals is also discussed, in order to explain how specific types of information are extracted by the sensory system. Most of the experiments presented in detail are behavioral ones, for these are unequalled in their ability to show the value of a given sensory system for an animal.

The book emphasizes peer-reviewed studies published in well-established scientific journals. Articles were chosen based on their scientific importance and clarity of exposition and are representative of a larger body of work in each research area. From these key sources, cited in the text and listed in the References, readers who want to delve more deeply into particular topics can readily gain access to the wider literature.

DO ANIMALS HAVE CONSCIOUS EXPERIENCE?

A final note: We can learn about an animal's sensory abilities and discover the biological basis of those abilities, but we cannot directly experience

an animal's sensations. We do not know, for example, whether the sky looks blue to a bird or a frog, much less to a mantis shrimp. It is likely that the sensations of apes are similar to ours, but beyond that we should reserve judgment. In fact, we don't know for certain that animals far removed from us on the tree of life have subjective experiences at all.

It is clear to me that the squirrel who watches me through the window of my study as it climbs the dogwood tree on its way to my attic knows what it is doing. But some have doubted this sort of attribution. The most influential of these was René Descartes (1596–1650), the French philosopher who proposed a sharp distinction between body and mind: Cartesian dualism. People have both a body and a mind, Descartes reasoned, and these clearly interact in some way, since when we stub a toe, we feel pain. He hypothesized that interactions between body and mind occur in the pineal body, a small structure roughly in the center of the head, which is now known to be involved in the control of circadian rhythms.

Since Descartes saw considerable overlap between the mind and the soul, he decided, partly for religious reasons, that animals have neither. He thought that dogs, for example, were automata with no conscious experience. While the details of Descartes' philosophy were not universally accepted, many agreed that there was an abrupt discontinuity—or no connection at all—between humans and animals.

This all changed with Darwin (1859, 1872), whose detailed descriptions and analyses of animal behavior convincingly indicated that feelings, especially emotional ones, are not the exclusive province of humans. The ensuing Victorian era saw great popular, as well as scholarly, interest in the mental life of animals.

But at least among psychologists, the pendulum swung back again at the beginning of the twentieth century with the rise of *behaviorism*, a school of experimental psychology founded by John B. Watson (1913). Watson sought neither to confirm nor to deny animal consciousness; his interest was in denying the usefulness of this concept in studying the behavior of any animal, human or otherwise. The result, especially in the United States, was a 40-year period in which it was widely considered inappropriate to acknowledge animal sentience in a scientific context.

However, as the study of animal behavior has deepened in recent decades, many who explore the details of animals' lives have concluded that trying to explain behavior entirely in terms of reflexes and conditioned responses is often less fruitful than attributing it to perception and/or cognition. There is now an emerging consensus (Balcombe, 2016; Crook, 2021; Elwood & Appel, 2009; Godfrey-Smith, 2020; Koch, 2019; Sneddon, 2019) that *sentience*—subjective awareness of one's surroundings—is a likely feature of all animals with a complex nervous system, including mammals, birds, fish, and some invertebrates. Plausible arguments have even been made for sentience in bees (Buchmann, 2023), although not everyone is convinced.

But there is a difference between complex processing (which has been amply demonstrated in a variety of animals) and sentience (which can be neither detected nor ruled out by current scientific methods). So when in this book I say things like "The octopus sees the shark," bear in mind that this is an *interpretation* of its behavior. Readers can decide for themselves whether the octopus's perception of the shark is just a metaphor or an important and consequential reality.

2 Vision in Vertebrates

Vision means using light to perceive the environment. All light is composed of elementary units called photons—infinitesimal packets of energy. A ray or beam of light contains billions of photons. What makes light so informative is that it generally travels in straight lines. For example, light travels in straight lines from the sun to the Earth, so the direction of these rays contains information about the sun's location in the sky. When this light hits an object, such as a rock, some of it is reflected and sets off in a new direction, now carrying information about the *rock's* location.

An eye is a structure that captures information about the direction (and other properties) of light. This information is of great survival value, not so much for what it tells an animal about light itself, as for what it reveals about the locations, movements, and other properties of seen objects. So it is not surprising that most animals complex enough to have a nervous system—and that includes vertebrates, arthropods, and molluscs—have eyes of some kind.

In the present chapter we will consider the visual systems of vertebrates, with an emphasis on the structure and function of their eyes and those parts of the brain that process visual information. We will start by describing the basic structure of the vertebrate eye, which has stayed remarkably constant over the course of evolution, but will then consider the variations that make it possible for animals to see in different environments (water and air), at different distances, and under different levels of illumination.

Next, we will describe visual components of the brain. Paramount in cold-blooded vertebrates is a midbrain structure, the optic tectum; in warm-blooded animals, forebrain structures (the visual cortex in mammals and the Wulst in birds) take on a major role, often overshadowing the tectum. The chapter concludes with an example of how light pollution threatens the survival of some animals.

The visual systems of arthropods and molluscs will be described in Chapter 3. Our discussion of vision will then continue with deep dives into two components of vision: color vision (Chapter 4) and visual space perception (Chapter 5).

THE VERTEBRATE EYE

HOW THE RETINAL IMAGE IS FORMED

HUMANS AND MOST OTHER MAMMALS

Because of its familiarity, we will start by describing the human eye. It is nearly a sphere, about 2.5 cm in diameter, with a bulge on the front called the *cornea* (see Figure 2.1). The transparent cornea is continuous with the *sclera,* the white of the eye. Together, the sclera and the cornea form a sturdy outer covering of the eye. The cornea admits light into the eye and bends it almost enough to bring it to a focus on the *retina,* a thin neural layer lining the inner surface of the back of the eye. Sandwiched between the retina and the sclera behind it is a highly vascularized

DOI: 10.1201/9781003362319-2

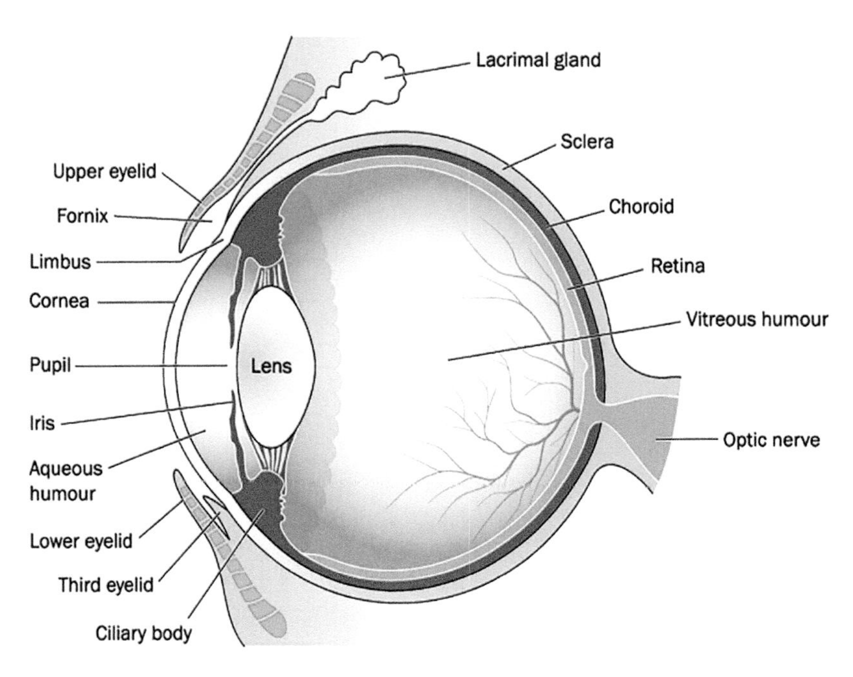

Figure 2.1 Drawing of the human eye.

Source: Used with permission of Shutterstock. Vector Contributor, Blamb.

layer called the *choroid*, which supplies nutrients to the back of the retina.

Assisting in the focusing of light is the lens, inside the eye, which is suspended in the path of the light and held in place by ligaments. The ligaments are anchored in a ring of muscle attached to the sides of the eye; when this muscle contracts, the ligaments go slack, allowing the lens to change shape, becoming thicker from front to back and thus optically more powerful. In this way the focusing power of the eye is fine-tuned to adjust for the distance to the viewed object, a process called *accommodation*. Reflex circuitry in the brain automatically triggers these adjustments. Just before reaching the lens, incoming light passes through the *pupil*, an opening in the center of the *iris*, which gives people their individual eye color—brown, blue, etc.

Two fluids, termed humors, fill most of the space inside the eye. The fluid in the front of the eye, between the cornea and the lens, is the *aqueous humor*. This watery liquid is constantly replenished, bringing nutrients to, and removing waste products from, the cornea, which has no direct blood supply since blood vessels would interfere with the passage of light. The region of the eye behind the lens is filled with another fluid, the *vitreous humor*, which is more gelatinous that the aqueous humor and is not replenished.

The cornea, lens, and aqueous and vitreous humors are collectively referred to as the *ocular media*. They work as an integrated system to produce a sharply focused image of the *visual field* (the part of the environment viewed by the eye) on the retina.

It was not always understood that this inconspicuous sheet of neural tissue is the eye's most critical component. It took an ingenious experiment by seventeenth-century philosopher and scientist René Descartes to establish this fact. He reasoned that since we see objects clearly, light must trigger vision at some location inside the eye where there is a clear image of the viewed scene. Descartes sought to locate this image by dissecting ox eyes obtained from a slaughterhouse. When he cut away some of the sclera and choroid from the back of the eye, so that he could view the translucent retina from behind, he saw on it a miniature, upside-down replica of the scene toward which the eye was aimed. No such image occurs anywhere else in the eye. Descartes had discovered that the process of vision starts in the retina.

OTHER VERTEBRATES

Much of this description of the human eye applies to other vertebrates as well. There are, however, important differences among vertebrate classes. One difference between fish and terrestrial vertebrates is that the two groups accommodate differently. For animals that live out of the water, light refracts significantly as it passes from air into the curved cornea, because of the difference in density between these two media. In other words, most of the work of bending light to create a retinal image is accomplished just as the light enters the eye. Fine adjustments of the eye's focus for objects at different distances are then easily handled by changes in the shape of the thin and supple lens.

But in fish, light enters the cornea from water, not from air. Since the cornea has about the same density as water, little refraction occurs at this interface. Instead, the optical burden of focusing the light onto the retina falls on the lens, which therefore needs to be a large, sometimes almost spherical structure. Because of its bulk, changes in its shape are not possible. So accommodation is accomplished instead by movement of the lens. It moves forward toward the cornea for near vision and backward for seeing more distant objects. In different groups of fish, different muscular arrangements are used to produce these movements (Walls, 1967).

A nearly spherical and inflexible lens is also characteristic of whales and some other marine mammals, and accommodation is achieved, as it is in fish, by forward movement of the lens (Mass & Supin, 2007).

Anableps. A few vertebrates have eyes specialized to see all parts of their environment clearly without the need for accommodation. One such is *Anableps*, the "four-eyed" fish of Central and South America, which swims at the surface of the water, with the top half of its eyes in air and the bottom half submerged (Zahl & O'Neill, 1978). It squirts water at flying insects and eats them when they drop into the water. Its name derives from the fact that each eye has two pupils, one above the water line and one below (see Figure 2.2).

Figure 2.2 *Anableps anableps*, the four-eyed fish. Each eye has two pupils, one just above the water line and the other just below. Its egg-shaped lens forms sharp retinal images of both airborne and underwater objects.

Source: **Used with permission of Shutterstock. Photo Contributor, gallimaufry.**

The challenge for Anableps is to simultaneously achieve sharply focused retinal images of both underwater and airborne objects. Anableps is able to keep both the upper and lower halves of its visual field in focus because its lens is shaped like an egg, angled downward toward the lower pupil. This gives the lens great optical power to bend rays of light from underwater objects, focusing them on the upper retina even though no refraction occurs at the water–cornea interface. But light coming through the upper pupil, already refracted at the air–cornea interface, needs only a little additional bending to be imaged on the lower retina. Just the right amount is provided as the light traverses the thinner dimension of the egg-shaped lens.

Pigeon. A less extreme optical asymmetry in the eye of the pigeon, *Columba livia*, allows near and far objects to be in focus simultaneously on different parts of the retina, so there is little need for accommodation. Its retina is composed of a larger yellow portion (the *yellow field*) and a smaller *red field*. The red field occupies the top third of the retina and therefore views the lower part of the visual field, that is, the ground; the yellow field receives light from more distant surroundings and the sky. The geometries of the lens and eyeball are such that faraway objects are generally in focus on the pigeon's yellow field. The red field, in contrast, is nearsighted, so that it receives focused images of objects on the ground, near the beak.

Moreover, the colors of the pigeon's red and yellow fields indicate the wavelengths of light that are reflected—and therefore not absorbed—by photoreceptors in those regions. In other words, red and yellow are the *opposites* of the colors these fields are most sensitive to. The pigeon is thus well adapted to viewing bits of food in the grass, while monitoring the blue sky for a hawk.

CELLS OF THE RETINA

RECEPTORS

A defining characteristic of the retina is the presence of visual *receptors*, specialized cells that produce an electrical response when they absorb light. These receptors are very numerous: For example, there are some 126 million of them in each human eye. They are elongated cells, tightly packed together, with their long axes parallel to rays of light reaching them from the pupil. At the back end of each receptor, closest to the choroid, is a part called the outer segment; it is here that light is absorbed and triggers an electrical signal, a process called *transduction*. Transduction occurs because the outer segments contain molecules that change shape when they absorb a photon. These molecules of *retinal* (emphasis on the last syllable) are made up of a ring of carbon atoms, with additional carbon atoms extending from the ring in a chain. In the dark, this chain has a bend in it at a certain place, but when a photon is absorbed, the bend straightens out: In chemical terms, retinal changes from the *11-cis* form to the *all-trans* form. Light isomerizes the molecule.

This change is important because the retinal molecule, sometimes called the visual pigment's *chromophore*, is tightly bound to a much larger protein molecule called *opsin*. Together, they make up a molecule of visual pigment. The 11-cis form of retinal fits together perfectly with the opsin, but the all-trans form doesn't. So when a photon is absorbed, the chromophore and the opsin separate, triggering a series of chemical and electrical events that can lead to vision. Before absorbing light the pigment is richly colored, but when the chromophore and opsin separate, they become colorless; so the effect of light on the pigment is called *bleaching*. Afterwards, opsin binds with new 11-cis retinal, and molecules of visual pigment are *regenerated*.

Each receptor cell contains millions of molecules of visual pigment. Vision occurs when a number of visual pigment molecules absorb photons at the same time, causing the receptor cells in which they are contained to initiate messages that travel to other layers of the retina.

OTHER CELL TYPES

The retina actually contains a number of cell types in addition to the receptors. Most important are the *ganglion cells*: These neurons receive information about light absorbed by the receptors and send that information to the brain via their axons. There are about a million ganglion cells in each retina, each with a single axon, and those axons make up the optic nerve.

The layer of ganglion cells is actually in front of the receptors, just behind the vitreous humor. This surprising arrangement means that light has to pass through the nearly transparent ganglion cells in order to reach the receptors and be absorbed; it is therefore said that vertebrates have an *inverted retina* (see Figure 2.3). This inversion allows the choroid to be close enough to the receptors to supply their demand for nutrients and 11-cis retinal, without blocking their view.

Several other types of cells are located between the receptors and the ganglion cells. *Bipolar cells* receive messages from the receptors and pass them along to the ganglion cells. *Horizontal cells* influence the transmission of signals from the receptors to the bipolars so that spatial differences in light intensity, such as between areas of sunlight and shadow, are emphasized. For example, a cloudless patch of sky will evoke a strong response from bipolar and ganglion cells if it includes a black speck that may indicate the presence of a predator high overhead. This contrast enhancement results from *lateral inhibition*, a process described in Chapter 3.

Amacrine cells influence the transmission of signals from bipolars to ganglion cells, by emphasizing movement or other temporal changes: They increase ganglion cells' responses to the black speck if it moves or grows larger. In other words, both horizontal and amacrine cells participate in the *processing* of visual information.

While these general principles apply to almost all vertebrate eyes, the circuitry in some retinas is more complex than in others. In frogs, for example, there are ganglion cells that will not respond to a field of dots moving back and forth but will respond vigorously if one of the dots moves with respect to the others (Lettvin et al., 1959). In everyday life, such a cell would respond to a bug crawling across a flower but not to the flower as a whole swaying in the breeze.

In contrast, the ganglion cells of cats, primates, sheep, and most other mammals have simpler properties, responding to a spot on a contrasting background with little regard for what is happening elsewhere in the field of view. In these animals with a well-developed cortex, detailed analysis of the visual scene appears to be postponed until the "data" reach this computationally powerful structure.

ADAPTATIONS FOR VISION IN BRIGHT AND DIM LIGHT

A remarkable feature of many vertebrate eyes is their ability to function over a very wide range of illumination levels. Light from the sky is 100 million times more intense on a sunny day than on a moonless night, yet we humans are able to see under both of these conditions and at all levels in between (although we see best in moderate daylight). Cats too can see both day and night, but in their case, much better at night. Animals that can see better in daylight are said to be *diurnal*, and those that see best at night are *nocturnal*. Even more extreme are strictly diurnal animals such as eagles and strictly nocturnal animals such as bats, but most vertebrates fall somewhere in between. A number of factors contribute to the ability of some vertebrates to see in daylight and others to see at night and of still others to adjust to a wide range of illumination levels.

THE PUPIL

In humans and other mammals, the pupil is a dynamic structure that helps the eyes adjust to different light levels. In many cases the pupil is round, and its size depends on ambient illumination. In humans, for example, it enlarges in the dark to about 8 mm diameter and constricts to 2 mm diameter in bright light. This change partially compensates for the change in external illumination: When a spelunker emerges from a cave into sunlight, the fraction of light admitted by the pupil will be reduced by a factor of 16. Illumination outside the cave may, however, be a million times greater than in the cave, so pupil changes compensate for only part of the increase. Pupil size also depends on emotional state (being larger during periods of excitement) and on the distance to a viewed object, constricting during near vision to optimize depth of field.

In some other mammals, such as the domestic cat, the pupil is round when dilated but in bright light becomes a slit that admits very little light. This indicates that the cat's eye is not equipped to function well in direct sunlight. But whatever its shape, the pupil of most mammals is highly mobile. Many birds, reptiles, and amphibians, and a few fish also show pupil changes.

PHOTOMECHANICAL CHANGES

Another method of regulating the amount of light reaching the receptors is widely used by fish, amphibians, and birds but never occurs in mammals (Walls, 1967). This method involves the *pigment epithelium*, a layer of cells between the choroid and the retina. These cells are flat on the side that presses against the choroid, but on the retinal side, they have thin processes that extend into the retina, between the outer segments of the receptors. These processes deliver nutrients from the choroid to the receptors. The pigment epithelium derives its name from the granules of melanin that it contains. Melanin is not a visual pigment—it is not bleached by light and does not generate a visual signal. But, as in the skin, melanin granules can move about within the cells that contain them. In a skin cell, melanin granules spread out in response to light, resulting in a tan. Similarly, in the eye's pigment epithelium, melanin granules can, in bright light, migrate into the processes of the cell that extend between receptors, shielding the receptors from excess light, just as sunglasses would. In the dark, the melanin granules migrate back into pigment epithelium cell bodies, so the receptors are better able to catch what little light reaches the retina.

In some non-mammals, the receptors themselves also undergo mechanical changes—growing longer or shorter—in response to light. In bright light they lengthen, so that the outer segments extend between the processes of the pigment epithelium and are partially shielded by them. In dim light the receptors shorten, enabling them to capture more of the light arriving from the front of the eye.

THE TAPETUM

In many nocturnal animals, sensitivity to dim light is increased by a shiny layer called the tapetum, on the front surface of the choroid. Light entering the eye reaches the retina, where some of it is caught, but a fraction passes through the retina and pigment epithelium and bounces off the tapetum. As these reflected photons pass forward through the retina, receptor cells have a second chance to catch them, thus doubling the eye's sensitivity. Those photons that are not caught continue forward, exiting the eye through the pupil and producing the bright eyeshine that is seen when a cat is caught in a car's headlights at night. In animals with a tapetum, the pigment epithelium cells in front of it do not contain pigment, which would render the tapetum useless by absorbing the reflected light (Tansley, 1965).

NEURAL ADAPTATION

When an animal goes from a dimly lit area to a bright one, as when it emerges from dense forest into a clearing, the increase in light induces chemical changes in receptors and other retinal cells that make the eyes less sensitive. For example, if the light is very intense, it will bleach away visual pigment faster than it can regenerate, and fewer photons will then be caught because there are fewer pigment molecules to catch them. The resulting drop in the sensitivity of receptors is called *photochemical adaptation*.

But reductions of sensitivity also occur in response to smaller increases in ambient illumination. These are due to decreases in the *gain* (amplification) of signal transmission within the retina and are referred to as *neural adaptation*.

These adjustments occur in reverse when illumination decreases, as when the animal goes from the clearing back into the forest. The resulting increase in sensitivity can be studied psychophysically in the laboratory by tracking the gradual drop in detection threshold of a small test light after other brighter lights have been extinguished. Measurements in the pigeon by Blough (1956) are a classic example.

RODS AND CONES

The tapetum helps to optimize visual performance in dim light, while pupil mobility, photomechanical changes, and photochemical and neural adaptation enable animals to adjust to different levels of illumination. But there is another factor more remarkable than any of these: the existence of two types of visual receptor cells, *rods* and *cones*, that are specialized for vision in dim and bright light, respectively. Their names derive from the fact that rods' outer segments are cylindrical, while those of cones taper toward the back (see Figure 2.3).

Cones are not sensitive enough to function in moonlight, while rods are dazzled in daylight, so there is a gradual transition from cone vision to rod vision as darkness falls. Having both types of receptors allows humans and many other vertebrates to see over a much wider range of ambient light intensities than we would be able to with only one type or the other.

The different roles of rods and cones were first realized when Max Schultze, a nineteenth-century German anatomist, studied the retinas of different animals under a microscope. He found that strongly nocturnal animals have retinas in which rods predominate, whereas strongly diurnal animals have a preponderance of cones. Vision using rods is called *scotopic vision*, and cone vision is also known as *photopic vision*.

Differences in the rod-to-cone ratio can exist even between related groups of animals, depending on their habitats. For example, some snakes are diurnal and have mostly cones, while others are nocturnal and have mostly rods (Walls, 1967). Similarly, both squirrels and rats are rodents, but squirrels are diurnal and have cone-dominated eyes, while rats are nocturnal and have mostly rods. Fish living near the surface

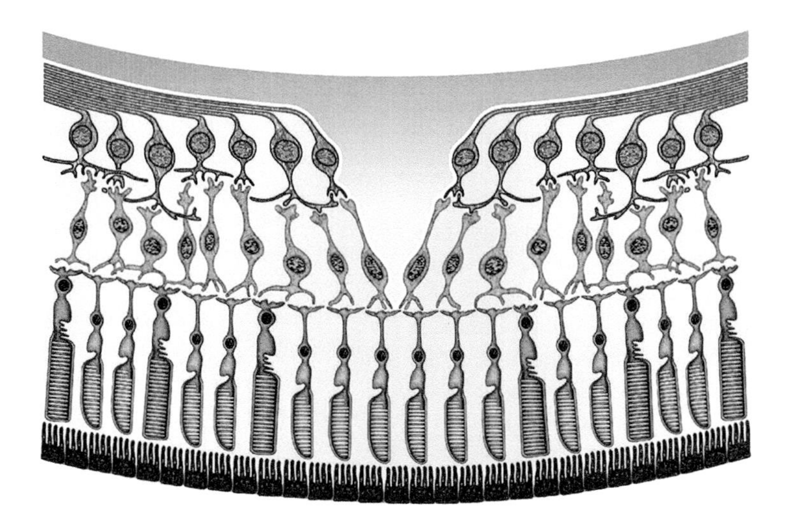

Figure 2.3 Simplified diagram of a portion of retina including the fovea. The back of the retina is at the bottom of the figure. Here, the two types of visual receptors, rods and cones (brown), absorb light and transmit visual signals to ganglion cells (purple) by way of bipolar cells (green). The fovea (center) is a small area of acute vision, found especially in primates and birds, where anterior neural elements are swept aside, allowing light unfiltered access to the receptors. Horizontal and amacrine cells are omitted from the figure.

Source: **Used with permission of Shutterstock. Illustration Contributor, ilusmedical.**

of the water have mostly cones, while those living in the murky depths have mostly rods.

Some overall trends are seen across classes of vertebrates. Most fish, regardless of the depths at which they swim, have a considerable number of both rods and cones; a similar balance is seen in amphibians. But as reptiles emerged and diversified, a tendency for their retinas to be dominated by cones occurred in many species. With few nocturnal predators, they could afford the luxury of eyes optimized for daytime hunting. The same predominance of cones is also seen in birds, except for those who hunt or migrate at night. And it is widely believed that the same was true of dinosaurs, who evolved from earlier reptiles and in turn gave rise to birds.

Life on Earth took a dramatic turn 66 million years ago, when a 50-km asteroid crashed into the sea close to the Yucatan Peninsula. The impact hurled millions of tons of dust and debris into the air, cooling the planet and darkening skies worldwide for several years. With their strongly diurnal eyes, dinosaurs were poorly equipped for life under these conditions; it may have been a factor in their extinction.

The eyes of mammals, however, had evolved along a different path, with rods vastly outnumbering cones in most species. Early mammals were small nocturnal creatures who were able to scurry about freely under cover of darkness, invisible to diurnal predators. Perhaps their rod-dominated retinas proved even more of an advantage in the wake of the asteroid, helping set the stage for the rise of mammals to a position of prominence among the vertebrates.

WHAT MAKES ROD VISION AND CONE VISION DIFFERENT?

What is it about rods and cones that enables the former to see well at night and the latter to see well in daylight?

First, consider rods, whose job it is to make the most of what little light is available at night. Several factors are involved in their ability to do so. First, an individual rod contains more visual pigment than an individual cone, which means more pigment molecules are able to capture arriving photons. Second, because of the shape of their outer segment, rods (unlike cones) are sensitive to light reaching them from any direction, such as light bouncing back through them after hitting the tapetum found in many nocturnal eyes.

Third, rods are connected up differently to the ganglion cells, whose axons, the fibers of the optic nerve, are the only way in which visual signals can reach the brain. There are about 100 times as many rods as ganglion cells, so a single ganglion cell receives signals from a large number of rods. The absorption of even a single photon by a rod enables it to generate a tiny signal that is sent forward to other retinal layers. This means that rods have achieved the ultimate in sensitivity—the theoretical maximum of which a visual receptor is capable (Hecht et al., 1942). One of these tiny signals is not enough to spur a ganglion cell into activity, but a group of simultaneous signals from different rods makes the ganglion cell fire, sending to the brain action potentials that can trigger a visual sensation.

Taken together, these specializations make scotopic vision about 1,000 times more sensitive than photopic vision. But there is a cost. The fact that rods can absorb light from any angle and that their signals are pooled makes rod vision quite blurred. Acuity is poor. For example, by moonlight, you can't read this book. Nevertheless, what counts for creatures active at night is vision good enough to discern predators and prey, and rods provide that.

Cones are specialized in the opposite way. They are adapted for optimal functioning in daylight, when light is plentiful. Their job is to achieve sharp, richly detailed vision, and they do this partly by absorbing only photons that approach them end-on, from the center of the lens, which is the portion with the highest optical quality. Cones' tapered outer segments shunt away light coming from other directions, so that it never passes through their visual pigment.

Furthermore, pooling of cone signals by ganglion cells is much less than the pooling of rod signals. It would be an exaggeration to say that each cone has a private ganglion cell to relay its signals to the brain, but the pooling of cone signals is like sharing a taxi to the brain, while rod signals, by comparison, are riding on a double-decker bus. The result of all these specializations is that visual acuity is higher with cone vision than with rod vision: An animal can see fine details using cones but only coarse features of an object using rods.

THE FOVEA

Like other mammals, humans have retinas in which rods vastly outnumber cones, and yet we are diurnal creatures in that we can see better, and are generally more active, by day than at night. How is this possible? The answer is that our retinas, and those of other primates and of birds, contain a remarkable structure called the *fovea* that enables our cones to play a role in vision out of all proportion to their number.

The fovea is a small central region of the retina where cones are tightly packed together, rods are excluded, and blood vessels and other retinal components are swept aside to give approaching rays of light unobstructed access to the cones, ensuring a sharp retinal image (refer again to Figure 2.3). When the image of an object, such as a ladybug on a leaf, falls on the fovea, we can see it in vivid detail. In other words, visual acuity is excellent in the fovea: Young individuals with normal vision can resolve details that subtend only 1 minute of arc (a 60th of a degree) at the eye. For example, they can tell the difference between this C and this O from about 3 m away.

However, cones are by no means restricted to the fovea; most are outside it, scattered among the much more numerous rods in the peripheral retina. This has allowed researchers to compare the acuity of rod vision and of cone vision in the same region of retina. They have found that peripheral cone acuity is better than rod acuity, although it falls far short of foveal acuity.

But of what practical use is the tiny fovea, if we must wait until the images of interesting objects happen to fall there in order to see them clearly? The answer is that we don't need to wait. When an object, seen as a blur off to the side, captures our attention, we turn our eyes so that the retinal image of the object falls on each eye's fovea. This happens whether the shifting of attention is voluntary (looking at a paperclip on the desk) or reflexive (looking toward a sudden flash of light). A large amount of brain circuitry is involved in the control of these eye movements, which are called *saccades*.

Another way the brain contributes to the dominance of foveal vision is the large amount of brain tissue devoted to analyzing the information that comes from this small retinal region. Messages from receptors in all parts of the retina activate ganglion cells, which forward them to the brain via the optic nerve. In humans and other mammals, these nerve signals are relayed to the primary visual cortex, in the occipital lobe at the back of the brain. Here, they are distributed in a highly organized way, so that a detailed neural representation or "map" of the visual field is formed. In animals with a fovea, this map is distorted: Signals from the fovea occupy a disproportionally large percentage of it, and the cortical magnification factor drops steeply for regions of retina farther and farther to the sides. The map on visual cortex is more fully described later in this chapter.

The vast neural resources devoted to the fovea, in terms of both eye movement circuitry and the analysis of visual signals, enable the fovea to play so dominant a role in our vision that we are by nature diurnal: Most of us are active during the day and do most of our sleeping at night.

A few fish and reptiles have a fovea, but this structure is most common among birds and primates. Consider the fovea of the common pigeon, *Columba livia*, as an example. In keeping with most other birds, cones are in the majority, constituting some 80% of the pigeon's receptors (Querubin et al., 2009). As noted earlier, its retina is divided into a large yellow field and a smaller red field. In the center of the retina, within the yellow field, is the fovea, the retina's only rod-free area. It views objects off to the side of the pigeon—objects that are seen with one eye only, such as an approaching fox. However, the red field also contains a special region, the *area dorsalis*, where the density of cones is almost as high as in the fovea, but where there are also some rods. The area dorsalis is offset in the two eyes so that the pigeon can view objects binocularly with it. Behavioral experiments show that the pigeon's acuity is about the same in the fovea and the area dorsalis; it is less than half as good as human foveal acuity (Blough, 1971; Hodos et al., 1985, 1991; Roundsley & McFadden, 2005).

Some birds of prey, such as falcons and eagles, have two foveas in each eye. In front of these birds is a region of space that they can see with both eyes. This region is imaged on the part of each retina farthest from the beak. Within this region of retina, each eye has a fovea, so positioned that the bird can use both eyes to look directly at an object that is straight ahead. The other fovea in each eye is more centrally located within the retina and is positioned to view the part of space off to the side of the bird, which only one eye can see. Whichever fovea it is using, a bird of prey requires excellent acuity, to see a small animal from a great height. A kestrel, for example, surveying its domain from an elevated perch, can spot a mouse from more than two football fields away (Gaffney & Hodos, 2003). Its acuity is about 50% better than that of a human with normal vision.

In addition to providing good acuity, there is another way in which photopic vision is richer than scotopic vision: It is responsible for color vision, whereas rod vision is only in black and white. Color vision is such an important component of perception in both vertebrates and invertebrates that a separate chapter, Chapter 4, is devoted to it.

VISION AND THE BRAIN

Let's consider how the eyes and the brain work together to enable vertebrates to see where an object is located. This is one of the most important of all visual abilities.

The fundamental basis of this localization ability is that objects in different parts of the visual field are imaged on different parts of the retina. A hawk overhead is imaged on the lower retina; a tree off to the left is imaged on the right half of the retina; and so on. This correspondence is an invaluable source of information for animals, because of the way not only their eyes but also their brains are organized.

The vertebrate brain consists of three large regions: from front to back, these are the forebrain, midbrain, and hindbrain. Most visual processing occurs in the forebrain and midbrain. The roof or *tectum* of the midbrain is the primary visual area in the brains of fish, amphibians, and reptiles and continues to play a significant role in mammals (where the homologous structures are called the *superior colliculi*) and in birds. But in the latter two classes, forebrain structures—the cerebral cortex in mammals and the Wulst in birds—have evolved advanced circuitry that supports new processing algorithms. These structures make contributions to visual processing that are as important as, and in some cases overshadow, those made by the midbrain.

In the following sections, we will first consider the pathways over which signals from the retinas travel to the brain. After that, we will take up the brain structures that receive and process this information: the visual cortex, the optic tectum, and (in birds) the Wulst.

FROM RETINA TO BRAIN

Let's now consider the pathway from retina to cortex over which signals arrive. It starts with the optic nerves, thick bundles of fibers that are the axons of retinal ganglion cells. These nerves extend diagonally along the bottom surface of the brain, traveling both backward and toward the midline. They intersect at a midline structure called the *optic chiasm* because of its resemblance to the letter X. Here, at least half of the fibers from each optic nerve continue across the midline without changing direction, while the remainder turn away from the midline and stay on the same side of the brain. The two fiber bundles that leave the chiasm, the left and right *optic tracts*, are each composed of a mixture of fibers from the two eyes.

The proportion of optic nerve fibers that *decussate* (cross the midline) at the chiasm depends on an animal's lifestyle. In general, animals that are carnivores and hunt down their food have eyes on the front of their heads, as we do. This allows them to get a double dose of information about a prey animal they are stalking. These two neural representations of the visual field—one from the left eye and one from the right—can be most efficiently analyzed if they are brought together in the brain so that they can be closely compared.

What makes this possible in the case of front-facing animals is that roughly half of the optic nerve fibers decussate, while the other half stay on the same side of the brain. The axons sort themselves out so that those coming from corresponding parts of the two retinas are brought close together. All fibers carrying information about the left half of the visual field (regardless of their eye of origin) leave the chiasm in the right optic tract and, therefore, terminate in the right half of the brain. Similarly, information about the right half of the visual field is routed to the left half of the brain.

In contrast, prey animals like horses and rabbits, who must constantly be on the lookout for predators, usually have eyes on the sides of their heads. This allows them to monitor their environment in all directions. Since there is not much overlap of their two eyes' visual fields, there is little need to combine signals from the two eyes. Most of their optic nerve fibers therefore decussate at the chiasm; only a handful remain on the same side of the brain.

VISUAL CORTEX

A short distance beyond the chiasm, the two optic tracts enter the brain. In cold-blooded vertebrates, they go mostly to the optic tectum, of which more is said later. But in mammals, a majority extend to the *lateral geniculate nucleus (LGN)*, a component of the thalamus. The optic tract fibers synapse on LGN cells that convey visual information the rest of the way to the cortex.

The *cerebral cortex* is most of what one sees when looking at a mammalian brain from above or from the side. A defining characteristic of cortex is that it contains layers of cells parallel to the cortical surface. Evolution has driven the cortex to enfold upon itself, like ribbon candy. Folding is a way of squeezing as much cortex as possible into the limited space available in the skull. When we examine a human brain, what we see of cortex from its surface are only its superficial gyri; many folds lie hidden in fissures and sulci.

Different regions of cortex have different functions, and a number of these regions are involved in the processing of visual information. Chief among these is a large area of cortex at the back of the brain called the *primary visual cortex*. This area has a distinctive white layer that appears in cross section as a bright line parallel to the cortical surface; it is named the *stripe of Gennari* in honor of the eighteenth-century medical student who discovered it, and primary visual cortex is sometimes called *striate cortex* because of it. The stripe indicates the cortical layer where many thousands of axons—white because they are covered with myelin—enter the cortex, bringing visual messages that originated in the retina and were relayed by the LGN.

THE MAP ON VISUAL CORTEX

Primary visual cortex, also known as V1, is a large region. When LGN fibers arrive there, they distribute themselves so that some fibers go to one part of V1, while others go to other parts. In a sense, V1 is a *map* of the visual field, with different parts of it receiving information about different portions of the visual field. The map is slightly different in different mammals. In people, for example, the foveal representation, where information about the thing you are looking at is processed, is at the very back of the brain, whereas in monkeys, the foveal representation is some distance forward on the lateral surface of each hemisphere. In animals with a fovea, its representation is a disproportionately large part of V1, a fact referred to as *cortical magnification*.

The existence of the map has been demonstrated in several ways. The oldest way, which is a part of the practice of medicine, consists of examining the vision of a person whose primary visual cortex has been partially destroyed, for example by a battlefield injury. The anatomical location of the injury can be compared with the resulting impairment of the patient's vision. Destruction of part of V1 always results in a *scotoma*—an area of blindness—in a specific region of the visual field. For example, if the damage is to the most posterior part of V1, the scotoma will be at in the center of the visual field, and the patient will not be able to perceive an object that is straight ahead. Damage elsewhere in V1 will result in a scotoma in a different part of the visual field, consistent with the map.

A second set of methods involve neuroimaging. For example, functional magnetic resonance imaging (fMRI) can be used to determine which parts of V1 are active under a particular set of conditions (see Figure 2.4). When a human research participant looks directly at a target, activity is detected at the posterior pole of the cortex, whereas when the target is presented in the periphery of the participant's visual field, the neural activity detected by fMRI is farther forward in V1 (Dougherty et al., 2003).

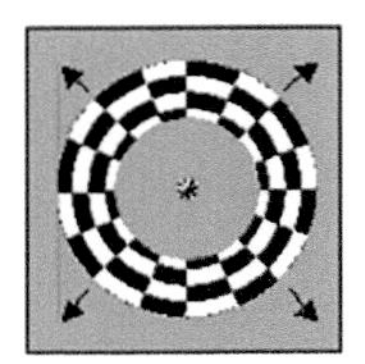

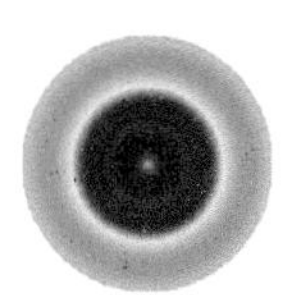

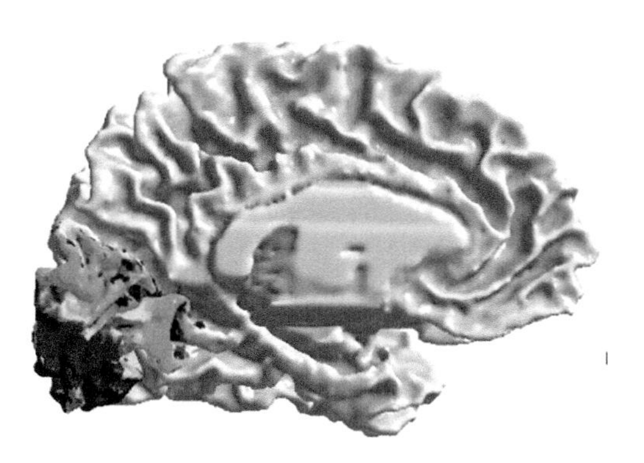

Figure 2.4 Computer image of the left hemisphere of a human brain, seen from the midline. fMRI study participants looked steadily at the center of the stimulus, a slowly expanding ring filled with contours (upper left). Colors indicate the portions of visual cortex successively activated as the ring expanded (see key in upper right). The foveal representation is at the posterior pole of the cortex. Reprinted with permission from Dougherty, R. F., Koch, V. M., Brewer, A. A., Fischer, B., Modersitzki, J., & Wandell, B. A. (2003). Visual field representations and locations of visual areas V1/2/3 in human visual cortex.

***Source: Journal of Vision*, *3*, 586–598. Used by permission of the Association for Research in Vision and Ophthalmology (ARVO); permission conveyed through Copyright Clearance Center, Inc.**

A third method, and the one most often used with animals, is neural recording from individual cortical cells. In these experiments an animal, a cat for example, is anesthetized, and a microelectrode is lowered into its cortex through a hole in the skull. Once the electrode is close enough to an individual cell to pick up its action potentials, the experimenters present visual stimuli on a screen in front of the unconscious animal. (Contact lenses are placed on its eyes to keep the corneas moist.) The experimenters then move the stimuli around on the screen until the cell responds. The region of the visual field to which a cell responds is called its *receptive field*.

RECEPTIVE FIELDS

Receptive fields of cortical cells can be used to delineate the map on V1, which closely corresponds to the map determined by other methods. But David Hubel and Torsten Wiesel (1962, 1977) learned surprising additional things about how visual cortex works by studying the receptive fields of cortical cells.

Their most fundamental discovery was that most cells in V1 respond best not to a spot, but to a *line*, such as the edge of an object. And to evoke the strongest response, the line must be in a specific orientation: horizontal, vertical, or something in between. There are different cells for different orientations.

The orientation specificity of cells in V1 was a surprise, because they receive their primary input from the axons of LGN cells, which (like retinal ganglion cells) have small, round receptive fields. Hubel and Wiesel's explanation for this change is that a cortical cell receives signals from a set of LGN axons whose receptive fields are *in a row* in the visual field. The row is vertical for some cortical cells, making them responsive to vertical contours, horizontal for others, and so on. Cortical cell receptive fields are thus larger than the receptive fields of cells earlier in the visual pathway and more specific in terms of the visual stimulus needed to activate them.

Cells having the same preferred orientation cluster together into narrow *orientation columns,* which extend all the way from the surface to the bottom of cortex. As Hubel and Wiesel lowered their microelectrode perpendicularly into the cortex, recording first from a cell near the surface and then from a series of cells at increasingly greater depths, the preferred orientation of each cell was the same.

However, if they advanced the microelectrode into the cortex at an angle, the preferred orientation of the cells they encountered gradually changed. For example, if the first cell they recorded from fired most vigorously to a vertical line, the second one might prefer a line tilted a few degrees clockwise from vertical, the third one a slightly greater tilt, and so on. Since the electrode was advancing diagonally, it was crossing from one orientation column to another. Eventually, they would reach another cell that, like the first, preferred vertical lines. Hubel and Wiesel called a complete set of adjacent columns an *orientation hypercolumn.*

Regardless of its preferred orientation, the receptive field of a V1 cell is just a small part of the overall visual field. The receptive fields of all the cells in an orientation hypercolumn overlap considerably and are in the same general region of the visual field. So a hypercolumn can be thought of as a cortical module that views a small region of the visual field and detects any lines present there, regardless of their orientation.

An object does not, of course, consist of just one line. Imagine a cat viewing a pigeon that is strolling nearby. The pigeon's silhouette consists of curves, but short segments of the silhouette are nearly straight lines. Cells in one hypercolumn will respond to the horizontal edge of the pigeon's back, cells in another hypercolumn will respond to the vertical edges of its neck, and so on. Building on these responses, the visual system is able to piece together the overall shape of the pigeon—a task that is accomplished not in V1 but in other cortical areas with which V1 is in communication.

The term receptive field means not just the location but the stimulus properties to which a cell responds. These properties can vary in ways other than preferred orientation: Some cells prefer a moving line to a stationary one, while others do not; some insist on a short line rather than a long one; and so on. In general, the cells with complicated properties are located near the top or bottom of cortex, while those with relatively simple properties are located about halfway down, near the stripe of Gennari where axons arriving from the LGN terminate. Cortical processing adds complexity to cells that are at some distance above or below this input layer. For example, most cortical cells close to the input layer can only be activated by signals from one eye, while cells close to the top or bottom of cortex respond best to simultaneous stimulation of both eyes.

PRE-STRIATE CORTEX

V1 is not the only cortical area with cells that respond to visual stimulation of the eyes. It occupies a large portion of the brain's occipital lobe, but in front of it is another large region that also contains visually responsive cells. This is not *primary* visual cortex, but it is visual cortex in a broader sense of the term. Because it lies anterior to striate cortex, it is called pre-striate cortex. It occupies the rest of the occipital lobe and spills over into the temporal and parietal lobes.

Pre-striate cortex receives its primary input from striate cortex. It is complex and somewhat mysterious, and its functions have been a matter of theoretical debate for well over a century. Early on, there was a tradition of considering pre-striate cortex to be responsible for more abstract or conceptual visual processing than striate cortex—for perception as opposed to sensation—so until about 1950, it was usually called visual association cortex.

A revolution in our understanding of pre-striate cortex began in the 1970s, when neuroscientist Semir Zeki recorded from single neurons in the pre-striate cortex of monkey and discovered that it is composed of a number of distinct areas that respond to different aspects of visual stimulation. The number of areas is quite large, perhaps running into dozens. For example, cells in one area, dubbed V4 by Zeki, are especially attuned to the color of a stimulus; cells in another area, V5, are most responsive to stimulus movement. It now appears that striate cortex sorts the visual input it receives from the LGN, sending different types of information to different pre-striate areas (Zeki, 1993).

These experimental discoveries in monkeys are consistent with clinical neurology, for it sometimes happens that a patient who has suffered a stroke that produces localized damage in a section of pre-striate cortex suffers from a specific sensory impairment, such as color blindness or an inability to see movement. These rare conditions reflect selective destruction of the pre-striate "module" responsible for the type of perception now lost.

So the cortex teases apart the different strands of visual information, presumably so that each can be analyzed more efficiently. But how are these separate strands put back together again, so that a meaningful overall percept—let's say, of a tree—can be experienced?

Several possible solutions to this *binding problem* have been considered (Treisman, 1999). One is that the various pre-striate areas send fibers to an integrative area still farther from V1, and that individual cells in the integrative area receive signals about multiple features of a stimulus. Suppose, for example, that a monkey is looking at a green snake moving across the ground. Nerve signals signifying the snake's shape, movement, and color reach V1 and are forwarded to three different areas of pre-striate cortex, where they are analyzed in detail. These three areas then send precise information about the object's properties to the integrative area, where neurons are wired to receive specific combinations of inputs. A few of these cells will be connected to inputs signaling wavy shape, undulating movement and emerald green and will therefore be triply activated, causing perception of the snake.

This idea sounds farfetched, but in fact, integrative areas have been found in the temporal lobe of monkeys. Cells here respond to objects with many features, such as the hand or face of another monkey (Gross et al., 1972; Perrett et al., 1982). It may be that activity of groups of these cells results in a perceptual experience of a particular object. Some connections may be strengthened by repeated activation, enabling recognition of familiar stimuli by the animal. Hypothetical cells that are activated by specific, meaningful stimuli are sometimes whimsically referred to as "grandmother cells" (Gross, 2002).

Another possibility, embodied in *feature integration theory* (Treisman & Gelade, 1980), is that *attention* is needed to unite the features of a perceived object. One piece of evidence for this theory is that people viewing several objects at once, such as a red circle, a green square, and a blue triangle, often combine their properties incorrectly if they are not paying attention. That is, they report seeing *illusory conjunctions*, such as a blue circle and a red triangle (Treisman & Schmidt, 1982). Feature integration theory is a persuasive contribution to cognitive psychology, but the underlying physiological mechanisms are unclear.

These two perspectives are not mutually exclusive, and each will probably contribute to an eventual understanding of how coherent visual perception is achieved. Theoretical and experimental work in this area continues (Zeki, 2015). The answer is certain to be complex and

to be consistent with principles of perceptual organization inferred from subjective experience (Koffka, 1935), a topic to which we will return later in the chapter.

OPTIC TECTUM

In fish, amphibians, and reptiles, the optic tectum is the brain region chiefly responsible for visual processing. It does not have the crisply layered structure characteristic of cortex, but its cells have receptive fields that are arranged in maps of the animal's visual field.

FROGS

There has been considerable research on the optic tectum of the frog. Its optic tracts largely cross at the chiasm, so each half of the tectum sees mainly through the contralateral eye. However, the region of the visual field close to "straight ahead" is visible to both eyes and is therefore represented in both the left and right tectal lobes.

Frogs respond promptly when they see a moving bug, worm, or other potential meal, by making a sudden movement toward it. Once the prey is within reach, they lunge forward and extend their tongue to seize it. To determine the role of the tectum in the orienting response, Kostyk and Grobstein (1987) dropped live mealworms at various locations within the visual field of leopard frogs (*Rana pipiens*) and noted the direction of the frog's initial movement. Intact frogs hopped very accurately in the direction of the mealworm.

In some frogs, the experimenters lesioned either the left or right lobe of the tectum. The operation was performed under anesthesia and the frogs were tested once they had recovered. They now responded only to mealworms that could be seen with the non-lesioned lobe of the tectum. Thus the tectum is essential for a frog's ability to orient toward prey.

While capturing prey is essential to the survival of frogs, escaping from an approaching predator is no less important. With this in mind, Yamamoto et al. (2003) measured the response of bullfrogs (see Figure 2.5) to a looming object. The frogs were positioned in front of a computer screen on which a black rectangle grew steadily larger, simulating an object approaching from a distance of 6 m. At some point, the frog hopped out of the way. To determine how the frog decides when to jump, the researchers made the rectangle expand at different rates, simulating approach velocities of either 2 or 4 m/s. If frogs are trying to escape the approaching object just before it smashes into them, they should hop away sooner if the rectangle is expanding faster, but they did not. Instead, they jumped when the rectangle grew to a certain size, subtending 20 degrees of angle in the frog's visual field—roughly the angle subtended by a soccer ball at arm's length. The neural computation performed by the frog is thus a relatively simple one.

Figure 2.5 An American bullfrog, *Rana catesbeiana*. Collision-sensitive neurons in its optic tectum prompt it to jump out of the way of a looming object.

Source: **Used with permission of Shutterstock. Photo Contributor, Ilias Strachinis.**

Recording from single cells in the frog's optic tectum, Nakagawa and Hongjian (2010) discovered *collision-sensitive neurons* that respond vigorously to a looming object. These cells are extremely selective: They give little response if the path of the approaching object will cause it to miss the frog. But if the frog is directly in the path of the object, the collision-sensitive neurons will fire more and more as it gets closer. Remarkably, their firing rate reaches a maximum when the object subtends just over 20° of visual angle, the same size found by Yamamoto et al. to trigger a behavioral response. Taken together, these findings make a strong case that the optic tectum plays a key role in the ability of bullfrogs to avoid approaching objects, such as a hungry snapping turtle.

SNAKES

An interesting feature of the tectum in many cold-blooded vertebrates is that, in deeper layers, it receives somatosensory inputs that complement the visual information reaching the top layers. An example is the tentacled snake, *Erpeton tentaculatus*, that lurks below the surface of Southeast Asian waters. This unique reptile is named for the two short tentacles that extend directly forward from its snout and are acutely sensitive to tactile stimulation.

While the snake is waiting for a fish to blunder within reach, its body extends upward through the water but its head bends back down, giving it the shape of an upside-down letter J (see Figure 2.6). When a fish comes within the curve of the J, the snake makes a feint with its body, causing the alarmed fish to flee toward the snake's jaws (Catania, 2010). Remarkably, the snake's strike exactly anticipates the trajectory of the fish's movement, indicating deep processing of movement information by its tectum.

Experiments show that the snake is capable of hunting in this way even when the fish is only a moving image on a computer screen, or when there is insufficient light for the snake to see. These results indicate that either vision or touch, by itself, is able to mediate the hunting behavior.

The snake's tectum receives both visual and tactile input, and each input forms a map on the tectum (Catania et al., 2010). The visual map is based on the locations of objects in space, whereas the tactile map is based on the part of the body stimulated, either by a direct touch or by vibrations of water caused by, say, an approaching fish. Importantly, the two maps are approximately in register, suggesting that they are able to send equivalent signals to motor centers that cause the snake to orient toward its prey.

The tectum also presumably mediates the remarkable visuomotor ability of a terrestrial snake, the spitting cobra, to hit its mark. These animals have hollow fangs with a hole in the front surface through which venom can be ejected with great force. Westhoff et al. (2005) studied the spitting behavior of two species of

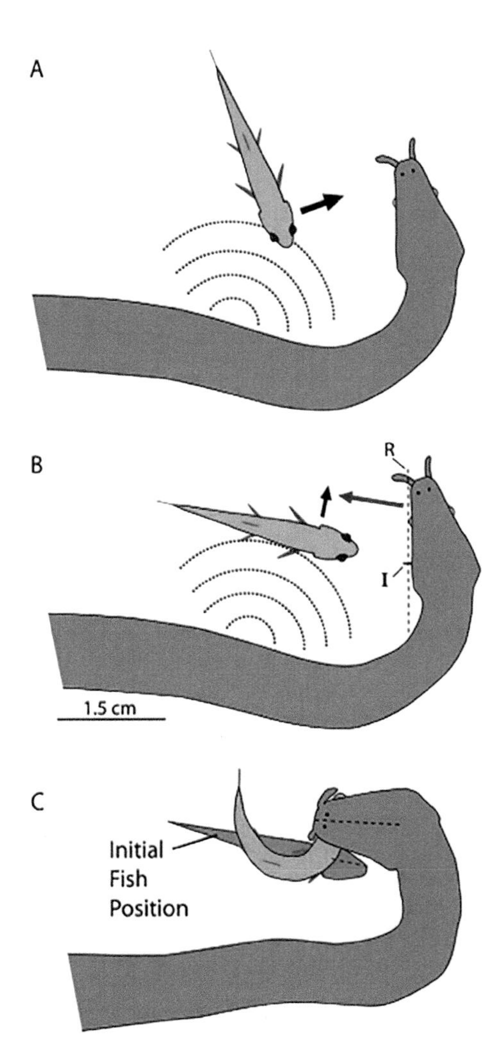

Figure 2.6 The tentacled snake, *Erpeton tentaculatus*, uses trickery to catch a fish. When it either sees or (in murky water) feels the approach of a fish, it feints (A), producing vibrations in the water that cause the fish to turn toward the snake's head (B), where it can be caught (C). In aiming its strike, *Erpeton* predicts the fish's movement, an ability that indicates sophisticated processing of movement information by the tectum.

Source: **From Catania, K. C. (2010). Born knowing: Tentacled snakes innately predict future prey behavior. *PLoS One*, 5, e10953. © 2010 Kenneth C. Catania.**

African spitting cobras of the genus *Naja*. The snakes were individually housed in wooden terraria, with a sliding panel that could be pulled away during an experimental trial so that the snake could see a visual stimulus, such as the face or hand of an experimenter. The cobras

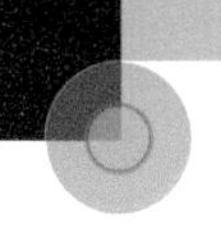

attentively watched the moving hand from the back of the terrarium, a distance of about 60 cm, but did not spit at it. In contrast, they frequently spat at the face, provided it was moving.

To both protect the experimenter and measure the accuracy of the snake, the experimenter wore a face shield when peering into the terrarium. The front of the shield was covered with a fine purple powder which became visible when hit by venom. The spray pattern included at least one of the experimenter's eyes some 80% of the time and often included the other eye as well. The occurrence and accuracy of the spits are quite adaptive, since the venom causes pain and corneal damage when it strikes an eye but has little effect on the hand or other parts of the body.

To determine how the spitting cobra achieves such accuracy, the experimenters (Westhoff et al., 2010) made precise measurements of the cobra's head movements before and during a spit and compared them with the irregular movements of the human target's head that elicited the spit. Before spitting, the cobra watches the person's moving face and moves its head to follow those to-and-fro "wiggles" of the face. Since it takes the snake about 200 ms to react, it is continually oriented to where the face was 200 ms earlier. If it spat while continuing to move in this way, it would miss the face. But as it begins to spit, the cobra's head movement accelerates, so that it catches up with the target's face. In other words, neural circuitry—presumably in the tectum—causes the snake to *anticipate* where the target will be when venom reaches it. This is a remarkable coordinated achievement of the spitting cobra's visual and motor systems.

SUPERIOR COLLICULUS

All mammals have both a visual cortex and a superior colliculus, a midbrain structure homologous to the tectum of cold-blooded vertebrates. Both cortex and colliculus contribute to visual behavior. Schneider (1969) found, in research on the Syrian golden hamster *Mesocricetus auratus* (see Figure 2.7), that these contributions are qualitatively different. He lesioned the visual cortex of some hamsters and the superior colliculus of others. After they had recovered from the surgery, he compared the visual behavior of these two groups. In one test, the animal stood before two panels, one with a mottled pattern and the other striped. One of these was designated the correct panel by the experimenter. If the hamster pushed on the correct panel, it swung open, allowing the animal to get a quick drink of water. Over trials, the animals with lesions of the colliculus learned which pattern was correct; however, the animals with cortical lesions were unable to do so. This indicates that the ability to distinguish patterns depends on the cortex.

Figure 2.7 A Syrian golden hamster, *Mesocricetus auratus*. Both visual cortex and optic tectum contribute to its visual abilities.

***Source:* Used with permission of Shutterstock. Photo Contributor, Alexruss.**

In another test, the experimenter held a sunflower seed near the animal's head, but without touching it. The position of the seed varied from trial to trial. Hamsters with a cortical lesion quickly turned toward the seed to get it, but those with collicular lesions froze, uncertain which way to turn. This finding indicates that the colliculus is important for visual orienting. An intact cortex enabled the hamsters to *identify* an object, whereas the colliculus enabled them to *locate* it.

The relative contributions of visual cortex and superior colliculus vary across mammal groups. This can be gauged by comparing the number of optic tract fibers traveling to the LGN (which sends visual information on to the cortex) with the number traveling to the colliculus. In rodents such as the hamster, a majority travel to the colliculus, whereas in primates, the LGN is the primary recipient. Even in primates, the colliculus has an important role in visual orienting, especially in the control of eye movements.

However, the primate cortex has taken on a role in object localization, overshadowing that played by the colliculus. It is now recognized that there is a spatial vision pathway from the striate cortex forward into the parietal lobe, complementing the object identification pathway from striate cortex into the temporal lobe (Mishkin et al., 1983).

Furthermore, in primates there is a major output from the parietal lobe of the cortex to the superior colliculus, suggesting that the two structures work together. The cortex appears to compute the location of an object, after which the colliculus orders eye and head movements to orient the animal toward the stimulus. In the deeper layers of the colliculus are cells that are activated by auditory and tactile stimuli. These cells respond best when multiple modalities are stimulated from the same location in space, as when an infant macaque in distress emits gecker calls, moves about, and tugs on its mother to attract her attention.

CENTRAL VISUAL PROCESSING IN BIRDS

THE WULST

Birds are unique in many ways. Like cold-blooded vertebrates, they have in their midbrain an optic tectum, which is the main processing region for visual information. However, the bird *forebrain* also contains well-developed visual centers that are more advanced than the homologous structures in fish, amphibians, and reptiles. The largest of these is the *Wulst*, a bulge on the top of the forebrain. Both tectum and Wulst receive retinal input: the tectum directly and the Wulst indirectly via a thalamic nucleus that is homologous to the mammalian LGN.

It was formerly believed that the Wulst lacked the precise anatomical organization of mammalian cerebral cortex, which is densely packed with *tangential* fibers parallel to the cortical surface and *radial* fibers perpendicular to the surface. Roughly speaking, the radial connections allow for detailed processing of information from a particular part of the visual field, i.e., within columns, while the tangential connections enable comparisons among stimuli in adjacent parts of the visual field.

Recently, Stacho et al. (2020) have examined the Wulst microscopically, using a new, three-dimensional imaging technique that uses polarized light. Many axons in the brain are myelinated, and the orderly arrangement of molecules in these myelin sheaths makes them *birefringent*, that is, having a refractive index that varies with the orientation of polarized light. Using a method that capitalizes on this property, the researchers discovered that fibers in the Wulst have a previously unsuspected organization resembling that of cortex, with a profusion of both horizontal and vertical fibers.

Birds are a diverse class of vertebrates, with many more species than there are in Class Mammalia, and so it is not surprising that there are differences in the structure and function of the tectum and Wulst across types of birds. Here, we will consider one example: the visual areas of the brain of the common pigeon, *Columba livia*, whose retina was described earlier.

VISUAL PROCESSING IN THE PIGEON'S BRAIN

Pigeons are social animals that can recognize one another. They can navigate over considerable distances, but this is normally in the interest of returning home rather than of migrating. They forage by eating bugs and seeds they find on the ground. They are preyed upon by falcons and other raptors, by foxes, snakes, and a variety of other animals, so it is important for them to keep an eye on their environment in all directions—which explains their laterally placed eyes.

Single-cell recordings from the pigeon's Wulst (Ng et al., 2010) show that the cells' receptive fields are arranged in a map and that, when tested with gratings, individual cells preferred some orientations to others. Curiously, more preferred vertical stripes than horizontal ones. Ng et al. speculate that this may be because when pigeons fly past objects, the retinal images of those objects will consist of horizontal streaks. By having few neurons that are activated by horizontal contours, the pigeon is not confused by these streaks.

It will be recalled that the pigeon has a retina that is divided into two parts: a yellow field that is monocular and looks off to the side and upward, and a smaller red field that is binocular and views objects that are straight ahead and close to the pigeon's face, like bits of food on the ground. Electrophysiological recordings have shown that the tectum receives information from both the red and yellow fields, while input to the Wulst is primarily from the yellow field.

The behavioral significance of this difference has been examined by lesioning the Wulst of pigeons and later testing them, using animal psychophysics, to determine whether vision is impaired (Budzynski & Bingman, 2004). The pigeons' task was to distinguish gratings of different orientations from one another. When the gratings were at a distance, the pigeon viewed them off to the side, using the fovea in its yellow field, but when the gratings were directly in front of the pigeon, in an operant chamber, it viewed them with its red field. Pigeons with lesions of the Wulst performed more poorly than control pigeons when tested with the far-away gratings. In contrast, their performance when using the red field was unimpaired. The implication is that the Wulst plays an important role in processing information from the yellow field but is not necessary for perception with the red field. So the Wulst appears well suited to monitor the pigeon's global environment—horizon and sky—for navigation cues and possible predators.

The tectum, meanwhile, appears to carry out a detailed examination of objects' color, form, motion, and other properties (Clark & Colombo, 2020). It is more advanced than the tectum of the frog: For example, in responding to an approaching object that is on a collision course with the animal, cells in the frog tectum code only for the angular size of the object—a rough measure of its proximity—while some cells in the pigeon's tectum compute the time to collision (Wu et al., 2005). This is a more sophisticated, and more useful, calculation (see Figure 2.8).

It should be emphasized that the pigeon's Wulst and tectum do not operate in isolation. They are part of a richly interconnected system that includes other nuclei (Clark & Colombo, 2020), and many puzzles remain to be resolved.

PERCEPTUAL ORGANIZATION IN ANIMALS

We have seen that in the visual systems of vertebrates, visual images are analyzed to extract information about edges and other simple features. The process starts with rods and cones, each of which responds to light in a tiny part of the visual field called its receptive field. As they are passed along to other types of cells, these neural signals interact with one another in various ways so that the cells respond only to specific features, such as lines with a particular orientation or movement in a particular direction. In later stages of processing, these features are somehow stitched together into perceived objects. As we saw earlier, the physiological basis of this binding process is not fully understood. However, from a phenomenological perspective, it can be described in terms of the *laws of perceptual organization* discovered a century ago by Gestalt psychologists (Koffka, 1935).

Image elements such as specks or short lines are more likely to perceptually combine into a larger unit if they are similar, and if they are close together. For example, adjacent spots more readily fuse into a larger patch if they are all red than if they are of different colors. This is the Gestalt *law of similarity*. And a series of diamond-shaped patches are more likely to be seen as belonging together if they are near one another than if they are in different parts of the visual field—the *law of proximity*.

FIGURE AND GROUND

The next step in perceptual processing is for some of these larger units to be seen as closer to you than other units. If you see a deer standing in front of some trees, the deer is, in Gestalt terms, the *figure*, and everything behind it is the *ground*. Figure/ground perception is crucial for animals, since they are more able to survive if they can perceive their prey as figure, while they themselves are "hiding in plain sight" by means of camouflage when predators look their way.

One component of camouflage is for the color and pattern of an animal to match its background. The value of background matching is

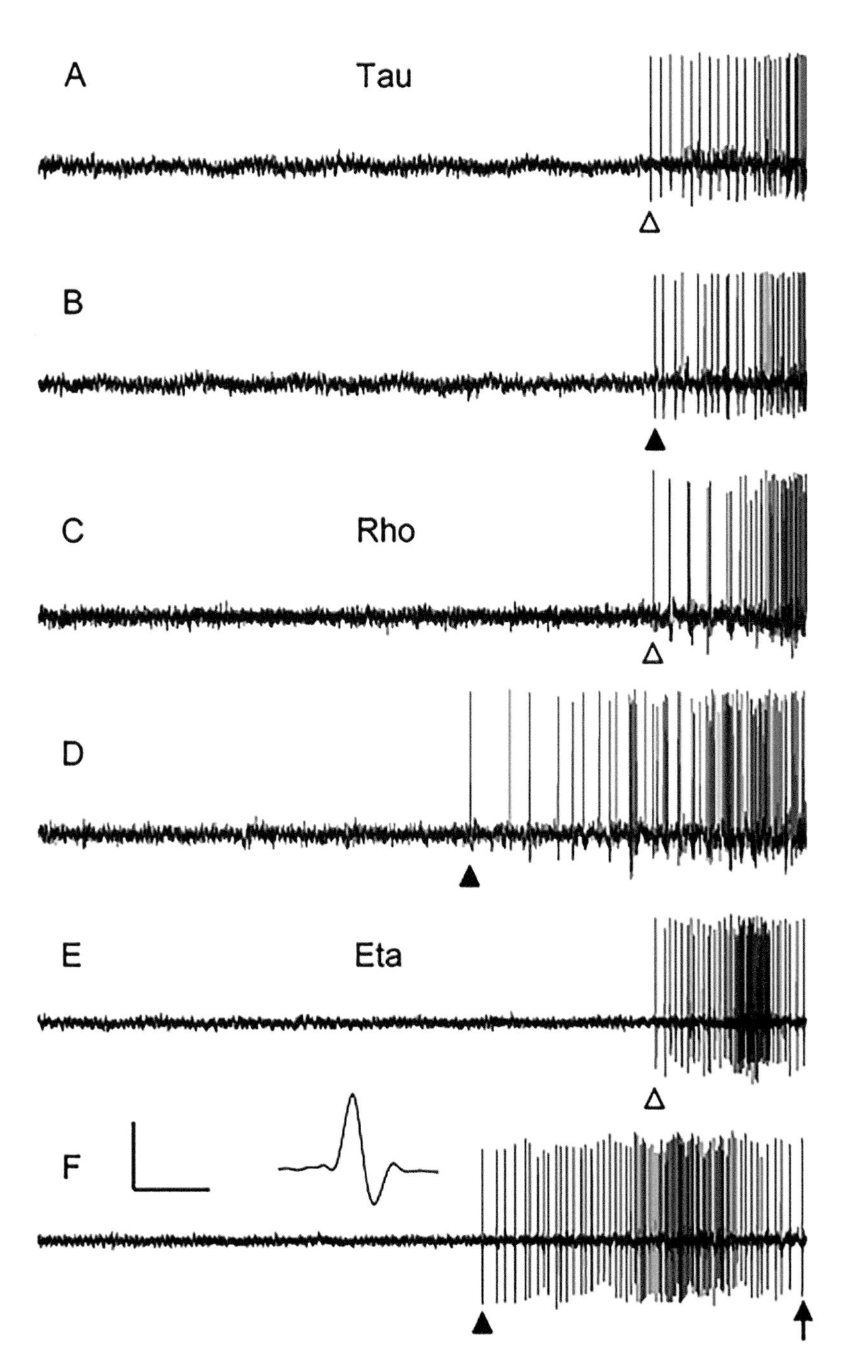

Figure 2.8 Responses of neurons in the tectum of an anesthetized pigeon to a looming object simulated on a computer screen. Vertical lines are action potentials, and each trace ends at the moment of (virtual) collision. Data from three neurons of different types (tau, rho, and eta) are shown. Each was tested with a small (A, C, E) and a large (B, D, F) object approaching with a velocity of 3 m/s. Rho and eta neurons began responding sooner to a large object than to a small one, as do frog tectal neurons. Tau neurons began responding at the same time regardless of object size, giving the pigeon a final warning to get out of the way.

Source: **From Wu, L.-Q., Niu, Y.-Q., Yang, J., & Wang, S.-R. (2005). Tectal neurons signal impending collision of looming objects in the pigeon.** ***European Journal of Neuroscience*****,** ***22*****, 2325–2331. Used with permission of John Wiley & Sons—Books; permission conveyed through Copyright Clearance Center, Inc.**

shown in a study of ground-nesting coursers and plovers in Zambia (Troscianko et al., 2016). These birds flee their nests when predators approach, leaving camouflage as their eggs' only protection. Visually, the eggs and their surroundings consisted mostly of interspersed patches of different shades of brown. The experimenters measured the contrast—that is, the difference between bright and dark areas—of both the eggs and their surroundings. They also noted which eggs had been broken open by predators, which included a mongoose and a baboon. Among the eggs that predators had not found, a common element was similarity between the contrast of the eggs and the contrast of their surroundings. This contrast-matching enabled the eggs to blend into the background so that predators did not perceive them as figure.

It strengthens camouflage if the prey doesn't just resemble the background but has borders between contrasting areas that intersect the actual outline of the animal. Cuthill et al. (2005) demonstrated this using butterfly-like targets consisting of a dead mealworm attached to a brown piece of card resembling wings. The targets were pinned to oak trees. Irregular black patches were printed on the cards, similar to the pattern on the trees; on some cards, these patches extended to the edge of the card, providing *disruptive coloration*, but on other cards, they did not.

The experimenters checked the targets over the course of 24 hours and made note of which mealworms had been eaten by birds. The mealworms attached to cards on which the pattern contours intersected the edge of the card were less likely to be eaten than those attached to cards on which the black patches did not extend to the card edge. This shows the value of disruptive coloration, which visually broke up the edge of the card and prevented birds from distinguishing the card as figure from the ground of the tree trunk.

AMODAL COMPLETION

However, the machinery of perception does not entirely ignore the ground. For example, when you look at a horse that is standing behind a tree, you perceive the horse as physically complete although part of it is blocked from your view. This process is called *amodal completion*; it shows that your visual system is hard at work filling in the "missing" parts of an object that is partially occluded and therefore part of the ground. Animals from mice to monkeys also experience this phenomenon.

AMODAL COMPLETION IN A BIRD

Amodal completion is not limited to mammals: some birds have it also. For example, Tvardíková and Fuchs (2010) showed that tits rely on amodal completion in avoiding predators. They studied several species, including the great tit, *Parus major*, counting the number of times the birds visited two widely separated feeding stations during a severe winter. Given that the feeding stations were on the ground in often snowy conditions, the distinctively colored tits had to be on the lookout for predators such as their nemesis, the Eurasian sparrowhawk *Accipiter nisus*.

The experimenters used a sparrowhawk dummy, perched at the edge of one feeding station, to see whether it would cause the tits to stay away. In fact, it did: Tits almost never visited this station. The crucial manipulation was at the other feeding station, where only part of a sparrowhawk was visible. This was either a truncated dummy consisting only of an upper half-body or a partially occluded dummy, its lower half concealed by foliage. The portion of the sparrowhawk that was visible was identical in the two cases, but their effects on the tits were very different. Tits showed some reluctance to feed near the half-hawk, but they strongly avoided the half-concealed hawk, which they apparently perceived as a complete hawk hiding in foliage. In other words, the tits showed amodal completion.

AMODAL COMPLETION IN A FISH

Even fish have now been shown to have amodal completion (Sovrano & Bissaza, 2008). Redtail splitfin (*Xenotoca eiseni*), colorful fish native to Mexico (see Figure 2.9), were trained and tested individually in a tank consisting of an inner chamber and an outer, surrounding portion.

Figure 2.9 A male redtail splitfin, *Xenotoca eiseni*. This endangered fish swims in the shallows of Mexican rivers and lakes, where its capacity for amodal completion may help it detect partly hidden objects.

Source: **Used with permission of Shutterstock. Photo Contributor, Pavaphon Supanantananont.**

During training, a fish was placed in the bare inner chamber, and worked to gain access to the surrounding chamber, a playground with conspecifics, plants, and food. Two doors connected the central chamber with the playground, and geometrical forms located below the doors signaled which door was the correct choice. It opened when pushed, while the other door remained locked.

Both patterns consisted of a red disk with a portion cut out and a black hexagon exactly corresponding to the missing portion of the disk. On one door, the red and black shapes were touching, giving human observers the impression of a disk partly covered by a hexagon; on the other door, the two shapes were slightly separated. Some fish were trained to choose the first pattern and others to choose the second pattern (see Figure 2.10).

Consider those fish who were trained to choose the door with the first pattern: What did they perceive? To answer this question during the testing phase of the experiment, the investigators dispensed with the hexagon and offered the fish a choice between a complete red disk and one with a missing portion. By two to one, they chose the complete disk, apparently indicating that during training, the missing part had been perceptually filled in by amodal completion. It is remarkable that this sophisticated, but clearly very adaptive, perceptual process is so widespread among vertebrates.

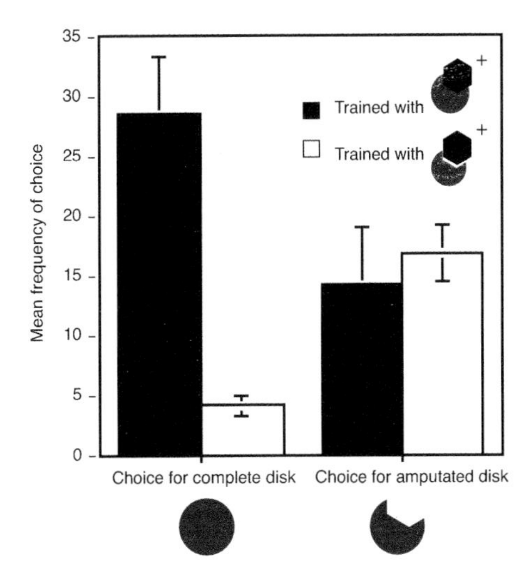

Figure 2.10 Fish were rewarded for choosing a pattern in which a partial red disk and a black hexagon were either touching or not. In a later test without the hexagon, fish trained to choose the touching shapes now selected the complete disk. This result suggests that they had experienced amodal completion during training.

Source: **From Sovrano, V. A., & Bisazza, A. (2008). Recognition of partly occluded objects by fish. *Animal Cognition*, *11*, 161–166. Used with permission of Springer Nature BV; permission conveyed through Copyright Clearance Center, Inc.**

EFFECTS OF ARTIFICIAL LIGHT AT NIGHT

We have considered light as a type of energy that enables animals to perceive objects and events in their environment. But perhaps the most universally experienced event signaled by light is the alternation of day and night. Most animals have some sort of circadian rhythm which prompts them to be active during the day, at night, or at twilight (crepuscular animals). Many also have annual rhythms corresponding to seasonal changes, including changes in the length of the day. In a stable ecosystem, the changes in light that are part of these rhythms help to regulate the interaction of many species of animals and plants. For example, turtle eggs hatch at night, when most avian predators are inactive, and many birds build nests in the spring when food is plentiful.

Humans have introduced a new factor into these equations with the invention and widespread use of artificial light, mostly at night. The

study of its effects is an active research field with a rapidly growing literature and its own acronym, ALAN, for Artificial Light at Night. In Chapter 5, we will consider ALAN's effects on navigation, especially in birds. Here, we will illustrate its effects on reproductive fitness in fish.

EFFECT OF ALAN ON FRESHWATER FISH

Anika Brüning and her colleagues (2018) at the Leibniz Institute of Freshwater Ecology and Inland Fisheries in Berlin studied the effects of ALAN on two species of freshwater fish: European perch (*Perca fluviatilis*) and roach (*Rutilus rutilus*). Fish were placed in large underwater net cages in a drainage ditch lined with streetlights, in an otherwise dark nature park. At night, fish in the experimental group were exposed to light from the streetlights, but fish in the control group were in a neighboring area with the streetlights off. The nighttime illumination of the water was some 80 times greater in the experimental group than in the control group. After a month, the reproductive statuses of the experimental and control fish were compared.

Spawning in fish is controlled by a hormonal "axis" extending from the brain to the ovaries and testes. A hormone produced in the hypothalamus descends into the pituitary gland, at the base of the brain, and induces it to secrete other hormones called gonadotropins, especially follicle-stimulating hormone and luteinizing hormone. These travel in the bloodstream to the gonads, where they stimulate the production of sex steroids including testosterone and estradiol, as well as the gametes themselves.

Brüning et al. found that both gonadotropins and sex steroids were greatly reduced in the experimental groups in both species of fish and in both sexes. It is very likely that these hormonal declines indicate a lessening of reproductive fitness and represent a threat to the population of these species where they are exposed to ALAN.

EFFECT OF ALAN ON A MARINE FISH

A different but equally dire effect of ALAN on reproduction has been discovered in a marine species, the common clownfish *Amphiprion*

Figure 2.11 *Amphiprion ocellaris*, the common clownfish, here sharing a symbiotic relationship with sea anemones. Research has shown that ALAN can interfere with this fish's reproductive success by blocking the hatching of its eggs.

***Source:* Used with permission of Shutterstock. Photo Contributor, blue-sea.cz.**

ocellaris (see Figure 2.11). It might seem that this reef-dwelling fish would be far enough from land to be spared any effects of ALAN, but substantial light pollution from coastal cities and developments, as well as from ships and offshore structures, reaches their habitat.

To determine the effects of disrupting the light/dark cycle of day and night, Emily Fobert and her colleagues (2019) carried out an experiment on mating pairs of clownfish in their laboratory at Australia's Flinders University. All fish received daylight-level illumination for 12 hours of the day. During nighttime hours, an experimental group received ALAN at a moderate intensity comparable to that present at an offshore reef, while control pairs were in darkness. The experiment continued for 60 days, during which spawning, egg laying, and hatching were recorded.

Both groups spawned with the same frequency, and fertilization success (determined from the appearance of the eggs) was also comparable in the two groups. However, there was a remarkable difference between groups in the likelihood of eggs hatching. In the control group, more than 80% of eggs hatched, but in the experimental group, none of them did.

A possible explanation for this result is that, in many species of reef fish, eggs hatch during the evening. The adaptive value of this behavior is that it serves to hide new hatchlings from predators. But if hatching is triggered by light offset, and the lights never go off because of ALAN, no eggs will hatch.

Taken in combination, these two studies indicate that ALAN can affect reproductive success in a variety of fish species and by way of different mechanisms. The strength of the observed effects suggests that the threat to fish populations, especially freshwater and coastal species, is serious. And as we will see in Chapter 5, ecological disruption by light pollution is not limited to its effects on fish or even on vertebrates. Its broader implications have been thoughtfully explored by Eklöf (2020).

SUMMARY

In this chapter, we have seen that the vertebrate visual system shows considerable evolutionary stability. Despite innumerable specializations, all vertebrate eyes from fish to primates are constructed on the same basic plan: ocular media that focus light onto a retina, where visual receptors absorb it and send neural signals to the brain. These signals contain information about the location and features of objects in the environment. Neural processing extracts these features and uses them to construct a meaningful representation of the animal's visual world, which is used in the selection and guidance of behavioral responses. New brain structures such as the mammalian visual cortex have emerged over the course of evolution, yielding increasingly elaborate signal analysis, but many visual phenomena experienced by humans, such as figure/ground perception, are demonstrable in other vertebrates as well, suggesting that some processing algorithms are strongly conserved.

The ways in which visual information is used are, of course, as diverse as vertebrates themselves, for vision is woven into every aspect of life. More examples will be given in later chapters, particularly Chapter 4, on color vision, and Chapter 5, on visual space perception.

First, however, we turn to invertebrates. What are the visual systems of arthropods and molluscs like, and how do these highly successful animals make use of the information their eyes and brains acquire? We will find that there are great differences between their vision and our own, but also surprising similarities.

Vision in Invertebrates

In this chapter, we will examine and compare the visual systems of two invertebrate phyla: Arthropoda and Mollusca. Their eyes and brains are very different from ours, and from each other's. But despite obvious differences in structure, there are surprising commonalities across phyla, such as the occurrence in all three groups of lateral inhibition, a neural process that increases sensitivity to edges in the visual field. Other shared characteristics—color vision (present in vertebrates and arthropods) and sensitivity to the polarization of light (present in arthropods, molluscs, and some vertebrates)—will be taken up in Chapters 4 and 5, respectively.

ARTHROPODS

We turn first to the arthropods, animals with an exoskeleton. Phylum Arthropoda is highly diversified and successful. As with vertebrates, there are arthropods that live on land, others that dwell underwater, and others that have mastered flight. There are vastly more arthropod species (certainly more than a million, although the actual total is unknown because many species no doubt remain to be discovered) than vertebrate or mollusc species. Why have arthropods been able to thrive since the beginning of the Cambrian Period, or perhaps even before? The answer has to do partly with their exoskeletal armor, which protects them from other animals; partly with their capacity for a modicum of learning, while relying primarily on a wide array of innate, stereotyped, and quickly activated behaviors; and partly with the short life cycle of many arthropods, which allows them to evolve quickly in response to changing circumstances. But another factor is their very effective eyesight: Their vision is not as keen as ours but is linked in surprising ways to specific behaviors.

SIMPLE AND COMPOUND EYES

Arthropods have eyes very different from ours. The eyes of vertebrates are extremely complex but are technically, and confusingly, called *simple eyes*. The term refers to the fact that the vertebrate eye is a single chamber, with just one retina, just one lens, etc. In contrast, most arthropods have *compound eyes*. Each compound eye is composed of hundreds or thousands of individual facets called *ommatidia*. An ommatidium is a functioning unit with its own optical system, consisting of a miniature cornea and underlying focusing element, the *crystalline cone*. The eye as a whole is convex, with the corneas of the individual ommatidia constituting tiny bumps on its surface.

The visual systems of only a small fraction of arthropod species have been studied in detail, but in those that have been examined, the general structure of compound eyes outlined above applies. Differences between compound eyes have to do partly with the number and types of receptor cells in an ommatidium. For example, in arthropods with color vision—the subject of the next chapter—different receptors in the same ommatidium contain different visual pigments.

DOI: 10.1201/9781003362319-3

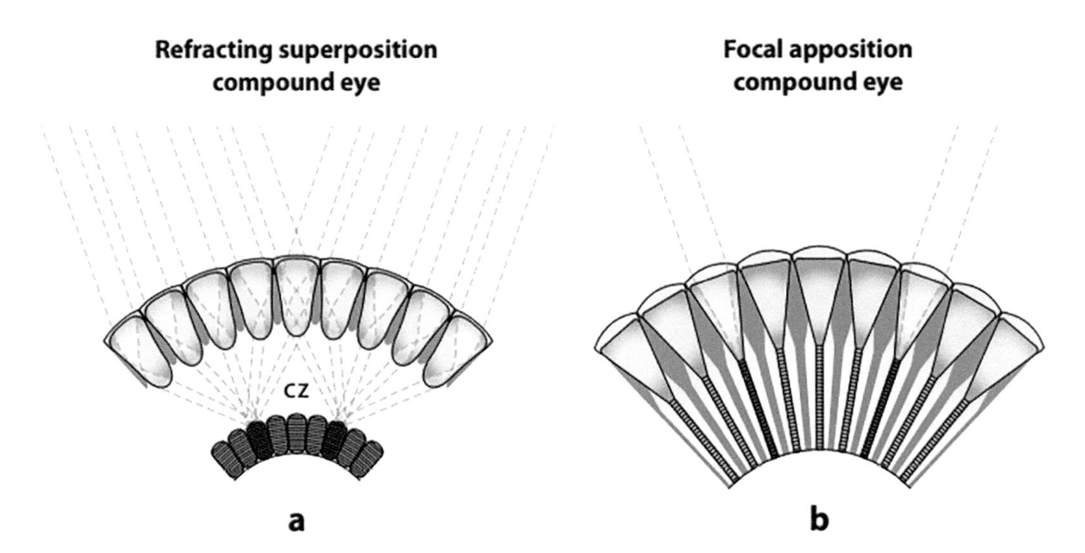

Figure 3.1 The two main types of compound eye. In an apposition compound eye (right), individual ommatidia are optically isolated from one another; their crystalline cones are in close proximity to receptor cells (brown). Superposition compound eyes (left) are found in many nocturnal insects because they use light more efficiently. A clear zone (cz) allows light to pass between ommatidia before reaching the receptor cells. The crystalline cones refract the light to form an image on the receptor cell layer.

Source: From Warrant, E., & Dacke, M. (2011). Vision and visual navigation in nocturnal insects. _Annual Review of Entomology_, _56_, 239–254. Diagrams courtesy of Dan-Eric Nilsson. Used with permission of Annual Reviews, Inc.; permission conveyed through Copyright Clearance Center, Inc.

INSECTS

Types of Compound Eyes

In the compound eyes of most diurnal insects, such as the bee, the receptor cells lie close below the corneas and crystalline cones, so that the light entering an ommatidium is largely absorbed by the receptors within that ommatidium. Eyes of this kind are called *apposition compound eyes* (see Figure 3.1). With each ommatidium receiving light from only a small region of the visual field, the bee is presented with what is essentially a mosaic. Each ommatidium channels the light it captures into a long translucent tube called a *rhabdom*, which is surrounded by receptor cells.

Each receptor cell in an ommatidium sends a *rhabdomere*—an array of microvilli—into the rhabdom. Each microvillus is packed with molecules of photopigment. The photopigment molecules within a rhabdomere are lined up with one another and locked in place; this is different from the arrangement in the rods and cones of vertebrates, where pigment molecules are free to spin within the discs that house them. As we will see in Chapter 5, this difference has implications for the ability to perceive the polarization of light.

When a receptor's visual pigment absorbs light, the cell sends signals along a fiber that exits the ommatidium. Signals from different receptor cells interact in a plexus, deep inside the eye, where connections are made between fibers.

A different structural arrangement exists in many nocturnal insects, such as the elephant hawk moth *Deilephila elpenor* (see Figure 3.2). In its *superposition compound eyes*, a large transparent gap exists between the superficial optics and the receptors, allowing light to pass from one ommatidium to another. The crystalline cones refract the light before it traverses this clear zone, forming an image at the level of the receptor cells. This arrangement enhances sensitivity by enabling receptors in one part of the array to capture photons that entered the eye at an angle through other ommatidia. The elephant hawk moth's sensitivity is so great that it has color vision even at night (Warrant & Dacke, 2011).

Figure 3.2 An elephant hawk moth, *Deilephila elpenor*. Its superposition compound eyes have such high sensitivity that the moth can see in color, helping it locate nectar-bearing flowers and conspecifics at night.

Source: **Used with permission of Shutterstock. Photo Contributor, HWall.**

Figure 3.3 A honeybee. Behind and slightly above its apposition compound eye is a small, round ocellus.

Source: **Used with permission of Shutterstock. Photo Contributor, Russell Marshall.**

Although compound eyes are the hallmark of arthropods, many have crude, simple eyes, called *ocelli*, in addition. In bees, for example, there are three ocelli on top of the head (see Figure 3.3). With high sensitivity but poor resolution, these ocelli have been shown to help bumblebees (*Bombus terricola*) navigate at dusk (Wellington, 1974). Larval insects have ocelli instead of compound eyes. An example is the larva of the sawfly, which can see well enough with its ocelli to crawl toward food and to spit at potential predators (Meyer-Rochow, 1974).

Visual Acuity

Even with their compound eyes, though, insects have much lower acuity than vertebrates. Average values are not of much use here, because acuity has only been measured in a handful of species, and it varies considerably from one species to another and as a function of testing conditions.

How can an animal's visual acuity be measured? The most common method involves gratings. A *grating* is a pattern composed of alternating light and dark stripes. Usually, all the stripes in a grating are of the same width. If the stripes are thick they are easy to see, but as the experimenter makes them thinner and thinner, they get harder to see. The stripes also get harder to see if the grating is placed farther from the animal.

Researchers usually report stripe thickness in terms of *visual angle*, the angle that the stripe subtends at the eye. Expressing the dimensions of a stimulus in degrees of visual angle has the advantage that it takes into account both the physical size of the stimulus and its distance from the eye. For example, the full moon, viewed from Earth, subtends a visual angle of 0.5°.

In a behavioral experiment, an animal can be presented with a choice between a grating and a uniform stimulus of the same average luminance and reinforced with a food reward for approaching the grating rather than the uniform stimulus. By learning this task, the animal shows that it can resolve the stripes of the grating. (Alternatively, it can be trained to discriminate between horizontal and vertical gratings.) It can then be tested with finer and finer stripes, until its performance falls to a chance level. This transition point defines the animal's visual acuity.

To give an example, honeybees can resolve gratings composed of stripes that are at least 2° wide (Srinivasan & Lehrer, 1988). In comparison, domestic cats can resolve stripes that are only .1° wide (Blake et al., 1974), and humans can do so down to a mere .02°. That is, a person can make out details 100 times smaller than those a bee can see.

Perceiving Shapes

There is more to vision than acuity. Can a bee perceive and recognize everyday objects based on their shape? A carefully designed experiment was needed to answer this question, because objects in the bee's Umwelt differ in many ways other than shape, including color and smell, on which a bee might base its responses.

To measure the effect of shape alone, Howard et al. (2021) gave bees the opportunity to choose between pictures of different flowers. The pictures were in black and white, eliminating color as a factor, and had no telltale floral aroma. The flowers were either species that birds pollinate or others that are pollinated exclusively by insects; on every trial, a bee was presented with two pictures, one of each type. Testing was carried out in a round arena with sloping sides, against which the photographs were propped up side by side; bees indicated their choice by climbing up the side of the arena and onto one of the pictures.

Remarkably, the bees consistently chose pictures of flowers that insects pollinate. The participants, members of the genus *Lasioglossum* that live on their own rather than in communities, were captured as adults and inducted into the study, so familiarity with flowers may have been the basis for their choices; inborn preferences may also have been a factor. But in either case, the results prove that bees can discriminate objects (flowers in this case) on the basis of their shape.

Honeybees can even discriminate between different human faces. Avarguès-Weber et al. (2010) allowed bees, tested individually, to fly toward a screen on which pictures of faces were posted. One face was designated the correct choice for some bees, and the other face for other bees. A small landing platform beneath each face provided either a drop of sugar water (for the rewarded face) or quinine water (for the non-rewarded face). The bees quickly learned this discrimination.

The researchers then asked whether bees who are trying to discriminate complex stimuli such as faces don't just attend to a particular feature such as the nose, but to several features at once. To answer this question, Avarguès-Weber and colleagues presented the bees with altered versions of the original photographs in which the hair and ears had been removed; and in a third test, hair and ears were restored but eyes, nose, and mouth were deleted. In both of these tests, the bees performed above chance, but not as well as they had with complete faces. This shows that they were basing their decisions on more than one feature.

But was it the mere presence of multiple features, or their configuration, that was important? To find out, the experimenters presented jumbled versions of the original pictures in which the parts of each picture were rearranged. Now, the performance of the bees fell to a chance level, indicating that they engage in configural learning.

For social insects, learning to discriminate the faces of different people would not be a very useful skill. But learning to recognize one another surely would be. To see whether such recognition actually occurs, Tibbetts (2002) studied paper wasps (*Polistes fuscatus*), who have distinctive yellow markings on their faces that vary from individual to individual (see Figure 3.4). The investigator applied tiny amounts of yellow or black paint to the faces of selected individuals in such a way that it either did or did not alter their facial appearance. She then returned the painted wasps to their nests and observed the results.

When a wasp's appearance had been altered, her nestmates lunged at her and bit her,

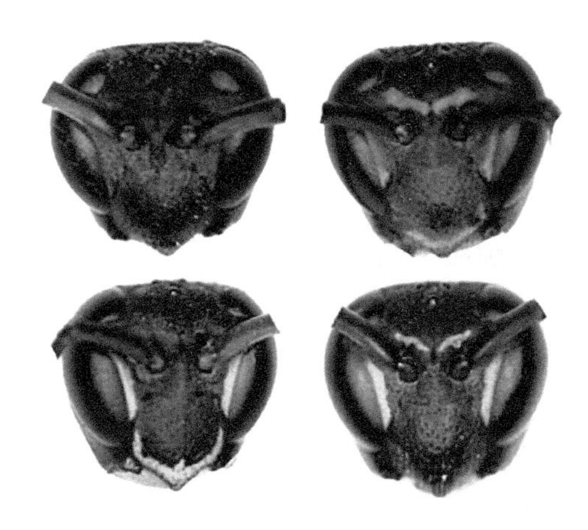

Figure 3.4 Faces of four paper wasps, showing individual variability in appearance.

Source: **From Tibbetts, E. A. (2002). Visual signals of individual identity in the wasp *Polistes fuscatus*. *Proceedings of the Royal Society of London B*, *269*, 1423–1428. Used with permission of The Royal Society (U.K.); permission conveyed through Copyright Clearance Center, Inc.**

aggressive responses indicating nonrecognition. Aggression declined as familiarity increased, suggesting that other wasps were learning to recognize her. Control wasps, painted in ways that did not alter their appearance, received much less aggression. In neither case was the aggression of the murderous type meted out to intruders. This suggests that the painted wasps were accepted as members of the nest, although not recognized as individuals (Tibbets, 2002).

Color vision and the perception of space are other important components of insect vision. These topics will be covered in Chapters 4 and 5.

SPIDERS

Spiders and other arachnids have simple rather than compound eyes, which in many cases are too large and complex to be called ocelli. Spiders are carnivores, and many rely on acute vision as they hunt for their next meal (see Figure 3.5). The most studied are jumping spiders, members of the family Salticidae. Most have four pairs of eyes: a pair of principal eyes that are directed straight ahead, and three pairs of smaller, secondary eyes that are primarily used to detect movement.

In a classic study of the red-backed jumping spider, *Phidippus johnsoni*, Land (1969a) found that its principal eyes are tubular in shape and that their retinas have four layers of receptors. These tiers ensure that some receptors will receive a focused retinal image regardless of the viewed object's distance from the eye (Blest et al., 1981). On anatomical grounds, it is estimated that with these large eyes (nearly a mm from front to back), *P. johnsoni* has an acuity about one-tenth that of a person, which is unusually good for an arthropod; and some jumping spiders have even better acuity (Cerveira et al., 2021). An unusual feature of *Phidippus* is that its retina can move about while the front of the eye remains stationary, allowing the spider to direct its keenest vision to objects of interest (Land, 1969b).

Figure 3.5 This red-backed jumping spider, *Phidippus johnsoni*, has used its keen vision to capture a katydid.

***Source:* Used with permission of Shutterstock. Photo Contributor, Ernie Cooper.**

Their good vision enables jumping spiders to adjust their predatory strategy depending on the nature of their prey. For example, *Phidippus* hunts caterpillars and flies in different ways (Jackson & Pollard, 1996). It approaches a caterpillar slowly, circling to face it head on, and creeping close before leaping onto the prey and pinning its head to the ground. In stalking a fly, in contrast, it makes haste to capture the prey before it can escape, jumping on it from a greater distance and from any angle and tearing at its thorax to disable its wings.

Experiments with another jumping spider, *Yllenus arenarius*, show just what visual cues it uses (Bartos & Minias, 2016). The researchers displayed simplified pictures of insects on a tiny screen in front of the spider and studied its response. *Yllenus* struck at what it perceived to be the head of the prey, based on features such as eyes and antennas. But it also responded to the prey's direction of movement in judging the location of its head. When these two cues were in conflict, the spider's behavior was guided by the stronger cue, for example striking at the trailing end when it had salient headlike features. This study is important for demonstrating the potential usefulness of the "false-head" strategy used by many insects to confuse predators.

HORSESHOE CRAB

The compound eyes of some arthropods are so large that individual ommatidia are visible to the naked human eye. This is the case with the Atlantic horseshoe crab, *Limulus polyphemus*. This lethargic denizen of the seashore is not a true crustacean; it is on a separate, small branch

of the arthropod family tree and has remained relatively unchanged since Jurassic times.

Their numbers are dwindling now, for a variety of reasons. One of these is that their blue blood is used in tests for the presence of bacterial toxins in, for example, medicines intended for injection. The crabs are bled in large numbers, and some die in the process. Fortunately, a way of producing their blood's key ingredient by means of recombinant DNA has been developed and is increasingly being adopted (Zhang, 2018).

The large size of *Limulus's* ommatidia and the thickness of the axons that lead from them made it an ideal choice for physiological recording in the middle decades of the twentieth century. Each ommatidium gives rise to a single optic nerve fiber, and these fibers interact in a plexus deep within the eye. H. Keffer Hartline, who won a Nobel Prize in 1967 for his work with *Limulus*, showed (in collaboration with Floyd Ratliff) how these inhibitory interactions between neighboring ommatidia enhance the neural response to edges in the horseshoe crab's field of view (Ratliff, 1965). In doing so, they confirmed the mathematical insight of Ernst Mach (1838–1916) that *lateral inhibition* underlies the perceptual phenomenon of enhanced edges—Mach bands—in humans.

To understand how lateral inhibition enhances perception of edges, consider an experiment of the type carried out in Limulus by Hartline and Ratliff (1957). First, light is shined on an individual ommatidium while the rest of the eye remains in darkness. Recordings from this ommatidium's axon shows that its firing rate increases when the light is made more intense; a bright light produces a firing rate of about 30 action potentials per second. Next, the experiment is repeated with a second ommatidium that is near the first one, with comparable results. A dim light is chosen that elicits about 10 action potentials per second from this unit. Because only one ommatidium at a time is illuminated, lateral inhibition is not affecting the results.

Now consider what happens when both ommatidia are illuminated at once, the first with a bright light and the second with a dim light. The two ommatidia now inhibit each other. Since an ommatidium's ability to inhibit its neighbors is proportional to the light falling on it, the first ommatidium inhibits the second more than the second inhibits the first. So inhibition will cause the firing rate of the second unit to decrease a lot, say from 10 to 7, while the firing rate of the first unit will decrease only a little, say from 30 to 29. Lateral inhibition has caused the difference in firing rate between the two units to increase from 20 (i.e., 30–10) to 22 (29–7). We may imagine that for Limulus, the perceived difference in brightness between these two spots in the visual field has been enhanced.

Of course the situation is more complicated when the outline of a complex visual stimulus, such as an approaching predator, affects the illumination falling on a large number of ommatidia, but the rules are the same.

Demonstration of this parallel between physiology and perception advanced the field of sensory neuroscience by inspiring researchers to look for other correspondences between neural and psychophysical measures.

THE ARTHROPOD BRAIN

The early ancestors of today's vertebrates, arthropods, and molluscs had central nervous systems consisting of nerve cords running the length of the body. Clumps of neuronal cell bodies, called *ganglia*, were located at intervals along the cords. Here, sensory impulses from the periphery arrived, interneurons combined or compared them with one another, and motor impulses were sent to the periphery, triggering behavioral responses.

As evolution proceeded, the ganglia underwent changes, sometimes merging with one another. Often, the process resulted in a large complex of neural tissue, a *brain*, at the front end of the animal, its "tip of the spear" in interacting with the environment.

This concentration of neurons in the brain has been especially pronounced in vertebrates, for whom the brain is the sole locus of perception and cognition. In arthropods, the cephalization of neural tissue has not been so extreme. Their brains are larger and more important than their other ganglia but are not the sole locus of decision-making and action control. Cockroaches, for example, are able to survive and move about for days without their head (Choi, 2007).

Although there are differences among the brains of insects, arachnids, and crustaceans, the similarities are great enough that we can take the insect brain as instructive regarding Phylum Arthropoda as a whole. The three major components of the insect brain are, from front to back, the *protocerebrum, deutocerebrum, and tritocerebrum*. Of these, the protocerebrum is the main region that receives and processes information from the eyes. Two large, lateral extensions of the protocerebrum, the *optic lobes*, contain initial waystations for signals traveling from the eyes toward the central brain.

Angelique Paulk and her colleagues have studied these brain areas in bumblebees (*Bombus terrestris*), who make good research subjects because they have large brains and mild dispositions. The investigators recorded from neurons in both the optic lobes and the protocerebrum, determining the characteristics of receptive fields in both brain regions. At each level, receptive fields are arranged in one or more maps of the visual field, as they are in visual areas of the vertebrate brain. And as in vertebrates, the way information is organized changes as it moves from one level to the next.

At the first waystation in the optic lobe, cells have tiny receptive fields because they respond to signals from individual receptors. As neural activity travels medially through the optic lobe, however, signals from different receptors are compared with one another, receptive fields enlarge, and a gradual separation of color information and movement information occurs (Paulk et al., 2008, 2009a). Color information then travels primarily to the anterior part of the central protocerebrum, while movement information is sent to its posterior portion (Paulk et al., 2009b); both regions receive information about spatial features such as edges. Finally, pathways from the posterior region pass to motor centers, to help control the bumblebee's flight and other movements.

In contrast, much of the information from the anterior portion is sent to the *mushroom bodies*, specialized protocerebral structures known to be involved in both visual and olfactory learning (Dyer et al., 2011; Erber et al., 1980). Circuitry here is crucial to the bee's ability to remember which flowers offer the most nectar and pollen. It makes use of long-term potentiation (LTP), an activity-dependent modification of synaptic strength (Menzel & Manz, 2005) that also contributes to memory formation in vertebrates and molluscs. The mushroom bodies are probably also involved in the ability, described earlier, of bees to learn to discriminate between different human faces, which for them are arbitrary patterns (Avarguès-Weber et al., 2010), and of wasps to recognize one another by their appearance (Tibbetts, 2002).

In summary, the organizational and behavioral complexity of the visual system of the bee is remarkable, given its dimensions; and the functional similarities between the bee's visual system and our own, such as the late-stage separation of color and motion information, are a compelling example of convergent evolution. We will return to the subject of the color vision of bees in Chapter 4.

MOLLUSCS

EVOLUTION OF MOLLUSC EYES

Molluscs, like vertebrates, have simple eyes. But while the basic structure of the vertebrate eye is similar across all classes—fish, amphibians, reptiles, birds, and mammals—the eyes of molluscs are more diverse in their structure. It is always problematic to say that a sense organ in one animal is more "advanced" than the corresponding sense organ in another, since in each case the organ has helped the species to survive and is therefore well suited to it. Nevertheless, some mollusc eyes are clearly more elaborate and refined than others (see Figure 3.6). Each of these refinements may have occurred many times over the course of evolution, on different branches of the molluscan family tree.

EYECUPS

In order to qualify as an eye, a light-sensitive structure must capture some information about the location of objects in the environment, i.e., the direction from which rays of light are reaching the eye. An early stage in the evolutionary development of eyes occurs in some molluscs (as well as

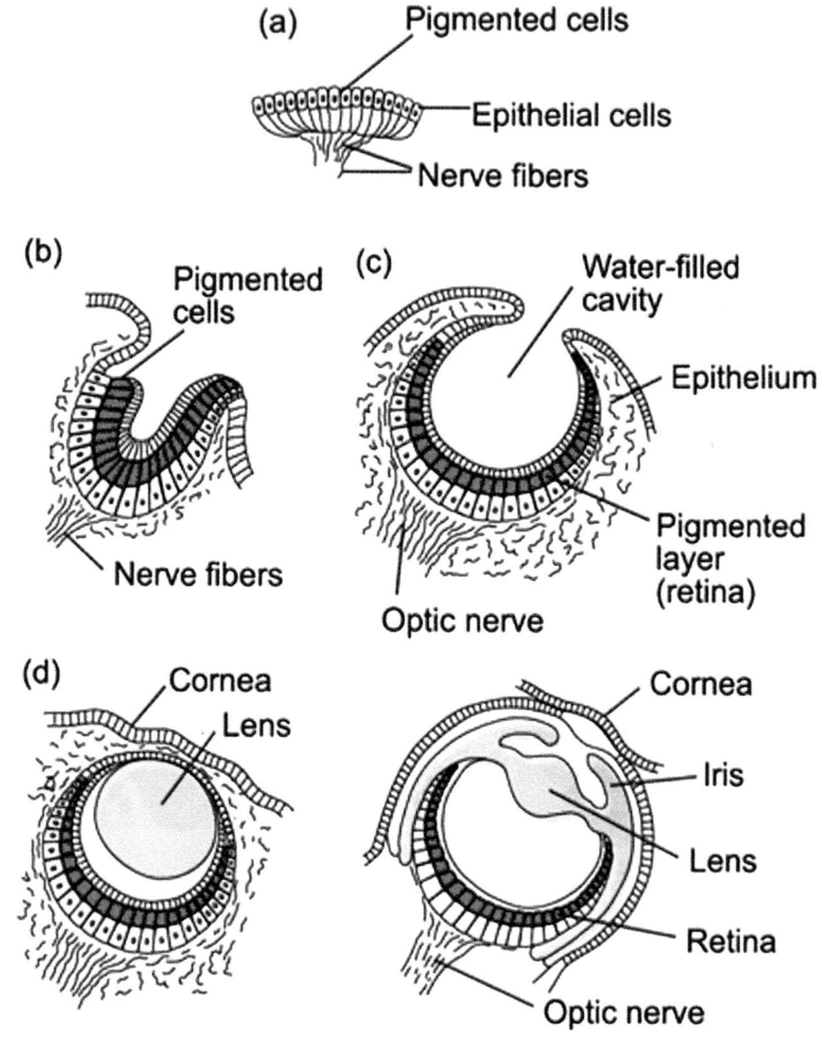

Figure 3.6 Presumed stages in the evolution of the eye in molluscs. Primitive molluscs may have had flat, light-sensitive eyespots (A). From these evolved eyecups (B), which captured some directional information. Increasing curvature gave rise to pinhole eyes, as in the nautilus (C). The next stage saw the appearance of a simple cornea and lens, as in some gastropods (D). The eyes of advanced cephalopods—octopus, squid, and cuttlefish—show additional refinements, such as the ability to accommodate.

Source: **Retrieved from https://en.wikipedia.org/wiki/File:Stages_in_the_evolution_of_the_eye.png. This work has been released into the public domain by its author, Remember the dot at English Wikipedia.**

members of other phyla such as flatworms and annelids) in whom eyespots are invaginated to form *eyecups*. In an eyecup, light from each region of the animal's environment falls on the opposite wall of the cup's inner lining, which we can call the retina. The abalone, a sea snail (genus *Haliotis*), has a pair of eyecups on stalks, which it moves about to capture visual information. In some species, the rim of this gastropod's eyecup is turned slightly inward, narrowing the opening through which light can enter (Land, 1984). Unfortunately for the abalone, their flesh is considered a delicacy by humans, and the iridescent inner surface, or nacre, of their shells is used to make jewelry.

Overharvesting and climate change are together threatening their long-term survival.

PINHOLE EYE

A major advance over the eyecup is found in the chambered nautilus (*Nautilus pompilius*). This mollusc lives in a coiled shell and might at first be mistaken for a snail; however, it is not a gastropod, but rather, a cephalopod, and thus related to octopus, squid, and cuttlefish. A chambered nautilus can live for many years, reaching a size of 20 cm or so. Like other molluscs whose shells are lined with nacre, the nautilus is extensively harvested and concerns for its future are growing.

The nautilus is called chambered because it only lives in a portion, or chamber, of its shell. As the animal grows, its shell grows with it, and previous chambers are sealed off, one by one.

The nautilus's eye is called a pinhole eye. It is like an eyecup in which the lip of the opening has turned farther inward, so that the aperture through which light can enter is small enough to be called a pupil (see Figure 3.7). This has the effect of greatly decreasing the blur of the retinal image, since from each point in the environment, only a small bundle of rays can pass through the aperture and illuminate a small region of retina. The result is something that can be reasonably described as a retinal image of the animal's visual field, thus making crude spatial vision possible.

Figure 3.7 A chambered nautilus, *Nautilus pompilius*. Note the pinhole eye. This mollusc is a cephalopod, but with a less complex nervous system than octopus, squid, or cuttlefish.

***Source:* Used with permission of Shutterstock. Photo Contributor, Teguh Tirtaputra.**

However, the pinhole eye has two serious drawbacks. One is that there is no covering—no cornea—on the front of the eye, so that seawater is what fills the eye. The second problem is that the pinhole permits only a small amount of light into the eye, and therefore, the retinal image is dim.

EYES WITH A CORNEA

These problems are avoided in the eyes of many snails, including the marsh periwinkle, *Littorina irrorata*. This gastropod lives in the intertidal zone, the stretch of shoreline that is dry at low tide but underwater at high tide. When the tide is out, the periwinkle crawls around finding food. When the tide comes in, bringing benthic predators (especially its nemesis, the Florida crown conch *Melongena corona*) with it, *L. irrorata* avoids them by climbing up plant stems (Hamilton, 1976).

Hamilton et al. (1983) carried out a detailed anatomical study of the periwinkle's eye, which is about 270 microns in diameter. Immediately behind the tiny cornea is an immobile and nearly spherical lens. Optical calculations showed that when the snail is in air, its retinal images are in focus, whereas when it is underwater they are blurred; in other words, the snail can see more clearly in air, where it usually is, than when it is briefly underwater as the tide comes in. But because the eye is small, the retinal image of an object such as another snail is likewise small and spread over only a few visual receptor cells. Visual details cannot be resolved unless they are imaged on separate receptors; so the dimensions of the receptor mosaic, rather than optical blur, are what set the limit on the snail's visual acuity.

Hamilton and Winter (1982) tested the periwinkle's visual behavior in the laboratory by placing it in the center of a circular arena. A target, consisting of a dark pattern, was placed against the white wall of the arena. The experimenter viewed the snail's shadow from below, through the translucent floor of the arena, to see what it would do. In general, snails moved toward any targets they could see. For example, they consistently crawled toward a black vertical stripe, regardless of its thickness (expressed in degrees of visual angle subtended at the center

of the arena). But the thinnest stripe presented by the researchers was 0.9°, so we know only that the snail's resolution limit is at least this small; they might have been able to see even thinner stripes. When the experimenters used horizontal instead of vertical stripes, the snails did worse. This difference in performance may reflect the importance of locating a plant stem if a hungry conch is approaching.

Gastropods vary widely in the ability of their eyes to form a retinal image and thus in their visual acuity. In general, terrestrial gastropods have poorer acuity than aquatic ones and use their vision mainly to avoid brightly lit areas where they could be seen by predators (Hamilton & Winter, 1984; Zieger & Meyer-Rochow, 2008).

MOLLUSCAN VISUAL RECEPTORS

Although molluscs don't have ommatidia, the visual receptors in most species, such as the octopus, are similar to the rhabdomeric receptors of insects and other arthropods. This similarity between mollusc and arthropod receptors suggests that the first cells with rhabdomeres date from before the branching that led to these two phyla. In fact, it was once held that animals with rhabdomeric receptors and those (like us) with receptors formed from specialized cilia represent two major, and largely distinct, evolutionary lines (Eakin, 1979). But this view has difficulty with the fact that both types of receptors are represented in molluscs. For while rhabdomeric receptors are the predominant form, some species, especially among the gastropods (snails and slugs) and bivalves, have ciliated receptors as well.

A striking example is provided by *Pecten*, the scallop, in which both types of receptors are found not just in the same animal, but in the same eye. This bivalve has something like 100 *non-cephalic* eyes, located not in the "head" region but scattered along the edges of the mantle. These are unique *mirror eyes* in which the retinal image is formed by light reflected by the shiny, concave tapetum behind the retina, and brought to a focus as it passes forward through the receptors (Land, 1965). There are two layers of receptors: a front layer of cells consisting of modified cilia and a rear layer of rhabdomeric cells. Hartline (1938) recorded from the optic nerve fibers emerging from both layers. He found that while the rhabdomeric receptors respond to the onset of light, cilia-based receptors are inhibited by light and respond when it goes off, a property they share with vertebrate photoreceptors.

A behavioral consequence of this physiological difference has been found by studying molluscs such as *Helix pomatia*, the Burgundy snail, in which the two types of receptors are found in different eyes. Helix's cephalic eyes contain rhabdomeric receptors, while its non-cephalic eyes have cilia-based receptors. When a moving shadow, such as might be cast by an approaching predator, falls on the cephalic eyes, it evokes no response, but when the shadow reaches the non-cephalic eyes, triggering off-responses, the animal withdraws into its shell (Land, 1984).

Regardless of the receptor type, visual pigment dynamics are different in molluscs than in vertebrates (Hamdorf, 1979). In the vertebrate eye, when a molecule of rhodopsin absorbs a photon, it changes briefly into an intermediate form called *metarhodopsin*, and then proceeds to break apart into all-trans retinal and opsin. In other words, the chromophore completely separates from the opsin, and the pigment bleaches, losing its color. In contrast, bleaching does not occur in the retina of molluscs. When a rhodopsin molecule absorbs a photon, it transitions to metarhodopsin and changes color slightly, but the chromophore and the opsin do not separate. It remains metarhodopsin until it absorbs another photon and transitions back to rhodopsin! This absorption of light by metarhodopsin does not produce a visual signal.

This molluscan method of replenishing the supply of rhodopsin has the advantage that the retina does not need to continuously acquire new molecules of 11-cis retinal, as the vertebrate retina does. But it has the compensating disadvantage that when the two pigments are mixed together, metarhodopsin absorbs some light that would otherwise be captured by rhodopsin and thus contribute to vision. The result is that visual sensitivity is reduced.

VISION IN THE OCTOPUS

Cephalopoda is a class of molluscs that includes not only nautilus but also octopus, squid, and cuttlefish. The latter three groups are the subject of intense research because of their intelligence and behavioral complexity. Their eyes, also, are highly developed and surprisingly similar to our own, at least in superficial ways. Here, we will consider the octopus eye.

STRUCTURE OF THE EYE

The common octopus, *Octopus vulgaris*, has an eye that is round but somewhat smaller than the human eye. It has a cornea, pupil, lens, and retina. Although functionally equivalent to the analogous structures in the vertebrate eye, they are different in important ways (Hanke & Kelber, 2020). For example, whereas the vertebrate cornea seals the eye off from the environment, the octopus cornea has a dorsal opening that allows sea water to enter and fill the space between cornea and lens—a vestige, perhaps, of the openness of the nautilus's pinhole eye. Like the lenses of fish, the octopus lens is nearly spherical and accommodates for near vision by moving forward rather than by changing shape. The octopus pupil constricts in bright light to a horizontal rectangle, something like that of a horse, but with more abrupt corners (see Figure 3.8).

The biggest differences between the octopus eye and that of vertebrates, however, are in the structure of the retina. In the vertebrate retina, it will be recalled, the receptors are at the back; other cell types, such as the horizontal and bipolar cells, are in front of them, so light must pass through these to reach the receptors, and electrical signals must pass forward from the receptors to reach the ganglion cells and their axons, the fibers of the optic nerve. The vertebrate retina, in other words, is *inverted*. In the octopus eye, in contrast, the receptors are in a more intuitive location: on the front. Moreover, the octopus's visual receptors have their own axons which pass out the back of the retina to form the optic nerve.

Each arrangement has its advantages. In the octopus, incoming light is not blurred by passing through other retinal structures before reaching the receptors. However, in the vertebrate retina, the receptors are in immediate contact with the pigment epithelium, a nutritive layer that helps the rods and cones maintain a high metabolic level that enables them to respond very quickly to light. This would not be possible if the retina were not inverted, for then the opaque pigment epithelium would be in front of the receptors, blocking the incoming light.

Given its overall similarity to the vertebrate eye, one of the most remarkable things about the octopus eye is that the visual receptors resemble those of arthropods more closely than they resemble ours. An octopus receptor has rhabdomeres—arrays of microvilli—extending from the cell body in two opposite directions. When

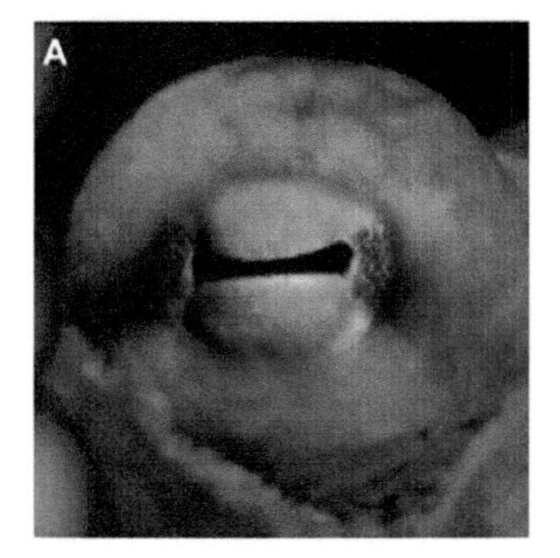

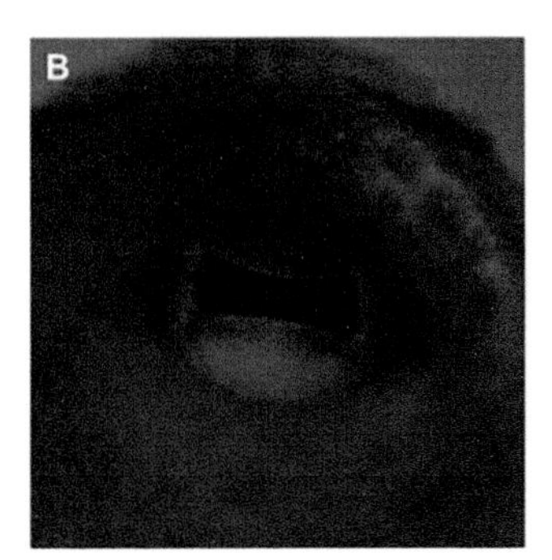

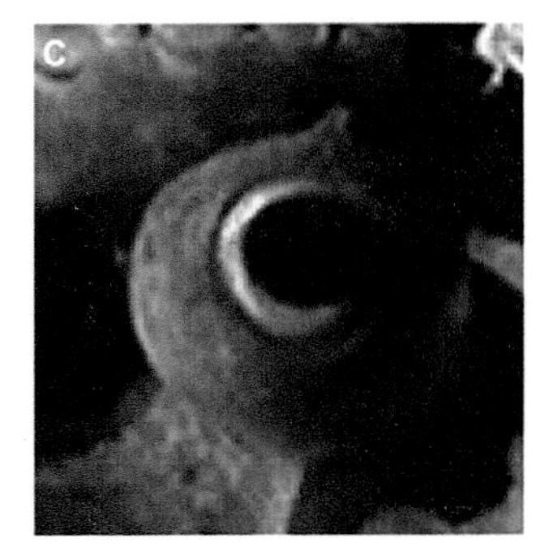

Figure 3.8 The pupil of the common octopus, *Octopus vulgaris*, is a horizontal slit in bright light (A) but expands and becomes round in dim light (C).

Source: From Hanke, F. D., & Kelber, A. (2020). The eye of the common octopus (*Octopus vulgaris*). *Frontiers in Physiology, 10*, Article 1637. © 2020 Hanke and Kelber.

the animal and its eyes are upright, these rhabdomeres extend either left and right, or up and down. Moreover, in an echo of the arrangement of retinula cells within an insect ommatidium, an octopus's receptor cells are arranged in a mosaic pattern in which microvilli from four neighboring cells form a square, called a rhabdom.

VISUAL ACUITY

How well can an octopus see fine details? To answer this question, Muntz and Gwyther (1988) measured the visual acuity of two species (*O. pallidus* and *O. australis*) that are similar to *O. vulgaris* but found in Australian waters. The animals were kept in individual tanks, and an inverted flower pot with the bottom cut off was placed in the center of each tank to serve as a "home" for the reclusive animal. Test stimuli, consisting of gratings (stripe patterns), were placed at the two ends of the tank, and the octopus was trained to attack a vertical grating rather than a horizontal one. The octopus hesitated in its home, looking at each end of the tank with one of its eyes, and then rapidly attacked one of the targets.

Once the animal had learned the task with coarse gratings, the stripes were gradually made thinner until the percentage of correct responses dropped to a chance level of 50%. The results indicate that the finest stripes the octopus can see are about .12 degree of arc wide. This is comparable to the acuity of some mammals, but less than that of primates or birds of prey.

However, octopuses can be reluctant subjects in laboratory experiments, and it is possible that they could sometimes perceive gratings that they did not bother to respond to. A more recent study on the gloomy octopus, *Octopus tetricus*, suggests that this may be the case. Nahmad-Rohen and Vorobyev (2019) used a more reflexive response measure, a darkening of the skin around the octopus's eye when it attends to a visual stimulus. In their lab at the University of Auckland, they presented the octopus with a uniform gray field that was periodically replaced with a vertical grating of the same average luminance. If its eye grew dark when the grating was presented, the experimenters knew that the octopus had seen it. The finest gratings that evoked a response had stripes that were only 0.04 degree wide, indicating good acuity.

Like most current researchers, Nahmad-Rohen and Vorbyev express the fineness/coarseness of a grating in terms of its *spatial frequency*, the number of cycles per degree of visual angle (cpd). A cycle of the grating means a dark stripe plus a light stripe. For example, a grating in which the stripes are 0.5° wide has a spatial frequency of 1 cpd. Expressed in these terms, Nahmad-Rohen and Vorobyev's octopuses had an acuity of about 12 cpd. This is not as high as a human's acuity (30 cpd) but greater than a cat's (5 cpd).

CONTAST SENSITIVITY

Nahmad-Rohen and Vorbyev continued their research, measuring not just the finest grating that the octopuses could see, but how *well* they could see coarser stripes. To do this, they adjusted the *contrast* of their gratings: the difference in luminance between the bright stripes and dark stripes. For each grating, whether fine or coarse, they determined the lowest contrast that evoked the eye-darkening response.

Data from three octopuses are shown in Figure 3.9. The horizontal axis of each graph indicates the different gratings used, from coarse to fine, in terms of their spatial frequency. The height of each point indicates the contrast sensitivity of that octopus at a particular spatial frequency: The higher the point, the lower the contrast needed for the grating to be visible to the octopus.

These contrast sensitivity functions resemble those measured in humans by earlier researchers, although the human functions are shifted to the right, i.e., to higher spatial frequencies, compared to those for the octopus. It is widely accepted that in humans, the peak is partly the result of lateral inhibition: It occurs when a light stripe is imaged on the excitatory portion of a visual neuron's receptive field, and adjacent dark stripes are imaged on inhibitory portions. This combination of strong excitation and weak inhibition results in a high firing rate of neurons, enabling a person to see the grating. Contrast sensitivity declines at high spatial frequencies mainly because detection is limited by acuity, but it declines at low spatial frequencies because

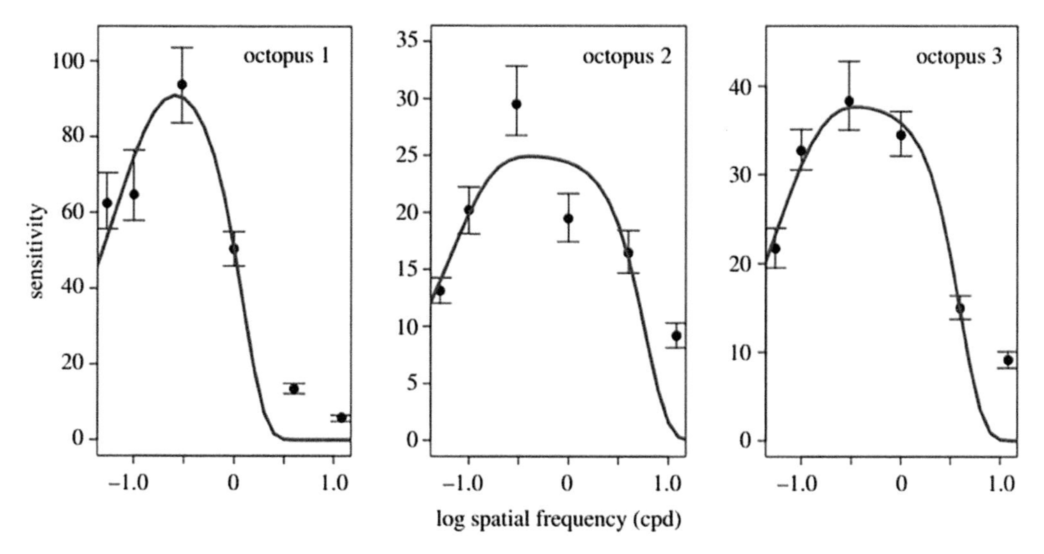

Figure 3.9 Contrast sensitivity functions of three octopuses. The lower the contrast at which an octopus responded to a grating, the higher the contrast sensitivity. Points represent the average of measured values at a given spatial frequency; error bars show the 25% and 75% quartiles; and the curves are theoretical functions fit to the data. Note that the X-axis shows the *logarithm* of spatial frequency. For example, the functions peak at −.5 log cpd, which equals 0.3 cpd. This spatial frequency is a "sweet spot" at which contrast sensitivity is boosted by lateral inhibition.

Source: **From Nahmad-Rohen, L., & Vorobyev, M. (2019). Contrast sensitivity and behavioural evidence for lateral inhibition in octopus. *Biology Letters*, *15*, 20190134. Used with permission of The Royal Society (U.K.); permission conveyed through Copyright Clearance Center, Inc.**

wide stripes do not fit neatly into the excitatory and inhibitory portions of receptive fields (De Valois & De Valois, 1990).

The fact that the octopus contrast sensitivity function falls off at low spatial frequencies implies that it too has lateral inhibition. We have now seen that this process contributes to vision in vertebrates, arthropods, and molluscs; and in fact, it is present as well in other sensory systems, including touch and hearing. It is one of the fundamental processes by which perceptual experience is constructed out of sensory information, as the physicist and sensory scientist Ernst Mach first proposed.

Incidentally, the octopuses in Nahmad-Rohen and Vorobyev's study consistently attended to the gratings over several weeks but eventually grew tired of the stimuli and refused to participate further, by hiding behind rocks or covering their eyes with shells. This is a remarkable example of autonomy in an invertebrate.

The horizontal and vertical dimensions of the visual field seem to be especially important to an octopus (Sutherland, 1957). The pupil, when constricted, is horizontal, and organs of equilibrium called *statocysts* keep it at that orientation regardless of the angle at which the animal may be slouching in its den. And there is a horizontal structure called the visual streak in the octopus retina, where receptor density is high and vision is thought to be keener than elsewhere. By analogy with a similar structure in the retinas of horses, rabbits, and other animals that are constantly on the lookout for predators, it probably helps the soft-bodied octopus to keep a close watch on the underwater horizon where water meets the sea floor. Behaviorally, the octopus is quite good at distinguishing horizontal from vertical contours, but not at distinguishing diagonal contours that are separated by 90°. When the animal's statocysts are removed, the eyes cease to be properly aligned, and vertical–horizontal discrimination is lost (Wells, 1960).

In everyday life, octopuses use their vision in sophisticated ways, for example camouflaging themselves at the approach of a predator to match the color and visual texture of their surroundings. They are able to recognize conspecifics,

based on appearance alone (Tricarico et al., 2011), and there are many anecdotal reports that they recognize individual humans as well.

LIGHT SENSITIVITY IN THE SKIN

Cephalopods do not have non-cephalic eyes, but the skin on some parts of their bodies is sensitive to light. Katz et al. (2021) trained octopuses (*Octopus vulgaris*) to reach through a tube in the opaque lid of their tank, and grope for pieces of food placed on the lid by the experimenter. When the tip of the foraging arm was briefly illuminated by an LED flashlight, the arm withdrew from the light, a *negative phototactic response*. Since the entire aquarium was covered in black plastic, the octopus could not see the light with its eyes but nevertheless reacted to it. This must mean that its skin contains some sort of photopigment. How the light is experienced by the octopus is unknown, but the authors are inclined to think that mechanoreceptors are involved, so that the octopus "feels the light." The withdrawal reaction may serve to protect the delicate tips of its arms from nibbling by fish or other animals.

Techniques including immunolabeling—the use of laboratory-produced antibodies that bind to biological molecules of interest—have revealed that opsin molecules are present not just in the eyes but also in the dermal tissues of a range of cephalopods (Kingston et al., 2015). They are found, for example, in *chromatophores,* small pigmented organs in the skin that contribute to the color changes of which many cephalopods are capable. Chromatophores expand when the skin in which they are embedded is illuminated, spreading out the nonvisual pigment they contain so that it is more noticeable. This light-induced expansion occurs even in isolated pieces of skin, showing that it is a local reaction. Ramirez and Oakley (2015) measured the spectral sensitivity of this process in the California two-spot octopus (*Octopus bimaculoides*) and found that it matches that of the animal's eyes, suggesting that the photosensitive pigment in skin and eyes is one and the same. Chromatophores are also responsive to nerve signals from the central nervous system, as when an octopus changes color during periods of excitement.

The bottom line is that light sensitivity in cephalopod skin serves at least two functions: protecting the arms by enabling them to be withdrawn in a well-lit environment and helping with camouflage or perhaps visual signaling by means of its effects on chromatophores.

THE OCTOPUS BRAIN

In molluscs, the central nervous system is even more distributed than in arthropods. In some, notably the gastropods (snails and slugs), there is not much of a brain—just a set of ganglia connected by nerve cords that forms a ring around the esophagus. However, the higher cephalopods (octopus, squid, and cuttlefish) do have a complex brain, formed by the coalescing of these ganglia to form a single, if highly subdivided, structure, with the esophagus forming a tunnel through its center. Its component parts are called lobes. In this section, we will consider the brain of the octopus, an animal that has been of sustained scientific interest because it is thought to be the most intelligent invertebrate.

The modern study of the octopus's brain began with the definitive anatomical work of pioneering British biologist J. Z. Young (1907–1997). But Young's initial contribution to cephalopod anatomy was actually in the squid, not the octopus: his discovery (1938) of the squid's giant axons, a pair of outsized motor nerve fibers—about 1 mm in diameter—that run from the brain down into the mantle, where they trigger the squid's jet propulsion through the water. Young's discovery set the stage for later researchers, notably Hodgkin and Huxley, who used the squid giant axon to learn how action potentials work.

Turning to the octopus, Young relied on a histological stain developed by Italian anatomist Camillo Golgi (1843–1926). When a thin section of neural tissue is treated with the *Golgi stain,* only a small percentage of neurons are stained. But the dark stain goes into all the processes of a neuron, so that the details of its anatomy can be seen under the microscope. The Golgi stain was quickly adopted and modified by Golgi's contemporary, the Spanish anatomist Santiago Ramón y Cajal (1852–1934), the founder of modern neuroscience. His realization that the stain did not travel from one neuron to another led him to propose the *neuron doctrine*, the idea that neurons are separate entities, not just nodes in a network.

With the goal of working out the cell types in the octopus brain and their interconnections, Young's anatomical study began with the octopus's eyes and progressed into its brain. In some ways, the brain of a cephalopod is reminiscent of an insect's brain: It consists of a central brain flanked by *optic lobes*, which are the direct recipients of optic nerve fibers.

OPTIC LOBES

In the octopus, as in other cephalopods, the optic nerve fibers are the axons of visual receptor cells. Having entered the optic lobe, these axons terminate on second-order neurons. The resulting transfer of information across these and later synapses is modulated by a large number of local interneurons, resembling the horizontal and amacrine cells of the vertebrate retina. In other words, early processing of visual information in the octopus occurs not in the retina, as it does in vertebrates, but in the optic lobe.

In the vertebrate visual system, it has been possible to record from individual neurons in both the retina and the brain and to determine those cells' receptive fields. This has proved very challenging in cephalopods, because it is difficult to stabilize the animal's soft body with the precision needed to position a microelectrode next to a neuron. But by looking at the shape of the dendritic trees of individual neurons in the optic lobes of *Octopus vulgaris*, Young (1971) was able to infer what their receptive fields are like. Some dendritic fields were round, but others were elongated, suggesting that the cells functionally resemble the edge-detecting neurons discovered by Hubel and Wiesel in mammalian visual cortex.

VERTICAL LOBE

Fibers leaving the optic lobes and entering the central brain go to a variety of motor centers, where they help to direct actions such as attacking prey or moving across the ocean floor. Some visual signals reach the *vertical lobe*, at the top of the brain, which is involved in learning and memory. The pattern of synaptic connections made here has been worked out using electron microscopy (Gray, 1970). Bundles of axons carrying sensory signals traverse the vertical lobe in parallel and deliver information to many closely packed small interneurons called *amacrine cells*. (Their name literally means "lacking a long axon," a characteristic they share with their namesakes in the vertebrate retina.) The amacrine cells in turn activate large neurons whose axons leave the vertical lobe and descend to motor centers. Different parts of the vertical lobe process information from different sensory modalities.

The incoming sensory fibers, running horizontally, and the outgoing axons of the large cells, traveling vertically, form a grid resembling the warp and weft of woven cloth. It is reminiscent of the grid seen in other brain structures where complex information processing occurs, such as the cerebral cortex of mammals, the Wulst of birds, and the mushroom bodies of insects. The grid layout may make it possible for the large neurons (1) to detect spatial patterns in the visual input by comparing activity across sensory fibers and (2) to detect temporal patterns by comparing activity at different points along a single sensory fiber. In other words, the grid layout of the vertical lobe may be what enables an octopus to perceive objects and events.

The vertical lobe also appears to be where the octopus's visual memories are formed, a physiological process studied by Binyamin Hochner and his colleagues at Hebrew University in Jerusalem (Shomrat et al., 2015). By electrically stimulating sensory fibers just after they enter the vertical lobe, the investigators triggered action potentials that traveled across the lobe, activating amacrine cells. They were able to record both the sensory fibers' action potentials and the resulting post-synaptic potentials (PSPs) in groups of amacrines.

Test recordings with a standard stimulus were made before and after a series of shocks delivered to the sensory fibers that caused an intense barrage of neural activity. Comparing the before and after recordings, the researchers

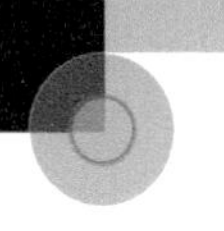

found that while sensory fiber action potentials were unchanged by the barrage, PSPs were four times larger in the final recording than in the initial one. This increase, called *long-term potentiation (LTP)*, is a well-established process that is crucial for learning. A stimulus strong enough to evoke a barrage of action potentials in sensory fibers—such as the sudden appearance of a shark—is probably one worth remembering, and so the octopus's nervous system is modified accordingly.

To evaluate the role of vertical lobe LTP in learning and memory, Shomrat et al. (2008) divided octopuses into groups. In one group, an electrode was used to deliver a barrage of shocks to several sites within their vertical lobe, inducing widespread LTP. Octopuses in a control group had sham surgery that did not induce LTP. All surgical procedures were carried out under anesthesia.

After recovering from the anesthesia, all the octopuses were trained *not* to attack a red ball. The animals in whom LTP had been induced learned this avoidance task more quickly than those in the control group. The results confirm that the LTP that occurs in the vertical lobe is important for learning and memory.

But in the everyday life of the octopus, when does LTP occur? It would not be adaptive for the octopus to form memories of objects or events that were irrelevant to its survival. The answer involves fibers that enter the vertical lobe from other brain regions, bringing information about the emotional significance of current stimulation. The origin of these input fibers is unknown, but they have been shown (Shomrat et al., 2010) to contain *serotonin*, a neurotransmitter often involved in emotional signaling. Its release in the vertical lobe greatly augments the LTP produced by sensory inputs.

So the octopus brain is beginning to yield up its secrets. But surprisingly, less than half of the octopus's nervous system is in the brain. A majority of neurons are found in or near the arms, in the service of sensory, motor, and perhaps even cognitive functions. This is a subject to which we will return in Chapter 6.

CAMOUFLAGE IN CUTTLEFISH

Although octopuses are generally acknowledged to be the most intelligent molluscs (and indeed, the most intelligent invertebrates), it is another cephalopod, the cuttlefish, that is the most adept at camouflage. It can rapidly change its appearance and frequently does so, causing it to blend in with its environment. Hanlon and Messenger (1988) found that the cuttlefish's camouflage displays fall into three groups: uniform, mottle, and disruptive. The animal displays a uniform (or lightly stippled) pattern when on a background, such as sand, that lacks large features. The mottle display has small light or dark patches distributed across the body, with a repetitive pattern sometimes being visible. The disruptive display is characterized by large feature elements such as a prominent white square in the center of the mantle. The salient contours of these elements presumably serve to distract other animals from the edges of the cuttlefish itself.

The cuttlefish changes its display by the expansion or contraction of chromatophores in the skin, which are controlled by motor neurons in its brain. Barbosa et al. (2008) showed that there are many variations within each display category. When a cuttlefish was placed on checkerboard patterns differing in size and contrast, it adjusted its own appearance in nuanced ways that presumably serve to optimize its concealment. A recent study (Woo et al., 2023) further demonstrated the complexity of the camouflage process, showing statistically that the displayed patterns vary along a large number of dimensions and that the cuttlefish tweaks its appearance in small steps until a close match to the background is achieved.

The cuttlefish must use vision during this process, somehow comparing its appearance with that of the background. This is a remarkable ability and no one is quite sure how it is done.

Most analyses of cuttlefish camouflage, including that of Woo et al. (2023), are based on the spatial and intensive properties of stimulus and response patterns rather than their color. Cuttlefish are, of course, known for their vivid and changeable colors, but the contribution of these to camouflage is uncertain, given that most if not all cephalopods are believed to be color blind. We will take up this issue in the next chapter.

SUMMARY

In this chapter, we have examined the visual systems of arthropods and molluscs. Most arthropods, including insects, have compound eyes composed of individual facets called ommatidia, each with its own lens and receptor cells. In many diurnal insects, the ommatidia are optically isolated from one another, but in most nocturnal species, light is more efficiently used, passing internally between ommatidia and forming an image on the layer of receptor cells.

Another difference between arthropod and vertebrate eyes is in the rhabdomeric nature of arthropod visual receptors: their visual pigment molecules are locked in place rather than being free to rotate. This gives arthropods the ability to detect the polarization of light, a subject we will consider in Chapter 5. In addition to their main eyes, many arthropods have small accessory eyes called *ocelli,* used mainly to monitor light from the sky.

Molluscs have a wide variety of simple (i.e., single-chambered) eyes, ranging from the eyecups of some snails and the pinhole eyes of the nautilus to the large, high-acuity eye of the octopus, which resembles in many ways our own eye. Yet molluscs' visual receptors, like those of arthropods, are mostly rhabdomeric.

As arthropod and mollusc eyes evolved, so did their brains, substantial portions of which contribute to the processing of visual information. One commonality across phyla is the presence of lateral inhibition, an algorithm that enhances outlines in the visual field, increasing the ability to perceive objects. Another widely occurring neural process is long-term potentiation, a synaptic strengthening that occurs following salient events. It supports the laying down of visual (and other) memories in specialized brain areas, the mushroom bodies of arthropods and the vertical lobe of advanced cephalopods. As a result of convergent evolution, representatives of all three phyla covered in this book share complex visual abilities, such as the ability to recognize individual conspecifics based on their appearance.

In the next chapter, we will consider the subject of color vision, an ability that is widespread among vertebrates and arthropods, but for which there is little evidence in molluscs.

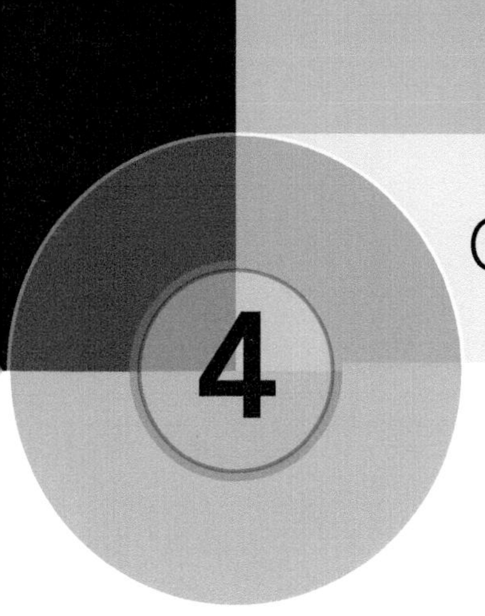

Color Vision

Our visual experience is complex. We recognize objects, see where they are, whether they are moving, and so on. One especially vivid component of vision is color—the blue of the sky, the green of grass, and the red of strawberries. Color is such an immediate and compelling sensation that is easy to assume that animals have color vision like ours.

Yes, most animals do have color vision, but not all color vision is the same. Some animals have richer color vision than others. And even the laws of color vision may differ between, say, vertebrates and arthropods.

In this chapter, we will explore these differences, taking as our first example the dog. Like most other mammals, it has a simple form of color vision that obeys easily understood laws. Next, we will apply those principles to the more complicated case of humans and other primates and then to the even richer color vision of most fish, birds, and other vertebrates. Later in the chapter, we will examine the equally diverse forms of color vision in butterflies, mantis shrimp, and other arthropods. The chapter concludes with a careful examination of the evidence for the widely held view that all molluscs are colorblind.

But let's start by talking about light. Light consists of elementary, zero-mass particles called *photons,* which travel in straight lines, but vibrate sideways as they do so. The vibration cycle repeats at regular, very short intervals, and the distance the photon travels in a cycle is called its *wavelength.* Photons of visible light have different wavelengths, ranging from about 400 to 700 nanometers (nm), distances somewhat less than a micron. We can't see a single photon, but we can see a beam of 400 nm light or of 700 nm light, or anything in between. Remarkably, these look different to us. The 400 nm light looks violet, 500 nm looks green, 575 nm looks yellow, and 700 nm looks red.

What is going on physically to cause these various color sensations? Remember that photons contribute to vision only when they are absorbed by molecules of visual pigment, which is confined to visual receptor cells in the retina. The most numerous receptors in the human eye are the rods, so named because the portion of the cell containing visual pigment, called its outer segment, is rod shaped, resembling corn on the cob. Each rod contains millions of molecules of a visual pigment called *rhodopsin,* and all the rhodopsin in all the rods is identical.

Rhodopsin does not absorb all the photons that strike it. Some photons pass right through the rhodopsin molecule and have no effect on it, whereas others are absorbed, activate the molecule, and contribute to vision. The likelihood of being absorbed varies with the wavelength of the photon: Photons with a wavelength of about 500 nm have the best chance of being absorbed, but this chance drops off gradually for photons with wavelengths differing more and more from 500 nm. In other words, the *absorption spectrum* of rhodopsin peaks at about 500 nm. A beam of 500 nm light that stimulates rods will therefore look brighter than a beam containing the same number of photons of a different wavelength, like 400 or 700 nm.

DOI: 10.1201/9781003362319-4

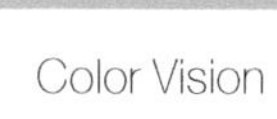

However, vision mediated by rods isn't in color: It is in black and white, like an old movie. It is the other type of retinal receptor, the cones, that mediate color vision. Cones require more light than rods to work well, which is why you don't see colors by moonlight.

How are cones able to mediate color vision whereas rods cannot? The key is that most vertebrates have more than one type of cone. Cones of different types contain visual pigments with peak absorption in different parts of the spectrum.

COLOR VISION IN MAMMALS

DOGS, CATS, AND OTHER NONPRIMATES

For example, the dog's retina has two types of cones. In one type, the visual pigment's absorption spectrum peaks at a short wavelength (429 nm), whereas the pigment in the other cone type absorbs maximally at a midspectral wavelength (555 nm). Since the ability of each pigment to absorb light drops off gradually on both sides of its peak, each wavelength of light will be absorbed to some degree by both pigments and will therefore activate both types of cones. However, the ratio of activations of the two cone types will vary with the wavelength. The color the dog experiences depends on this ratio.

An experiment carried out by Jay Neitz and his colleagues (1989) at the University of California at Santa Barbara illustrates this principle. They trained dogs to view an array of three translucent panels, illuminated from behind, and to select the one panel that was different from the other two by touching it with their nose. Two of the panels, randomly chosen on each trial by the experimenter, were illuminated with white light. This light was a mixture of all the wavelengths in the spectrum in roughly equal proportions, and so it activated the two cone types equally.

The third panel was illuminated with a single wavelength of light. To understand the dog's task, imagine that you are looking at a blue panel and two white ones; it would be easy to choose the one that isn't white. For most wavelengths, in fact, the dogs could easily tell that one of the panels was not white (see Figure 4.1). But when the wavelength of the odd panel was at the point in the spectrum where the absorption curves of their two cone pigments cross, the dogs made many mistakes. This means that this test wavelength looked white to them.

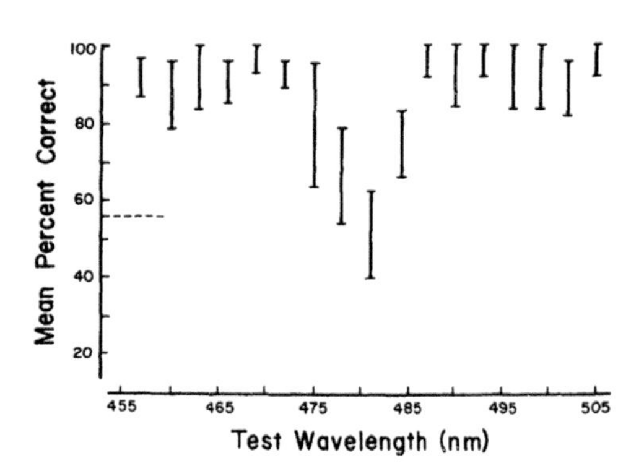

Figure 4.1 Results for one canine participant, a female toy poodle, in a color discrimination experiment. The dog's task, on each trial, was to tap the one panel out of three that was transilluminated by a colored test light. White light was presented in the other two panels. Several 300-trial sessions were run for each of the 17 test wavelengths indicated on the X-axis. A percent-correct score was determined for each session, and the range of these values for each test wavelength is shown by a vertical bar. The dog's performance was excellent at most test wavelengths but fell nearly to a chance level (33% correct) at about 480 nm, indicating the presence of a neutral point. The intensity of the test light was varied from session to session, to ensure that dogs were identifying it on the basis of color, not brightness.

***Source:* From Neitz, J., Geist, T., & Jacobs, G. H. (1989). Color vision in the dog. *Visual Neuroscience*, *3*, 119–125. Reproduced with permission of Cambridge University Press.**

This wavelength of light (about 480 nm) was a *neutral point* for the dogs. It occurred because at this wavelength, the two types of cones absorb equal amounts of light, just as they do when stimulated by daylight. The experiment shows that color perception in the dog is based on the *ratio* of stimulation of their two types of cones. For reasons that will be explained later in the chapter, it is widely believed that dogs see wavelengths shorter than their neutral point as blue and wavelengths longer than the neutral point as yellow.

Color vision like that of the dog is also found in domestic cats. Cats are less enthusiastic than dogs about participating in perception experiments, but Clark and Clark (2016) were able to tempt them to undertake a few trials a day, using tuna treats. It turns out that cats also have a neutral point, although at a somewhat longer wavelength (505 nm) than the dog.

Color vision based on two types of cones, called *dichromacy*, is widespread among mammals (Jacobs, 1993, 2009). It has been demonstrated behaviorally in animals ranging from the tree shrew to the horse, and is presumed to occur in many other species as well, based on spectrophotometry showing the presence of two cone pigments. Usually, one of these pigments has an absorption spectrum that peaks in the blue or violet part of the spectrum; the other pigment's absorption spectrum peaks in what to us is the green or yellow part of the spectrum.

Rats and mice are dichromats, but unusual ones. Their short-wave pigment peaks not in the blue-violet part of the spectrum, but at an ultraviolet wavelength (360 nm) that is invisible to humans. This presumably enables them to see flowers, seeds, and perhaps urine markings, that strongly reflect UV light (Chávez et al., 2003).

HUMANS

The average person's color vision is richer than that of a dog, cat, or mouse. For example, light at the dog's neutral point is seen by most humans not as white, but as a highly saturated blue. This is because we have three cone pigments, rather than two like the dog. In technical terms, most people are *trichromats*. Trichromats don't have a neutral point because there is no wavelength in the spectrum where all three absorption curves cross.

The basic rules of trichromatic color vision were worked out more than two centuries ago by Thomas Young (1773–1829), a British scientist who proved that color sensations depend on the ratio of absorption by different receptor types. His psychophysical work was far ahead of contemporary anatomy and physiology, and so it was not known at that time that these receptors were what we now know as cones.

Young's most important experiments made use of *additive color mixture*. This means combining lights of different colors, for example by shining different beams of colored light onto a screen at the same time. (Mixing paints is a more complicated process, called subtractive color mixture, which we will not discuss.) In some experiments Young used a red light, a green light, and a blue light. He found that the appearance of the combined light depended very much on the proportions of the three components, or *primaries*. For example, a mixture of a lot of red light and blue light, with only a little green light, looks purple, whereas a mixture of mostly red and green light, with a little blue, looks pale yellow. In fact, almost any color can be produced if the observer can adjust the individual intensities of the three primaries.

What these adjustments are doing physiologically is to change the relative amounts of light absorbed by the three cone pigments, and therefore the relative activity levels of the three types of cones. And the relative activity levels of the three types of cones determine the colors that you see. Each cone is, by itself, colorblind, and sends to the brain only a signal indicating the number of photons it absorbed, not their wavelengths; color is created in the brain when the three cone signals are compared.

In the last century, it became possible to measure the absorption spectra of the three cone pigments in the human eye (Brown & Wald, 1964; Dartnall et al., 1983). This is done by shining a narrow beam of light through individual cones and measuring the fraction absorbed, a procedure called *microspectrophotometry*. One pigment's absorption spectrum peaks at about 420 nm, another's at about 530 nm, and the third at about 560 nm (see Figure 4.2). The cones containing these pigments are often called the S (for short-wavelength), M (medium-wavelength), and L (long-wavelength) cones, and we will also refer to them this way. However, each pigment absorbs light across a wide span of wavelengths, so that it is not possible to activate just one type of cone in a normal human observer.

Molecules of visual pigment are proteins and are constructed based on DNA in our chromosomes. It turns out that the genes for the M and L cone pigments are on the X chromosome, which is one of the sex chromosomes. Women have two X chromosomes, whereas men have one X and one Y chromosome. This means that women normally have two copies of the L gene and two

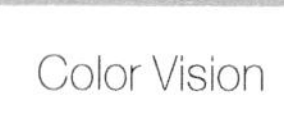

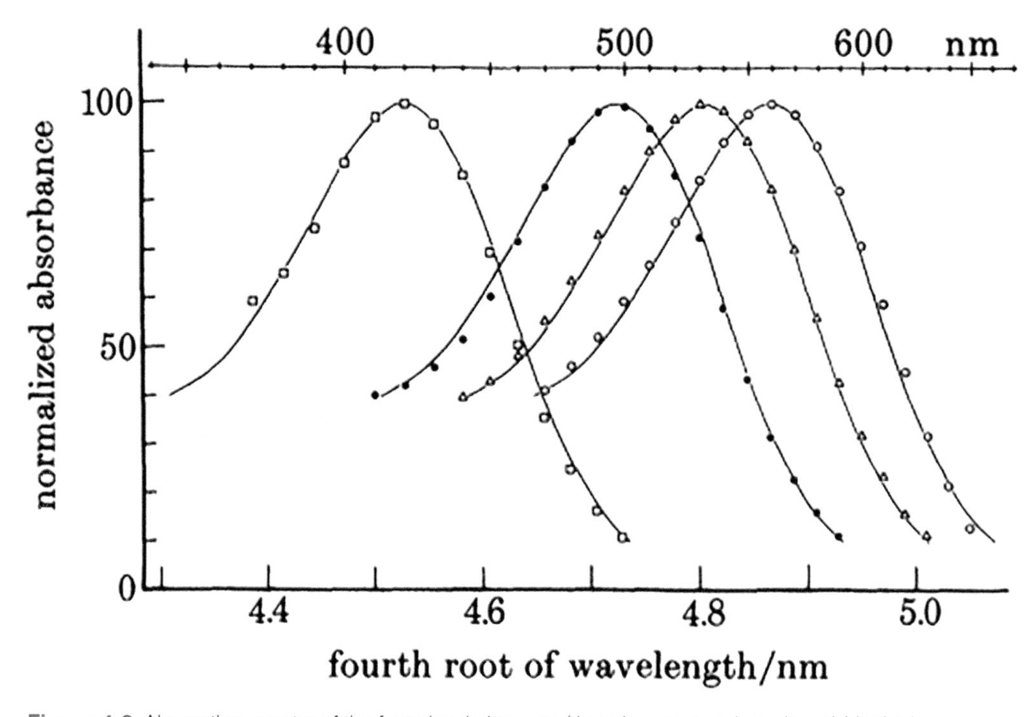

Figure 4.2 Absorption spectra of the four visual pigments (three in cones and one in rods) in the human eye. From left to right, they are the pigments in the S cones, rods, M cones, and L cones. Wavelength in nm is indicated by the horizontal axis at the top of the graph. The functions have been normalized so that they all have a peak value of 100. Plotted here on a transformed scale (bottom), all four spectra have the same shape.

Source: From Dartnall, H. J. A., Bowmaker, J. K., & Mollon, J. D. (1983). Human visual pigments: Microspectrophotometric results from the eyes of seven persons. _Proceedings of the Royal Society of London B_, _220_, 115–130. Used with permission of The Royal Society (U.K.); permission conveyed through Copyright Clearance Center, Inc.

copies of the M gene, while men have only one of each. The gene for the S cone is on another chromosome, and both men and women have two copies of this gene.

Partial color blindness. Some people are missing one cone type, and so are dichromats. They are usually said to be "color blind" but, in fact, have considerable color vision, comparable to that of a dog. The first person to give a subjective description of dichromacy was John Dalton (1766–1844), the English chemist who proposed that everything is made of atoms. His eyes were preserved after his death and kept at the Manchester Literary and Philosophical Society. After 150 years, the retinas were subjected to DNA analysis. It revealed that Dalton lacked the M type of cone but had S and L cones (Hunt et al., 1995). This form of dichromacy is called *deuteranopia*. Dalton had a neutral point, which he described as a narrow white zone in the spectrum, flanked by shorter wavelengths that he said looked blue, and longer ones that he said looked yellow.

Deuteranopia is much more common in men than in women. The reason is that men have only one copy of the M-pigment gene, and a defect in that gene will block the production of the M pigment. A woman has two copies of the gene, so if one is nonfunctional, she will still have enough of the M pigment (produced by the other gene) to get by. But she will be a carrier of deuteranopia and may have sons who are deuteranopes. The same is true for protanopia, an absence of the L cones. In contrast, tritanopia—absence of the S cones—is very rare, and equally so in men and women, since it requires two defective genes for the S pigment.

But let us stop for a moment and consider in more detail the color vision of a woman who is a carrier for protanopia or deuteranopia. Careful study of the color vision of small regions of her retinas, using a tiny beam of light, shows that they are a patchwork of dichromatic and trichromatic areas (Born et al., 1976). In everyday life, this abnormality goes unnoticed because most objects are large enough to project to trichromatic as well as dichromatic patches of retina.

However, it very occasionally happens that one eye follows the instructions of one gene with respect to the pigment in question, while the other eye follows the instructions of the other gene. When this occurs, the woman will be a trichromat in one eye, but a dichromat in the other eye. Only a handful of people with this extremely rare condition have been studied in the laboratory.

What makes these individuals especially interesting is that they can describe the colors they see with their dichromatic eye, using color names appropriate for trichromatic vision. For protanopia and deuteranopia, the types of human color blindness most similar to the dog's dichromacy, the answer is that unilateral dichromats see light at their neutral point as white, light of shorter wavelengths as blue, and light of longer wavelengths as yellow (Graham & Hsia, 1958). This is exactly what Dalton predicted.

We can't know for certain what color sensations dogs, cats, and other dichromatic mammals experience, but perhaps they also see short wavelengths as blue and long ones as yellow.

OTHER PRIMATES

The first primates appeared about 80 million years ago and quickly began to spread and diversify. At that time, Pangaea, the supercontinent that comprised most of Earth's landmass, was in the process of breaking apart into today's continents. As the Americas pulled away from Africa and Eurasia, primates in the "New World" were separated from their cousins in the "Old World," and both groups continued to evolve, but independently. Advanced forms called *monkeys* emerged in both hemispheres, but Old World monkeys and New World monkeys were—and are—quite different. One of the ways they are different has to do with their color vision.

Most species of New World monkeys (such as squirrel monkeys, tamarins, and marmosets) have an S pigment and an L pigment. But there are different versions of the latter pigment in different individuals within a species, a diversity called *polymorphism*. For example, there are three versions of this pigment in squirrel monkeys (Mollon et al., 1984), with peak absorptions at 537, 550, and 565 nm.

To understand polymorphism, remember that a visual pigment molecule is made up of a chromophore (11-cis retinal) and a large opsin molecule. It is the chromophore that absorbs light, but the way the chromophore is held by the opsin influences which wavelengths are most easily absorbed. Slight variations in the gene that codes for the opsin can modify its structure and so cause a shift in the peak of the pigment's absorption spectrum. The details of color vision, such as the location of the neutral point, can therefore vary from one individual (or species) to another. The reason this type of polymorphism is so common among New World monkeys is not known.

As in humans, the gene for a squirrel monkey's L cone pigment is on the X chromosome. Males only have one X chromosome, so whichever of the three versions of the gene is present, the monkey will still be a dichromat. However, females have two X chromosomes, so if they have two different versions of the L pigment (one from each chromosome), they will be trichromats. The overall result is that there are six different forms of color vision—three types of dichromacy and three types of trichromacy—in the squirrel monkey (Jacobs & Nathans, 2009)!

In Old World monkeys, as well as apes and people, a significant mutation made trichromacy possible for almost all individuals (Nathans et al., 1986). The X chromosome site where the L-pigment gene resides duplicated itself, so there were then two sites, close together, each of which could direct the construction of a visual pigment. These continued to evolve independently, with one eventually housing a gene for an L pigment and the other a gene for an M pigment. The result is that the average person, chimpanzee, or macaque (leaving aside the minority who are "color blind") has three types of cones—S, M, and L.

Old World (and some New World) primates are thus the only mammals—other than a few

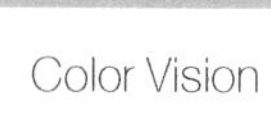

marsupials (Ebeling et al., 2010)—who have trichromacy. In other words, dogs and cats have only one color dimension (blue vs. yellow) while we have an additional dimension: red versus green (Hurvich, 1981).

Has the addition of the red/green dimension helped primates to survive? It has long been suspected that trichromacy is adaptive because it enables primates to locate ripe fruit against a background of green leaves. To evaluate this possibility, Caine and Mundy (2000) measured the ability of marmosets to spot corn puffs, dyed orange to resemble ripe fruit, in a complex environment of grass, weeds, and other plants. The species of marmoset used, *Callithrix geoffroyi* (see Figure 4.3), is native mainly to Brazil, but the study focused on 14 individuals living at the San Diego Wild Animal Park. Some of the animals were dichromats, and the rest were trichromats. They were released in groups (each a mixture of trichromats and dichromats) and the experimenters noted how many puffs each individual found within 10 minutes.

The trichromats found significantly more puffs than the dichromats. Puffs dyed green were also used, as a control; orange and green look different to a trichromat (that's why we give them different names!) but are indistinguishable to a dichromat who does not experience the sensations of red and green. The trichromatic marmosets found more orange puffs than green ones because orange contrasted more with their environment, but the dichromats found equal numbers of the two colors. The results of this study support the idea that trichromacy provides a selective advantage in the monkeys' everyday life.

Figure 4.3 A white-headed marmoset, *Callithrix geoffroyi*. Trichromacy may enhance the ability of some members of this species to spot ripe fruit.

***Source:* Used with permission of Shutterstock. Photo Contributor, Edwin Butter.**

COLOR VISION IN NONMAMMALIAN VERTEBRATES

It was long believed that the trichromatic vision of people, apes, and Old World primates represented the best color vision that nature has to offer. It is now realized, however, that many nonmammalian vertebrates, including fish and birds, have four cone visual pigments and, as a result, have color vision that is in some ways better than ours. Animals whose color vision depends on comparing absorptions in four sets of receptors are called *tetrachromats*.

Of their four cone pigments, three are similar, although not identical, to ours, and we may refer to them descriptively as S, M, and L cones. In addition, most cold-blooded vertebrates and birds have a class of cones that are maximally sensitive in the ultraviolet. (It will be recalled that rats and mice have ultraviolet-sensitive cones, but in their case, these are *instead* of cones with a peak in the violet, i.e., S cones, rather than in addition to them.)

FISH

Tetrachromacy evolved first in fish and was passed along to reptiles and birds. But why did it evolve? If we find three classes of cones to be entirely satisfactory, why do fish need more? According to one view (Sabbah et al., 2013), the reason is that color vision underwater is more complicated than color vision on land. The survival value of color vision lies in the fact that it enables us to detect or recognize objects based on the spectral distribution of the light that is reflected by those objects into our eyes. But this distribution depends both on the amount of each wavelength in the light illuminating the object and on the ability of the object to reflect each of those wavelengths.

By comparing signals from cones viewing an object with signals from neighboring cones that are viewing surrounding objects, the visual system is able disentangle the color of the illuminant from the reflectance of the object (Hurvich, 1981). For example, you can identify the color of objects in your living room whether the lamps have "soft white" or "daylight" bulbs. Animals need to do the same in their natural environment.

For terrestrial animals, the spectral distribution of overhead illumination changes from time to time, such as when the sun rises or sets or passes behind a cloud, or the animal emerges from shadow into a clearing. But for a fish, such changes in the illuminant occur constantly, as a result of the fact that light is significantly absorbed by water, and increasingly so as the wavelength of the light increases. Waves on the surface of the water mean changes in the amount of water through which the illuminant must pass, and therefore changes in its spectral composition; and the same is true when the fish moves up or down in the water column. Moreover, light from the sky is refracted when it passes into water, sending different wavelengths in different directions.

Factoring out changes in the illuminant is thus more complicated for an underwater animal than for a terrestrial one, and many fish may need a fourth cone type to perceive object reflectance with the same accuracy that humans do with three cone types (Sabbah et al., 2013).

However, fish are an extremely diverse group, and the number of cone pigments they have, as well as the absorption spectra of those pigments, show considerable variation. While many have four classes of cones, fish swimming in deeper water where there is less light may have three, two, or only one class of cones (Levine & MacNichol, 1982). Furthermore, the spectral locations of the cone pigments' absorption spectra can vary depending on the color of the water that a fish species normally inhabits, clear water being bluer than water containing organic matter. Variations in the structure of the opsins account for much of this variation (Carleton et al., 2020).

One teleost that has been especially well studied is the goldfish, *Carassius auratus*. Behavioral work (Hawryshyn & Beauchamp, 1985) shows that it can see UV light, and the existence of four cone pigments—UV, S, M, and L—was confirmed by microspectrophotometry (Bowmaker et al., 1991). But do signals from the UV-sensitive cones actually contribute to the goldfish's color perception?

Neumeyer (1992) carried out color-mixture experiments to answer this question. He trained goldfish to view two patches of light, side by side. One was a mixture of primaries, and the other was daylight white. The fish were trained to select the mixture rather than the white light to obtain a food reward.

As Thomas Young showed, people can adjust a mixture of violet, green, and red lights, or primaries, to look identical to (i.e., to *match*) daylight. But this didn't happen with the goldfish: A match was possible only when a fourth primary, UV light, was added. With most settings of the primaries, the fish were able to tell which patch of light was the mixture. Only when all four primaries were adjusted to exactly the right intensities did goldfish perceive the mixture as identical to daylight. Neumeyer's experiment showed that goldfish are genuine tetrachromats.

However, studies of other fish that have four types of cones suggest that the UV cones may sometimes have additional, special functions. For example, damselfish (*Pomacentrus amboinensis*) recognize members of their own species by facial markings that are only visible under UV light (Siebeck et al., 2010). And as we will see later in the chapter, UV cones are sometimes involved in detecting polarized light (Hawryshyn, 2000).

Despite the benefits of UV sensitivity, it also has drawbacks. Light scatters more as its wavelength decreases, so blue light scatters more than green, yellow, or red. That's why the sky is blue; what we're seeing is the blue component of sunlight scattered in many directions by the atmosphere. Scattering also occurs inside the eye, causing visual acuity to be worse in blue light than in light of other colors. And UV light scatters even more. So if humans were sensitive to UV light, we wouldn't see objects as clearly as we do now.

In fact, the lens in our eye, unlike a fish's lens, absorbs UV light, greatly reducing the amount of it that reaches the retina. This property of the human lens is an important one, because UV light can, over time, damage the retina. This has little effect on survival in animals that live only

a few years, but in long-lived species like ours, it would be more of a problem. The bottom line is that UV sensitivity comes with some trade-offs; it is just as well that we do not have it.

AMPHIBIANS

Amphibians were the first vertebrates to emerge onto land. Their name refers to the fact that they live in water early in life but on land as adults. As tadpoles, they breathe by means of gills, but at metamorphosis, they undergo a change to the adult form, and switch to breathing with lungs.

It is difficult to train amphibians to cooperate in vision experiments involving arbitrary tasks, such as color matching. But their strong tendency to move toward certain colors, called the *phototactic response*, has given researchers a way to study their color vision and its development.

It has been known for over a century that adult frogs, placed on a small pedestal and encouraged to jump off it, will jump toward blue light rather than light of other colors. Muntz (1963) decided to study the development of this behavior be testing tadpoles at various stages of development. He constructed a simple maze in the shape of the letter Y, filled with enough water for the tadpoles to swim. They were placed at the base of the Y, and swam forward until reaching the choice point, after which they ventured into one or the other of the two arms of the maze. Muntz tested tadpoles who were in an early stage of development and others who were well into the process of metamorphosis, so that they had rudimentary limbs. He found that the youngest tadpoles preferred green to blue, but that with advancing age, they gradually came to prefer the blue. Hunt et al. (2020) confirmed the green preference of young individuals, and proved, by testing tadpoles who had been reared in the dark, that it is independent of visual experience.

It is generally believed that the shift from a green to a blue preference is adaptive, enabling young tadpoles to seek the shelter and nutrition provided by greenery, but prompting adults to more effectively escape predators by jumping into open water. But how does this shift come about? The answer has to do with the frog's photoreceptors.

Adult *Xenopus* and other frogs have three types of cones (Witkovsky, 2000), rather than the four found in many fishes. However, frogs also have two types of *rods*: one type contain a green-sensitive pigment (rhodopsin) like the rod pigments of other vertebrates, whereas the other type has an absorption spectrum peaking in the blue. This means that in amphibians, rods as well as cones may be capable of color vision, which always involves a comparison of signals from different classes of receptors. In fact, it has recently been shown that frogs can distinguish colors even when evening is so far advanced that only rods are active (Yovanovich et al., 2017). The blue-sensitive rods are not present in young tadpoles, but develop during metamorphosis, so they may be responsible for reactions to blue that emerge as the tadpoles transition to adults (Parker et al., 2010).

REPTILES

Ancestral reptiles play a key role in the story of vertebrate evolution. The descendants of early amphibians, they in turn gave rise to dinosaurs (and through them, to birds) and to mammals. Despite their descent from amphibians, however, reptiles have retinas more like those of fish than of today's frogs and toads.

Best studied of the reptiles are freshwater turtles. Their retinas contain both rods and cones. Like many fish, they have UV, S, M, and L cones (Grötzner et al., 2020), and behavioral color-mixture data indicate that they are tetrachromats (Arnold & Neumeyer, 1987).

A major innovation of the turtle retina is the increased role in vision played by cone oil droplets. Some fish and amphibians also have an oil droplet within each cone, just in front of the outer segment that houses the visual pigment. The transparent oil has a high refractive index, so the droplets serve as miniature lenses that concentrate light onto the visual pigment, increasing visual sensitivity. In fish and amphibians, the droplets are clear, or occasionally yellow. In turtles, however, there is a greater range of colors, with red, orange, and yellow droplets, as well as clear ones (Walls, 1967).

The significance of this diversity is that the droplets can change the shape of the cone visual pigments' absorption spectra. Each colored droplet acts as a filter that allows long wavelengths to pass, while absorbing short-wavelength light before it can reach the outer segment and contribute to vision. A red droplet, for example, passes red light but absorbs light of other colors. In the slider turtle *Pseudemys scripta elegans*, these red droplets are found in L cones. Without the droplet, the visual pigment in the L cones would absorb not only red light but also some light of other colors. The droplet absorbs these other colors, thus preventing the L cones from responding to them. The same thing happens in an M cone with a yellow oil droplet; S and UV cones have clear or almost clear oil droplets and so are not affected. An overall effect of the oil droplets is to narrow some cone sensitivity functions (Neumeyer & Jäger, 1985), improving wavelength discrimination.

BIRDS

Cone vision reaches its apogee in birds. They are heirs to the tetrachromacy of fish and reptiles and make full use of oil droplets. In addition, they continue the trend toward greater reliance on cones and the near-elimination of rods in many species. And some, especially birds of prey, have two foveas in each eye, allowing them to see clearly straight ahead of their body, as well as off to the sides. Add to that their ability to fly, and birds were, for a time, unrivaled victors in the struggle for survival. Mammals became successful as well, not by outdoing the superb sensory and aerial abilities of birds but by playing a different game: relying on rod vision to carve out a largely nocturnal niche.

As we saw in Chapter 2, birds can be readily trained to carry out tasks requiring them to make visual discriminations. Goldsmith and Butler (2005) used these operant conditioning procedures to study color vision in the budgerigar (*Melopsittacus undulatus*), a parakeet native to Australia and popular as a pet. In the laboratory, the birds quickly learned to approach a food-reinforced color rather than one that was not rewarded. Trained to fly to spectral yellow (which substantially excites both L and M cones), the birds were able to distinguish it from most mixtures of red light (which excites mostly L cones) and green light (which excites mostly M cones). But there was a specific combination of green light and red light which excited the M and L cones in the same proportions as did the spectral yellow light, and the birds were not able to distinguish this mixture from the yellow. This color-mixture experiment implies that the signals from the M and L cones combine to create different color sensations (orange, yellow, chartreuse, etc.) depending on their proportions, just as they do in humans.

But are signals from UV cones compared in the same way with signals from other cone types, or are they functionally isolated so that they don't contribute to color vision? To answer this question, Goldsmith and Butler repeated their earlier experiment, but now using UV, violet, and blue lights. The bird could distinguish violet light from a mixture of UV and blue, except when the mixture excited the UV and S cones in the same proportions as the violet light did. The results imply that signals from the budgerigar's UV and S cones combine to produce a series of intermediate color sensations, just as signals from the M and L cones do. Taken as a whole, the data show that the budgerigar is a genuine tetrachromat.

The same is true of most birds. Nocturnal owls are an exception. Their cones, though greatly outnumbered by rods (Fite, 1973), are sufficient to mediate some color vision (Martin, 1974). However, they lack UV cones and are therefore not tetrachromats (Höglund et al., 2019).

Given birds' outstanding color vision, it is not surprising that color plays important roles in their lives, from prey and predator detection to camouflage to sexual selection. As an example, let us consider the case of bowerbirds, a family comprising 20 species of polygynous birds that live mainly in Australia and New Guinea. To attract females, males of most species build special structures called bowers, which they decorate with leaves or other brightly colored objects. Females may visit more than one bower before choosing a mate, basing her choice on some combination of the male's colorful appearance, courtship display, and the quality of the bower.

Best studied is the satin bowerbird *Ptilonorhynchus violaceus*. The male's plumage is a deep but shiny blue; the female's is a more muted combination of green and brown. To determine whether the male's coloration plays a role in the female's choice, Savard et al. (2011) used automated video equipment to record the activity of birds in the bowers—where all copulation occurs—during the mating season, and they were thus able to tally each male's mating success.

Photometric measurements had previously been made on each male, to record the percentage of light reflected by his plumage at each of a series of wavelengths across the spectrum. By comparing these light measurements with the behavioral records, the investigators were able to see that only reflectance in the blue part of the spectrum (405–480 nm) was significantly associated with mating success.

Is the importance of blue for satin bowerbird sexual selection confined to the male's plumage, or does it apply also to bower decorations? To answer this question, Borgia and Keagy (2006) made use of the fact that males will search their environment for decorative objects with which to make their bowers more attractive (see Figure 4.4). The researchers placed an assortment of plastic squares of different colors a short distance from each bower. Males quickly reacted to the availability of these potential decorations, moving blue squares toward their bowers and pushing red ones farther away.

What makes blue special for these birds? Borgia and Gore (1986) proposed that its importance derives from the fact that there are few blue objects in the animals' environment, so males must compete for them, stealing them from one another's bowers. By displaying many blue objects in his bower, a male signals that he is aggressive and dominant, traits that females may find attractive.

Figure 4.4 A male satin bowerbird decorating his bower with blue objects.

***Source:* Used with permission of Shutterstock. Photo Contributor, Ken Griffiths.**

But could satin bowerbirds simply have a preference for all things blue, rather than one tied specifically to sexual selection? To test this possibility, Borgia and Keagy (2006) tested the preference of both male and female satin bowerbirds for food (pieces of breakfast cereal) dyed different colors. The answer was clear. The birds hurried to the pile of red cereal and ate it first and then proceeded to eat the orange, yellow, and green, in that order, before—if they had room—turning to the blue. When choosing food, their color preferences were the opposite of those in play when females choose a mate. This reversal is adaptive for these fruit-eating birds, since warm colors generally indicate ripe fruit.

The overall message is that birds use their color vision in context-specific ways to help them make choices with survival value.

COLOR VISION IN ARTHROPODS

INSECTS

BEES

Many insects also have color vision, including the honeybee. In a series of classic experiments beginning more than a century ago, Karl von Frisch (1967) trained bees by placing a dish of sugar-water on a card of a certain color. Nearby were other cards that were of various brightnesses, but all were gray. Later, von Frisch tested the bees by setting out new cards identical to the originals, but without the dish of sugar-water. The bees continued to go to the colored card. Von Frisch reasoned that if the bees were colorblind and responding only to the brightness of the card, they should have confused it with a gray card of similar brightness. Since they did not do this, they must be able to perceive color.

Later research has shown that bees have three classes of photoreceptors, each with its own visual pigment. Their spectral sensitivities were examined by maneuvering a microelectrode into individual receptors and recording the electrical activity elicited by light of different wavelengths (Menzel & Blakers, 1976). The sensitivities of the three classes of receptors were found to peak in the ultraviolet, blue, and green parts of the spectrum; we will refer to them as the UV, B, and G receptors, respectively. So bees are trichromats, like humans, but their visible spectrum is shifted to shorter wavelengths than ours: They can see UV light but are nearly blind to red light. Their UV sensitivity gives bees the ability to perceive aspects of the environment, such as UV-reflective patterns on flowers, that we cannot see.

Although their visual pigments are different from ours, the rules of color vision (like the ratio principle) appear to be similar in bees and humans (see Menzel & Müller, 1996; von Frisch, 1967). For example, for humans, there are pairs of *complementary colors*, such as blue and yellow, that can be combined to produce a mixture that looks white, although it is physically different from daylight. It turns out that bees also have complementary colors, although the wavelengths involved are different.

ANTS

Ants, like bees, have color vision. However, they appear to be dichromats rather than trichromats (Aksoy & Camlitepe, 2012), with UV and G receptors. This is not surprising, given that ants spend much of their time underground and have less need than bees to identify objects at a distance based on their color.

BUTTERFLIES

Some insects have more than three types of receptors, giving them very keen color vision. The Asian swallowtail butterfly *Papilio xuthus*, for example, has at least six types of receptor cells (Arikawa, 2003). These were discovered by recording the electrical responses of individual receptors, while *monochromatic* lights (lights of a single wavelength) were shined on the eye. This butterfly has UV, V (violet), B, G, and R (red) receptors. In addition, they have a broad-band receptor that shows sensitivity across the spectrum; its spectral sensitivity curve is wider than that of any known visual pigment, indicating that this receptor cell contains a mixture of visual pigments. A final complexity is that the butterfly has three different types of ommatidia, each containing a subset of the receptor types.

To determine the behavioral significance of these receptoral complexities, Hisaharu Koshitaka and his colleagues (2008) at the Graduate University for Advanced Studies (Sokendai) in Hayama, Japan, measured the ability of swallowtail butterflies to distinguish one wavelength (λ) of light from another. Lights that are of very different wavelengths are easily discriminated by an animal with color vision, but lights of similar wavelengths are harder to distinguish. The smallest difference between two wavelengths that is just detectable is called the $\Delta\lambda$.

To provide context, let's briefly consider examples of wavelength discrimination in humans. A typical person with trichromatic color vision can just distinguish 540 nm light from light of either 535 nm or 545 nm, so this person's $\Delta\lambda$ at 540 nm is 5 nm. In contrast, the same person may have a $\Delta\lambda$ of only 1 nm at 480 nm, a part of the spectrum where a small change in wavelength produces a big change in the ratio of M-cone signals to S-cone signals. Human wavelength discrimination is equally good at 580 nm, where the ratio of L-cone to M-cone signals changes rapidly. In fact, a graph of $\Delta\lambda$ as a function of wavelength is W-shaped in humans, with minima (regions of best wavelength discrimination) at about 480 nm and 580 nm. In general, wavelength discrimination functions dip at places in the spectrum where there is a transition from one cone type to another as to which is more strongly activated.

Now let us return to Koshitaka et al.'s (2008) experiment with swallowtail butterflies. These insects have a delicate snout, called a *proboscis* (see Figure 4.5). It is normally curled up against their face, but they can extend it to suck up nectar. The experimenters first trained butterflies to feed on nectar in the presence of a light of a certain wavelength and then tested them by allowing them to choose between two wavelengths. During the test, the tethered butterfly

Figure 4.5 An Asian swallowtail butterfly, *Papilio xuthus*. Note the insect's proboscis.

Source: Used with permission of Shutterstock. Photo Contributor, feathercollector.

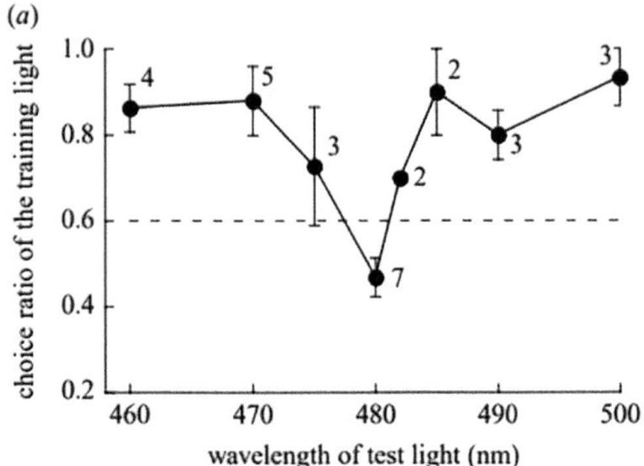

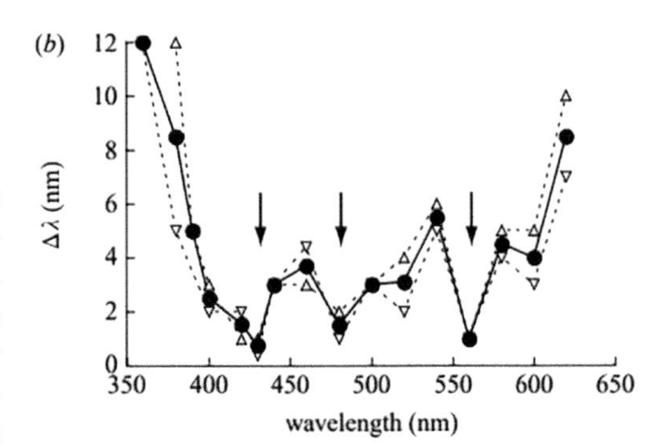

Figure 4.6 Wavelength discrimination in the swallowtail butterfly. (a) Method of determining Δλ at one wavelength. Discrimination of a training wavelength (480 nm) from test wavelengths plotted on the horizontal axis was excellent except when the test wavelength was close to the training wavelength. Δλ is the wavelength difference that supports 60% correct responding (dashed line). Numbers near points show the number of butterflies tested. (b) Δλ for the different training wavelengths plotted on the horizontal axis. The presence of three minima (arrows) indicates tetrachromacy.

Source: From Koshitaka, H., Kinoshita, M., Vorobyev, M., & Arikawa, K. (2008). Tetrachromacy in a butterfly that has eight varieties of spectral receptors. *Proceedings of the Royal Society B*, *275*, 947–954. Used with permission of The Royal Society (U.K.); permission conveyed through Copyright Clearance Center, Inc.

was brought close to two translucent panels, side by side, lit from below by two different wavelengths. One was the same wavelength used earlier in training, and the other was a different wavelength; they were adjusted to be equally bright for the butterfly. The question of interest was which wavelength the butterfly would extend its proboscis toward.

If the butterfly consistently extended its proboscis toward the training wavelength, rather than the comparison wavelength, this meant that it could tell the difference between the two lights. In fact, the butterfly consistently chose the training wavelength when the comparison light was very different, but had a harder time discriminating them when the wavelengths were close together. Koshitaka and colleagues considered two wavelengths to be just discriminable when the butterfly chose one (the training wavelength) 60% of the time (see Figure 4.6).

Two remarkable results were obtained. First, at some points in the spectrum, the butterflies could discriminate wavelengths that were only one nanometer (nm) apart. This is astonishingly good wavelength discrimination—much better than has been found in any other arthropod, and fully equal to the performance of humans. The second major result was that the wavelength discrimination function had *three* minima, implying that four classes of receptors were being used.

In the wild, butterflies use their color vision to recognize favorite flowers, and members of their own species. A number of studies have found that altering the colorful wing patterns of female butterflies dramatically alters the tendency of males to mate with them. In one such study, Shapiro (1983) used a developmental manipulation to change the wing pattern and coloration of laboratory-raised female buckeye butterflies (*Junonia coenia*). When they were in the pupa stage, Shapiro lowered their temperature, resulting in some of them developing wings closely resembling those of another species with which buckeye butterflies do not mate.

Once the females emerged from the chrysalis, the experimenter took them to an open field where wild buckeye males were flying about, released them one-by-one, and observed the results. Males attempted to mate with females that retained the normal wing pattern and coloration, but not those who resembled members of another species. The findings suggest that vision, including color vision, plays an important role in butterfly reproductive success. The fact that males will mate with artificial butterflies made of paper helps to establish the sufficiency of visual cues (Silberglied & Taylor, 1978), although pheromones are also involved under natural conditions (Douglas, 1986).

Figure 4.7 A peacock mantis shrimp, one of the most colorful of the stomatopods. The pale horizontal belts of its eyes can be seen, along with additional dark markings. Below the antennas are clublike appendages that can be swung rapidly forward to smash prey.

Source: **Used with permission of Shutterstock. Photo Contributor, Maxfield Weakley.**

CRUSTACEANS

Color vision that is more than trichromatic is also present in some crustaceans, a diverse group that includes shrimp, crabs, and lobsters. They have more than four pairs of jointed appendages, a feature that sets them apart from insects (with three pairs) and spiders (with four pairs).

As an example, we will consider the mantis shrimp, which has an array of visual pigments and receptors so complex that we are only beginning to understand it. These unique creatures, also called stomatopods, are an order of marine crustaceans that includes several hundred species. They have a worldwide distribution but are most common in shallow, warm waters of the western Pacific. The mantis shrimp is not a true shrimp: It branched off the main crustacean line some 400 million years ago (Cronin et al., 1994).

They are aggressive predators. Many use clublike appendages to attack and smash their prey (see Figure 4.7). Their first name pays tribute to the praying mantis, another highly developed predator, although the two are not related.

The mantis shrimp has complicated eyes in which dorsal and ventral portions, resembling typical compound eyes like those of the horseshoe crab, are separated by a strip or belt across the middle. This *horizontal belt* consists of several rows of ommatidia, and the ommatidia in different rows contain different visual pigments. Moreover, in some of the rows, the individual ommatidia have a multi-tiered internal structure: Some of their receptor cells constitute a superficial layer or tier just below the cornea, while other receptors make up a deep tier farther down in the ommatidium. And the receptor cells in the superficial tier have a different visual pigment from those in the deep tier.

Recall that light travels down the rhabdom, a translucent channel running the length of the ommatidium, where it is absorbed by visual pigment in the microvilli of the receptor cells. A cell's complement of microvilli, called a *rhabdomere,* fills a segment of the rhabdom in either the superficial or the deep tier. Since light must traverse the superficial tier in order to reach the deep tier, the superficial visual pigment filters the light before it reaches the deep tier, absorbing some of it.

The effect is as if the receptors in the deep tier were wearing colored sunglasses. Because the filtering pigment absorbs some wavelengths more than others, the ability of the cells in the deep tier to respond to those wavelengths will be reduced—they can't respond to light that doesn't reach them. The superficial pigment's absorption spectrum peaks at a shorter wavelength than that of the deep pigment, so its filtering action will distort the spectral sensitivity curve of receptors containing the deep pigment, red-shifting its peak (Cronin et al., 2000).

To add to the complexity, additional colored pigments—not visual pigments, but inert ones like those in the oil droplets of turtles—are often present in the crystalline cone and other parts of the ommatidium. The overall effect of all this filtering is that the mantis shrimp's visual receptors have narrower spectral sensitivity curves than would be expected based on the absorption spectra of their visual pigments alone (Cronin et al., 2014).

The plethora of receptor types with narrowly tuned spectral sensitivity curves—a dozen or more—would suggest that the mantis shrimp has color vision far superior to our own. Surprisingly, however, their ability to distinguish between wavelengths is only about one-tenth as good as that of a human or a butterfly (Thoen et al., 2014). This has led some researchers to believe that the mantis shrimp bases its color vision on which receptors are most active, rather than on the precise ratios of their levels of activity. The fact that a mantis shrimp's moving eyes view an object with different rows of the horizontal band in sequence and that these rows contain different visual pigments suggests that some sort of spatial or temporal code for color may also be used (Thoen et al., 2014). Clearly, there is much more to learn about the visual system of this remarkable crustacean.

DO MOLLUSCS HAVE COLOR VISION?

In contrast to vertebrates and arthropods, molluscs are thought to have little or no color vision. Even the advanced cephalopods—octopus, squid, and cuttlefish—are apparently colorblind.

COLOR CHANGES IN CEPHALOPODS

This is puzzling in view of the ability of cephalopods to change their appearance, enabling them to blend in with—or sometimes to stand out from—their surroundings. One of the most remarkable of these dynamic events is the *passing cloud display*, in which dark areas sweep along a cephalopod's body. The function of this display is uncertain: One hypothesis (Mather & Mather, 2004) is that in octopus, it startles potential prey so that they reveal their location.

As for color changes, consider the deadly blue-ringed octopus (genus *Hapalochlaena*): Its venomous bite injects tetrodotoxin (the same nerve poison that pufferfish contain) that kills by preventing the victim from breathing. When this octopus is threatened, iridescent blue rings appear on its skin, warning away potential predators. (Look ahead to Figure 8.7.) The rings are an adaptation that has survival value, even if the octopus itself cannot appreciate their color. Similarly, some octopuses turn a dark red when threatening another member of the same species; is it only the darkening, rather than the redness, that the second octopus perceives?

Color also plays a role in camouflage. How accurately a cephalopod's colors mimic those of its natural surroundings has been examined by Hanlon et al. (2013). SCUBA-diving researchers approached giant Australian cuttlefish (*Sepia apama*) in shallow water, in an area with grasses and algae on rocky and sandy substrates. To measure color, the optic fiber probe of a spectroradiometer was pointed at small patches on the skin of the cuttlefish and at equivalent patches on the surface of nearby objects.

When the animals were in visually homogeneous surroundings, such as an area of sea floor carpeted with green algae, the spectral distribution of light from their skin closely matched that of their immediate environment. That is, it provided excellent color camouflage. Hanlon and colleagues do not commit to a particular interpretation of how the cuttlefish are accomplishing this. Perhaps it is mostly a matter of making their skin highly reflective, so that it literally mirrors their surroundings.

In more complex environments, local variations appeared in the color and luminance of the cuttlefish's skin that resembled variations in the background. Backgrounds with distinct areas such as a clump of algae or a patch of sand elicited disruptive patterns in which discrete areas of the animal's skin (such as the white square) matched the colors of some—but not all—nearby objects. Environmental colors in Hanlon's study were mostly green, gray and

Figure 4.8 A cuttlefish filmed at Thailand's Phi Phi Island.

Source: **Used with permission of Shutterstock. Photo Contributor, Kjersti Joergensen.**

white; the occasional clump of red algae was an exception, and one that the cuttlefish did not mimic. Overall, camouflage was excellent, but it is not clear how much of a role was played by active chromatic changes in the skin.

However, dramatic, evanescent colors and patterns often occur that are clearly not a form of camouflage, as in the cuttlefish shown in Figure 4.8. The reason for their occurrence is at present a mystery. If these animals are colorblind, they cannot be using color to signal one another. Perhaps the colors serve to attract prey or to frighten predators.

HOW CEPHALOPODS CHANGE COLOR

Color changes are brought about by miniature organs in the cephalopod's skin: the chromatophores, iridophores, and leucophores (Mäthger et al., 2009). The *chromatophores* are sacks of inert, i.e., nonvisual, pigment, with colors such as orange, yellow, and black. Muscle fibers under the control of the nervous system pull on the edges of chromatophores, expanding them, and this can occur independently for chromatophores of different colors. Beneath the chromatophores are *iridophores*, shimmering, highly reflective structures that can be influenced by hormonal state, as in the case *Hapalochlaena*'s blue rings. When chromatophores contract into tiny dots, more light reaches and is reflected by the iridophores, giving the cephalopod's skin a jewel-like appearance. And beneath the iridophores of octopuses and cuttlefish (but not most squid) is a layer of *leucophores*, white but not mirror-like reflectors. The combined action of these different structures in the skin makes it possible for cephalopods to rapidly change color.

THE CASE FOR COLORBLINDNESS

Given this ability, why is there a consensus that cephalopods are colorblind? A strong piece of evidence is that, with one documented exception—the deep-dwelling luminescent firefly squid studied by Michinomai et al. (1994)—cephalopods' retinas contain only a single visual pigment, similar to our rhodopsin.

There is also behavioral evidence that they do not have color vision. For example, Lydia Mäthger and her colleagues (2006) at the Woods Hole Marine Biological Laboratory used the cuttlefish's talent for camouflage to test their ability to distinguish colors. The investigators put the animal on a checkerboard made of blue and yellow squares. When the blue squares were much darker or much lighter than the yellow ones, the cuttlefish took on a patch-like appearance, causing it to blend in (except for color) with the checkerboard. However, when the blue and yellow squares were made equally bright for the cuttlefish, it assumed a uniform appearance, indicating that now it could no longer see the squares of the checkerboard. The fact that it couldn't tell the difference between blue and yellow squares implies that it was colorblind.

However, a discovery by Alexandra Kingston and her colleagues (2015) at the University of Maryland adds complexity to the story. As mentioned in Chapter 3, these investigators found, in the longfin squid *Doryteuthis pealeii*, that rhodopsin is present not just in the retina but also in the skin. There is even rhodopsin within the squid's chromatophores, which get their color primarily from the nonvisual pigments they contain. If rhodopsin is mixed with or located below these inert pigments, its absorption spectrum could be shifted by the filtering action of the inert pigments. Rhodopsin molecules in, say, red and yellow chromatophores could thus

have different post-filtration absorption spectra, creating the potential for some wavelength discrimination (Mäthger et al., 2010) via the skin. One senses that the last word on color perception in cephalopods may not yet have been written (Godfrey-Smith, 2016).

SUMMARY

Color vision means the ability to perceive differences between lights of different wavelengths. An individual receptor cell initiates a neural signal proportional to the number of photons it absorbs but cannot encode the wavelengths of those photons. Color vision becomes possible when an animal can simultaneously see with two or more types of receptors that absorb preferentially in different parts of the spectrum.

Most mammals, such as dogs, have two types of cones and therefore have some color vision, but it is limited compared to ours. They probably see short-wavelength light as blue and long-wavelength light as yellow, with a white neutral point in between. Old World primates (including humans) have three types of cones and therefore see reds and greens as well. Most fish, reptiles, and birds have four types of cones and therefore color vision with more dimensions than ours, including the ability to see ultraviolet light. Behavioral experiments in goldfish, turtles, budgerigars, and other vertebrates show that the colors they perceive are based on ratios of absorption in different classes of cones.

Although their receptor cells are anatomically different from those of vertebrates, arthropods also have color vision, a remarkable example of convergent evolution. Honeybees, for example, are trichromats like us. The most remarkable arthropod with respect to color vision is the mantis shrimp, a crustacean with about a dozen types of receptors.

Although cephalopods such as cuttlefish can quickly change color, most evidence indicates that they (and other molluscs) are colorblind. They have only one visual pigment. However, some of this is found in the skin, in and near chromatophores, where it may serve functions yet to be discovered.

5 Visual Space Perception

In this concluding chapter on the visual sense, we look at some of the ways in which vision contributes to animals' perception of where objects are located in their environment and where they themselves are going as they move through it.

A subject of much research is depth perception, the ability of animals to see how far away objects are. Accurate depth perception depends on the use of several types of visual information, called depth cues, in the field of view. Motion parallax, the relative shifting of objects across the visual field as the perceiver moves, is a powerful depth cue for almost all animals. It is often present in combination with pictorial cues (used by animals with complex nervous systems) and/or binocular disparity (available to animals with overlap of the two eyes' fields of view).

We then consider ways in which animals use vision to navigate. In traveling through familiar territory they often use landmarks, such as a particular tree or building. For longer journeys, such as migrations, animals typically use celestial cues such as the position of the sun or stars. One such cue is the pattern of polarized light in the sky, which indicates the location of the sun even when it is obscured by clouds. Invertebrates can readily detect the polarization of light because of the rigid ordering of visual pigment molecules in their receptors. But in many vertebrates (mammals excepted) workarounds have evolved that also enable them to detect polarization. The chapter ends with examples of how artificial light at night can interfere with animal navigation.

PERCEIVING EGOCENTRIC DIRECTION

We humans, and other animals, learn a lot about our surroundings through vision. We can see whether it is day or night, what the weather is like, what inanimate objects and living things are present, and what colors they are. Once we see someone or something besides ourselves, we want to know *where* it is: Crawling along the ground, or flying overhead? On our left or our right? The direction from a perceiver to a perceived object is called its *egocentric direction*. Fortunately, the location of an object's image on the retina can help provide this information, for the overall retinal image is a reversed, upside-down picture of the spatial layout of our surroundings. For example, a deer in the left half of our visual field is imaged on the right side of the retina. And importantly, when the deer's image is on the right half of the retina, *we perceive it to be on our left*.

The visual direction associated with particular retinal locations is innate, and only with difficulty can it be transiently modified in the laboratory. An accurate sense of visual direction is present in animals soon after birth—usually by the time they can move about under their

DOI: 10.1201/9781003362319-5

own power. We will see examples of this in the next section.

DEPTH PERCEPTION

In addition to seeing whether an object of interest is to our left or right, and whether it is in the sky or on the ground, we animals need information about how far away the object is. The ability to perceive the distance of objects from ourselves is called *depth perception*. Retinal images in the eyes of vertebrates and molluscs and composite images in the compound eyes of arthropods are two dimensional, but they contain many types of information about the third dimension. These sources of information are called *depth cues*. In this section, we will discuss the three main types of depth cues: motion parallax, pictorial cues, and binocular disparity.

MOTION PARALLAX

When you look out the side window of a moving car, you can see houses and trees drift past, opposite to the direction of the car's movement. Importantly, the closer the objects are to the car, the more rapidly they move across your field of view; faraway objects, such as the moon, may even seem to move with the car. This means you can tell how close an object is to the car by how fast it is moving past you, a type of depth information called *motion parallax*.

Motion parallax is the most widely used depth cue in the animal kingdom, as ubiquitous as movement itself. For example, consider a goldfish swimming in a pond, with the bank on its left, so that with its left eye, it continuously views the mud, pebbles, and roots lining the bank. Now consider a mosquito larva, floating in the water midway between the bank and the swimming fish. Because the larva is closer to the fish than the bank is, the fish can see the larva moving relative to the bank and eat it. Motion parallax has enabled the fish to locate the larva in depth.

VERTEBRATES

A clever experiment by Hataji et al. (2021) showed that pigeons use motion parallax to judge depth. The pigeons were trained to peck at a target, a white spot on a black screen, to obtain food. A peck is a rapid, ballistic event; before initiating it, the pigeon moved its head about, getting into position. The experimenters recorded these preliminary head movements and used them to simultaneously move the target to the left or right on the screen. Because these movements of the target caused motion parallax, they influenced the way the pigeon pecked. For example, if the target moved to the right while the pigeon was making a preliminary movement to the left, the pigeon pulled back before pecking, as if the target were in front of the screen. But if the target moved in the same direction as the pigeon's preliminary movement, the pigeon leaned forward before pecking, indicating that it perceived the target to be behind the screen.

Most vertebrates use motion parallax as a source of information about depth and are able to do so from an early age. This was demonstrated in a classic series of experiments by Eleanor Gibson and Richard Walk (1960), psychologists at Cornell University. They devised an apparatus called the *visual cliff*, consisting of a sturdy horizontal piece of glass forming the surface of a table. Half of the glass had patterned paper attached to its lower surface; for the other half, the patterned paper was far below. Animals were placed on the "shallow" half, and the researchers watched to see if they were willing to venture onto the "deep" half. They almost never did. As soon as the animals could walk or crawl (for some, this was on the first day of life), they showed a determination not to fall off the cliff! Control experiments showed that motion parallax was the most important source of information used by the animals to perceive a difference in depth between the shallow and deep sides and to avoid the latter.

INSECTS

Insects can also use motion cues to perceive depth. For example, Srinivasan (1992) carried out a number of experiments with honeybees (*Apis mellifera*) who had been trained to fly along a tunnel to receive a reward. Vertical stripes lined the two sides of the tunnel. In a baseline condition, the stripes were physically stationary;

they drifted steadily backward across each eye's visual field as the bee flew along. Bees in this condition consistently stayed in the middle of the tunnel, thereby avoiding collision with either wall.

But then Srinivasan repeated the experiment, making the stripes on one of the walls move slowly as the bee flew along. On some trials the stripes moved in the same direction as the bee, and in other trials they moved in the opposite direction. When the stripes moved in the same direction as the bee, this reduced the rate at which they drifted backward across her visual field, just as if the wall were far away. Because of this motion parallax, the bee flew closer to the moving wall. In contrast, when the actual movement of the stripes was opposite to the bee's flight, increasing their rate of backward drift across the visual field, she adjusted her path to be farther from that wall, suggesting that she perceived it to be close by.

This set of experiments shows that bees use motion cues to obtain depth information. But which visual receptors are they using to acquire this information? To answer this question, Srinivasan et al. (1989) did other experiments in which the colors of the visual stimuli were varied. In one study, bees in the laboratory were trained to land on an artificial flower, in an artificial meadow, to receive a reward. There were many flowers, at different heights, but only those at a certain height held a reward. Bees flying above the flowers were able to use motion parallax to identify flowers at the correct height. For example, tall flowers were close to the bee, and therefore moved rapidly across her visual field as she flew past them; short flowers moved more slowly.

Bees readily mastered this task when the flowers were black and the meadow was white, a sharp contrast that activated all three receptor types—the UV, B, and G receptors. But when blue flowers were presented on a yellow meadow, the results were more complicated.

By adjusting the intensity of the yellow meadow, Srinivasan and colleagues caused it to activate the G receptors either more or less than the blue flowers did, so that these receptors "saw" the flowers as darker or brighter than the meadow. Under these conditions, the bees were able to choose the correct flowers by means of motion parallax. However, when the meadow and the flowers stimulated G receptors equally, those receptors could no longer distinguish the flowers from the meadow. In that condition of the experiment, the bees were unable to tell which flowers were at the correct height, although they could still see them with other receptors.

In other words, the bee's ability to use motion parallax depends not just on its ability to see objects but also on the special ability of some receptors—the G receptors—to acquire movement information. Further research has shown that G receptors use movement information for carrying out a variety of other tasks as well, leading Srinivasan (1992) to describe them as part of a "workhorse" subsystem.

Another insect that derives depth information from motion is the praying mantis, that lurks in foliage as it lies in wait for its next meal to come within reach (see Figure 5.1). It remains

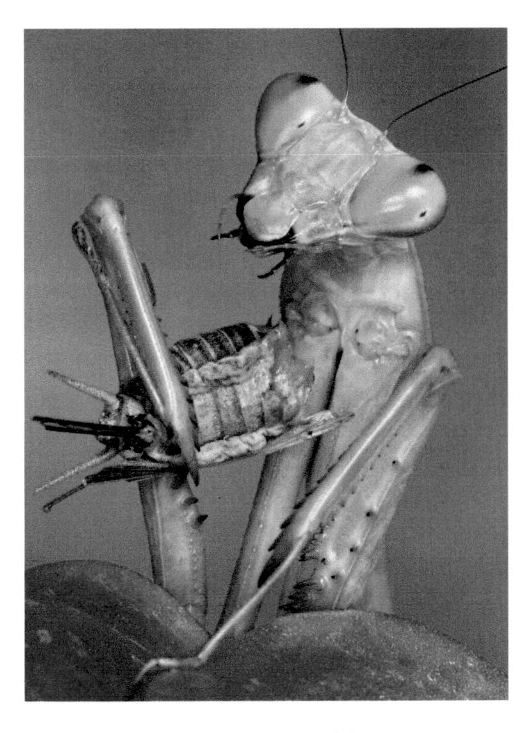

Figure 5.1 A giant Asian mantis (*Hierodula patellifera*) eating a cricket. Praying mantises typically peer at their prey before seizing it, moving their head from side to side. The resulting motion parallax enables them to accurately calibrate their strike.

***Source:* Used with permission of Shutterstock. Photo Contributor, Cathy Keifer.**

motionless while waiting, except for its head, which moves from side to side, a behavior called *peering*. With these movements the mantis produces motion parallax, enabling it to precisely determine the distance to its prey.

PICTORIAL DEPTH CUES

These examples demonstrate that motion parallax—movement of objects across the visual field, particularly when this is caused by movement of the observer—constitutes a powerful depth cue. But even in a stationary retinal image, information about depth may be present. During the Renaissance, artists learned how to use some of these features in their paintings to create an illusion of depth. As a result, they are called *pictorial depth cues*, although they are present not just in pictures but also in everyday life. For example, if one object partially blocks our view of another object, the first object is perceived to be closer to us than the second; this is the pictorial depth cue of *occlusion*.

Human infants aren't born with the ability to use pictorial cues but gradually acquire this skill by the age of five or six months (Kavšek et al., 2009). We know this from experiments using the *preferential reaching procedure* that rely on an infant's tendency to reach for a nearby object in preference to a faraway one. Two objects are simultaneously presented to the infant, at the same distance, but with pictorial cues being used to create the illusion that one is closer. Infants below the age of five months reach equally often toward the two objects, but beyond that age, they more often reach for the seemingly nearer object.

MONKEYS

A study by Gunderson et al. (1993) tested whether infant macaque monkeys could use pictorial depth cues. During their participation, the seven-week-old monkeys wore an eyepatch over one eye, to prevent the use of binocular depth cues. Two objects (small toys or dolls) were positioned at the same distance from the monkey but were attached to different parts of a large picture containing depth cues. For example, in one experiment, the picture was of a textured array resembling a tiled floor extending away from the viewer. One of the toys was lower on the picture than the other, so that it was superimposed on a coarser part of the background, making it appear closer if the monkey could make use of the depth cue called *texture gradient*. The monkeys consistently reached for the lower toy. Overall, the study suggests that even in infancy, macaque monkeys can use pictorial cues to capture information about depth.

PIGEONS

Experiments have shown that some birds are also able to use pictorial depth cues. In one such study, Cavoto and Cook (2006) let adult pigeons (*Columba livia*) view computer-generated scenes in which three objects (a sphere, a cube, and an icosahedron) were presented against a checkered background. Some of the scenes had pictorial cues, individually or in combination, that (for humans) made the three objects appear to be at different depths (see Figure 5.2). The cues were occlusion, texture gradient, and *relative size*. (Relative size refers to the fact that of two similar objects, the one with the larger visual angle appears closer.)

Pigeons were trained to peck a key only when presented with scenes in which the three objects appeared to be in a specific depth order, for example, sphere nearest and cube farthest away. They worked at this task over a long series of trials in which different cues were present, and the objects were in different depth orders. The researchers found that the birds' performance was better, the more pictorial depth cues were present; they were quite accurate when all three were present.

CUTTLEFISH

Sensitivity to pictorial depth cues is not limited to vertebrates, for there is evidence that cephalopods can also use them. Josef et al. (2014) placed cuttlefish (*Sepia officinalis*) in a tank where they were able to swim from one end to the other. The animals swam above a transparent platform most of which was covered with a checkerboard pattern. However, portions of the platform presented a different visual stimulus. In what we may call a "visual cliff" condition, these test

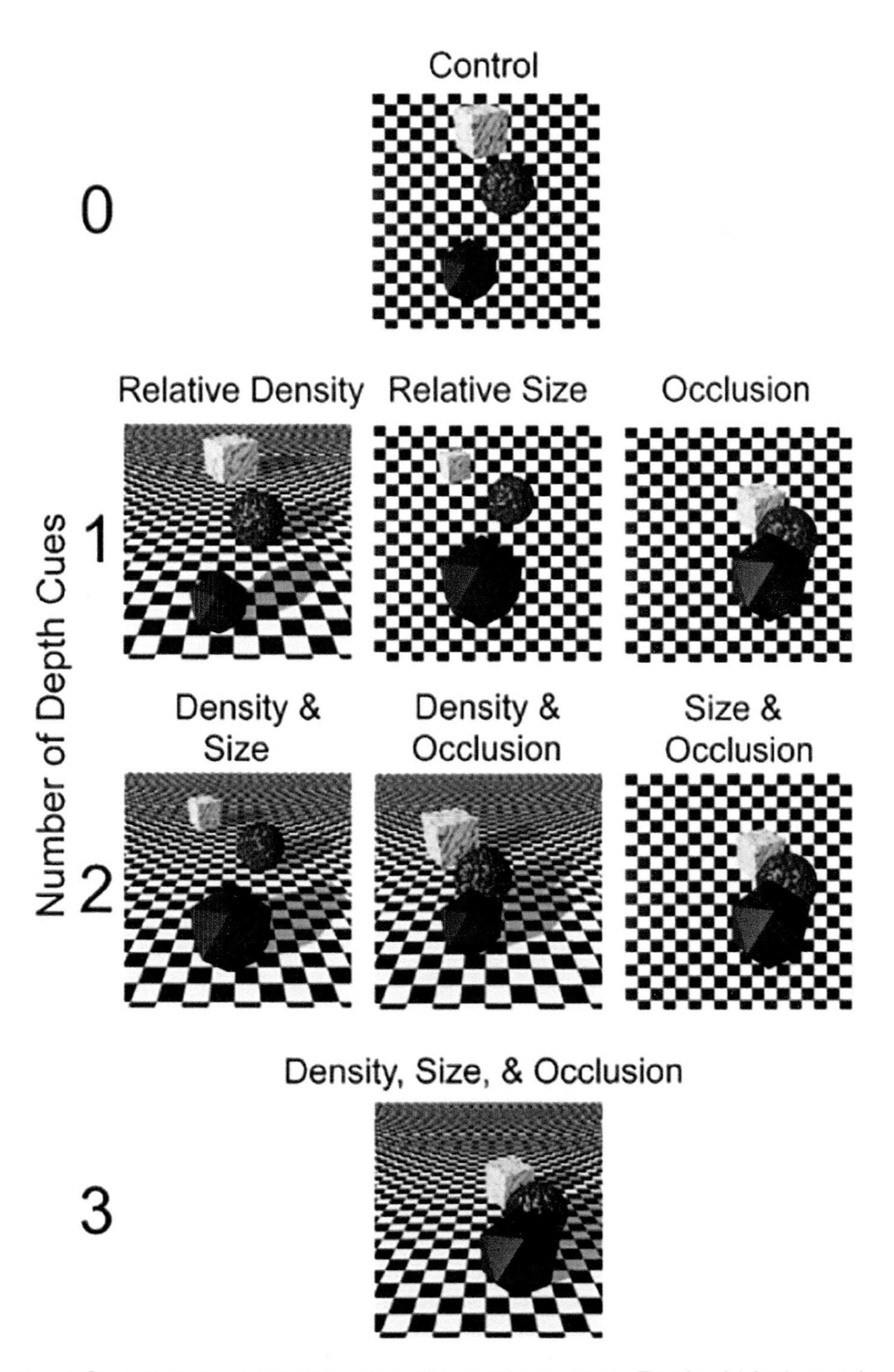

Figure 5.2 Some of the pictures used to study depth perception in pigeons. The pictorial depth cues of occlusion, relative size, and relative density (also called texture gradient) are present, singly or in combination, in the pictures. Their presence caused the pigeons to perceive the objects to be at different depths.

Source: From Cavoto, B. R., & Cook, R. G. (2006). The contribution of monocular depth cues to scene perception by pigeons. *Psychological Science*, *17*, 628–634. Used with permission of Sage Publications Inc. Journals; permission conveyed through Copyright Clearance Center, Inc.

areas were clear, allowing the cuttlefish to see through the platform to the floor of the aquarium far below. When cuttlefish swam along the tank in this condition, they hurried past these seeming drop-offs; as in other animals tested with a visual cliff, motion parallax probably made a major contribution to their depth perception here. But in another condition, a pictorial depth cue was presented: The test areas, instead of being clear, now had a distorted checkerboard pattern in which texture gradients gave the appearance of a deep crevasse. The cuttlefish avoided this as well, suggesting that they can use at least one pictorial depth cue.

BINOCULAR DISPARITY

Still another type of depth perception is *stereopsis*, the ability to use *binocular disparity* (small differences between the two eyes' views) to perceive depth (see Figure 5.3). Charles Wheatstone (1802–1875), inventor of the stereoscope, used it to prove that people can perceive depth based on binocular disparity, especially a horizontal disparity of vertical contours.

Since then, research has shown that many other animals also have this ability. Stereopsis of course requires that an animal be able to see the same objects with both eyes, as we can.

MAMMALS

Most animals have two eyes, but this contributes in different ways to the survival of different species. Those who must always be on the alert for predators generally have eyes on the sides of their head, enabling them to monitor their surroundings in all directions. The rabbit and the deer are examples. On the other hand, predatory animals that can afford—intermittently at least—to ignore everything except their prey generally have eyes that face forward and have stereopsis. For example, cats (Fox & Blake, 1971) and monkeys have been shown to have stereopsis. In fact, the stereoacuity of macaque monkeys—that is, the ability to detect small differences in depth

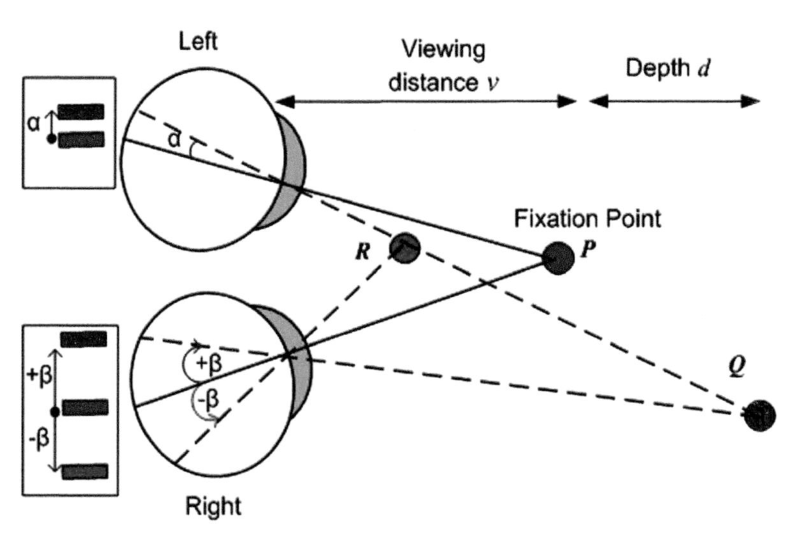

Figure 5.3 Binocular disparity, the stimulus for stereopsis. In this overhead diagram, a person or other animal is looking directly at point P, so it is imaged on the foveas of both eyes. Farther away and to the right of P is another object at point Q. Each eye forms an image of Q on the left half of the retina. However, the distance of this image from the fovea is smaller in the left eye (α) than in the right eye ($+\beta$). The difference between α and $+\beta$ is the binocular disparity of Q; it indicates that Q is farther away from the animal than P. Point R has a disparity of opposite sign, indicating its nearness.

Source: **From De Silva, D. V. S. X., Ekmekcioglu, E., Fernando, W. A. C., & Worrall, S. T. (2011). Display dependent preprocessing of depth maps based on just noticeable depth difference modeling.** ***IEEE Journal of Selected Topics in Signal Processing*****, 5, #2, 335–351. Used with permission of The Institute of Electrical and Electronics Engineers, Incorporated (IEEE); permission conveyed through Copyright Clearance Center, Inc.**

based on binocular disparity—is comparable to that of humans (Sarmiento, 1975).

Ambush predators like cats often have vertically elongated pupils, while some prey animals, such as sheep, have horizontally elongated pupils. Species differences in the importance of stereopsis may help to explain these differences in pupil shape (Banks et al., 2015). Light passing through the center of the eye's lens forms a sharp image on the retina, whereas light passing through its more peripheral regions blurs the image, a phenomenon called *spherical aberration*. An animal with a small round pupil is therefore likely to have retinal images that are sharp, but not very bright; if it has a large round pupil, the retinal images will be bright but blurred.

The cat's vertical pupil represents a good compromise between these extremes for a predator: It results in a retinal image with vertical contours that are both sharp and moderately bright, making good stereopsis possible. On the other hand, a horizontal pupil is adaptive for prey animals with lateral eyes and little or no stereopsis, because it ensures a sharp retinal image of the horizon, where predators may appear.

Cats, monkeys, and people have a well-developed visual cortex, and that brain region plays an essential role in their stereopsis. It will be recalled from Chapter 2 that the visual field is mapped onto visual cortex. Where the visual fields of the two eyes overlap, individual cortical neurons receive inputs from both eyes: That is, these neurons are *binocular*. In other words, a binocular cell in visual cortex has a receptive field—a small region of retina from which it receives information—in each eye. For many of these brain cells, the two receptive fields are in exactly corresponding locations on the two retinas. For example, they might both be 5° of visual angle to the right of the fovea of a person or monkey, or, in terms of the visual field, 5° to the left of the fixation point.

However, some binocular cells in visual cortex have receptive fields that are not in exactly corresponding locations in the two eyes; instead, one receptive field is slightly shifted to the left or the right with respect to the other receptive field. Cells like this contribute to the perception of depth. For example, consider a binocular cortical cell in a monkey's visual cortex that has a receptive field in the right eye's visual field that is 5° to the right of the fixation point, and a receptive field in the left eye that is 5.2° to the right of the fixation point. Now imagine that this monkey is looking at a piece of fruit in a tree, while just off to the right is a snake. If the snake is closer to the monkey than the fruit is, it will be imaged on the two retinas with binocular disparity. This will cause the snake to stimulate the cortical cell through both eyes at once, producing strong activation and contributing to a sensation of depth. In other words, the cell is sensitive to binocular disparity. The monkey will notice the snake and see that it is closer than the fruit.

We know this from an experiment by DeAngelis et al. (1998), who electrically stimulated clusters of cortical cells of the type just described, in an alert, behaving monkey. Before electrical stimulation, the money looked at a visual array having a certain disparity and accurately reported how far away it was. But when a cluster of neurons sensitive to a different disparity was activated by the experimenter, the monkey's perception changed: The stimulus array was now perceived to be at a different distance, closer to that coded by the electrically stimulated neurons. This experiment indicates that disparity-sensitive neurons play a crucial role in stereopsis.

BIRDS

A well-developed visual cortex is found only in mammals. However, other vertebrates process visual information with other brain structures, in ways that are sometimes similar to those found in mammals. A noteworthy example is birds: As we saw in Chapter 2, the roof of their forebrain follows organizational principles similar to those in cortex (Stacho et al., 2020). Does this imply that birds may also have stereopsis?

Fox et al. (1977) tested this possibility with a kestrel, a type of falcon (*Falco sparverius*). The bird was trained to sit on a perch and view two lighted panels 1.7 m away, at the ends of separate, parallel tunnels. Each panel contained a matrix of randomly arranged red and green dots. The kestrel wore a helmet with colored filters that allowed it to see only the red dots with one eye and only the green dots with the other eye. In one of the panels, the red and green dots were in register, but in the other panel, a group of red dots, forming a rectangle, was shifted with respect to the corresponding green dots. This introduced disparity, causing (in an animal with stereopsis) the appearance of a rectangle standing out in depth. If the kestrel flew to the 3D panel (which was sometimes on the left and sometimes on the right), it obtained a morsel of beef heart. The animal gradually learned to make the correct choice, showing that it had stereoscopic depth perception.

TOADS

To determine whether this sensory ability extends to cold-blooded vertebrates, Collett (1977) tested toads (*Bufo marinus*), who capture prey by quickly sticking out their tongue to the correct distance, showing that they have accurate depth perception. But does it depend on stereopsis? To answer this question, Collett placed prisms in front of a toad's eyes, then let it view a tasty mealworm. The prisms bent the light reaching the toad, shifting the mealworm to the right in the left eye's visual field, and to the left in the right eye's visual field, so that a binocular disparity was created. This wouldn't work in a person, because our eyes would reflexively converge so that they both looked directly at the worm, eliminating the disparity, but toads don't make eye movements. The result was that the toad's tongue undershot the mealworm, showing that disparity made the worm appear closer than it really was and, therefore, that toads have stereopsis.

PRAYING MANTIS

Remarkably, at least one insect—the praying mantis—has been proven to have stereopsis, with experimental methods similar to those used by Collett in toads. Mantises, like toads, suddenly capture their prey, but by grasping it with their forelegs rather than their tongue. Typically, the well-camouflaged mantis waits as an unsuspecting fly or other insect wanders closer. When it gets within reach (about 25 mm), the mantis suddenly lashes out. Rossel (1983) measured this triggering distance by attaching a fly to the end of a thin rod and moving it slowly toward the mantis. Then, the experiment was repeated, but with base-out prisms placed in front of the mantis's eyes to produce disparity that reduced the fly's optical distance. The effect of the prisms was that the mantis now attempted to seize the approaching fly while it was still out of reach but had an optical distance of 25 mm.

CUTTLEFISH

Among molluscs, the most likely possessors of stereopsis are the cephalopods, which have highly developed visual systems capable of examining their surroundings and then camouflaging themselves to blend in. To determine whether stereopsis is present in cuttlefish (*Sepia officinalis*), Feord et al. (2020) used blue and red anaglyph glasses to present separate images to the animal's left and right eyes. When they presented a stereogram of a walking shrimp, the cuttlefish positioned itself at an optimal striking distance before extending its tentacles to seize the prey. In the absence of disparity, this distance was about 12 cm, measured from the cuttlefish's eyes to the display screen. But when disparity indicated that the shrimp was in front of the screen, the cuttlefish backed off before striking; and when disparity indicated that the shrimp was behind the screen, the cuttlefish inched forward before lashing out. These adjustments prove that it is using stereopsis in hunting.

In summary, a wide range of animals, including many vertebrates, at least one insect, and a mollusc, have stereopsis. Unlike the pictorial cues, the use of which involves a certain amount of judgment and interpretation, the extraction of depth information from binocular disparity is a crisply defined mathematical process that uses computer-like algorithms. The overall perceptual construction of stereoscopic depth in a scene is more complicated in a monkey (Poggio, 1995) than in a mantid (Rosner et al., 2019), but in both cases the first step is the same: the comparison, by individual brain cells, of sensory signals from the two eyes.

VISUAL NAVIGATION

Many animals navigate, which is to say that they find their way to a goal. The squirrel who returns to the place where a nut is buried; the robin bringing a worm to its chicks waiting in the nest; the copperhead moving silently in the night toward a small, warm something hiding in the grass—all are navigating. When animals navigate over long distances, the process is called migration. Examples are the fabled return of the swallows to Capistrano, the flight of monarch butterflies from Canada to Mexico, and the circumnavigation of the ocean by young loggerhead turtles.

Navigation by animals has been compared (von Frisch, 1967) to the efforts of a ship's captain to reach port. While on the open sea, the captain (in the days before GPS) used compass and sextant to navigate on the basis of what we may call *global cues:* types of information that are available anywhere, such as compass direction and the position of sun and stars. But when the ship enters the harbor, the captain switches to the use of *local cues* which emanate from the goal itself such as the sight of the pier and waterfront buildings.

Animals also use local and global cues when navigating. They can obtain information about the distance or direction to the goal through multiple sensory channels—vision, hearing, smell, the feel of wind and water currents, the sensing of magnetic fields. In relying on their senses, animals generally use an inclusive approach, relying on whichever of their senses are useful at a given time.

Scientists have made many discoveries as to the types of sensory information particular animals use, and under what conditions, but much remains unknown. For example, a dog given to a new owner may break free and return to its original home miles away. Are familiar sights and smells sufficient to explain this phenomenon, or is there more to it? Research on such questions continues.

In this chapter, we will consider the use of *visual* information in navigation. The contributions of other senses will be considered in later chapters.

LANDMARKS

Vision provides navigational information in the form of landmarks, both large (mountains, seacoasts, and rivers) and small (trees and buildings). Although the use of landmarks is perhaps the most obvious potential navigation strategy when animals are in a familiar environment, clear evidence for it has been scarce until recently. The challenge is that several types of information may be available, making it difficult to establish which ones are being used.

But a study by Mora et al. (2012) using homing pigeons (*Columba livia*) is persuasive. Each pigeon was trained to return home from two release sites more than 6 km away. One return path led over territory with no distinctive landmarks, while the other path led near four highly salient wind turbines. When pigeons were taking the turbine route, they set off in a more accurate direction and flew more directly home than when they were taking the control route. This suggests that the turbines were functioning as landmarks that helped the birds find their way home.

The question is not whether landmarks are the *only* way a pigeon finds its way home (they are not), but whether they make an important contribution. To confirm that they do, Mora and colleagues carried out an additional experiment in which they interfered with another navigational tool, the pigeons' ability to home based on the sun's position in the sky. This ability, called *a sun compass,* is described in the next section; to interfere with it, the experimenters misled the birds as to the time of day. When the affected pigeons flew home on the turbine route, almost all of them eventually made it to the loft. When attempting to take the control route, however, they were thrown far off course and fewer than half of them made it home. This indicates that pigeons can use both a sun compass *and* visual landmarks: When there are salient landmarks, pigeons rely heavily on them, but in the absence of landmarks, there is stronger reliance on the sun. Overall, the study shows that homing pigeons use landmarks, when available, together with other types of sensory information.

Further evidence for the use of visual landmarks when homing pigeons are in a familiar environment has been obtained by Gagliardo et al. (2020). Each bird was first allowed to learn the way home from each of three release sites, wearing a GPS tracking device. The sense of smell was then inactivated in half of the pigeons, with a zinc sulfate nasal wash that caused olfactory neurons to degenerate. (They grew back after the experiment had ended.) Pigeons were then set free at each of the release sites, and their journeys home compared with the earlier ones. The anosmic pigeons stuck close to the exact route they had followed earlier, presumably relying on a series of landmarks, whereas the control pigeons extemporized more the second time, having olfaction available to help them choose their route. This study, like that of Mora et al., shows that visual landmarks are important in familiar environments, particularly when other navigational tools are compromised.

THE SUN COMPASS

One of the most important and widely used navigational cues is the position of the sun. The ability to extract and use this information is widely possessed by both vertebrates and invertebrates. It is the earliest global cue to be recognized and analyzed by scientists. We will begin by describing its presence in honeybees, which navigate over distances of a kilometer or two on a daily basis. Bees have aroused interest and curiosity since ancient times because of their usefulness to people in the form of crop pollination and honey, and their remarkable ability to communicate spatial information to one another. Foraging bees who find flowers containing nectar or pollen return home with a sample of the food, and let others know where to go for more.

How do they do this? Inside the dark nest, a returning forager clings to the side of a vertical comb while performing a stereotyped series of movements called a *dance*. The coded information in the dance was deciphered by Karl von Frisch (1967). He observed that, after returning from a distant food source, the forager dances along a straight line while "waggling" her abdomen from side to side and making a buzzing noise with her wings to attract the attention of nestmates. After about a centimeter, she loops around to the starting point and begins another waggling run; she continues in this way, looping alternately to the left and right. Other bees, attracted by the commotion, crowd around the forager, touching her to learn the orientation of the waggle run. This orientation symbolizes the direction to the food, relative to the sun. If the waggle run is upward, the worker bees must fly directly toward the sun. A tilted waggle run means the food is in a direction to the left or right of the sun. Only the azimuth of the sun (direction with respect to north, south, east, and west) is represented; the sun's elevation is ignored.

The tempo of the waggle dance indicates the distance to the food. When the food source is less than 100 m from the nest, the dance is simpler, serving mainly to alert nestmates to the presence of food nearby.

Honeybees are able to take the time of day into account, based on an internal clock. If, for example, a foraging honeybee returns to the nest just before sunset and waggles upward, indicating that food is in the direction of the setting sun, and then the nest is sealed overnight by the experimenter, bees setting out in the morning will fly *away* from the rising sun, and so reach the food.

Many birds, especially those who migrate, also use a sun compass. There is an extensive literature, going back more than a century, on the multiple factors involved in bird migration; and birds' use of the sun as a source of directional information was the first piece of this complex puzzle to be definitively established (Kramer, 1952).

Kramer carried out a series of experiments on starlings. In one experiment, a starling was kept in a circular cage, inside a small pavilion with six windows. Tested at the time of spring migration, the starling took up a north-facing stance inside its cage, whenever the sun (or a patch of sky close to the sun) was visible through one of the windows, and regardless of the time of day. But when the sky was heavily overcast, its orientation was inconstant and haphazard. This showed that the bird was using the sun to determine the correct direction for migration.

In a second experiment, Kramer placed a starling in a circular arena surrounded by feeders. The arena was covered by a canvas tent to block out the bird's view of landmarks, and a bright light was positioned just outside the tent to shine through the canvas and serve as an artificial "sun." The bird was trained, always at the same time of day, to obtain food at a certain feeder.

Finally, the starling was tested in the arena at different times of day, but with the artificial sun remaining in the same place. The bird shifted from one feeder to the next depending on testing time, compensating for its (now incorrect) belief that the "sun" was moving across the sky. This experiment indicates that the starling uses an internal clock to update its sun compass.

The idea of a sun compass is actually a complex one, for different birds, not to mention other animals, may use information about the sun's position in different ways (Guilford & Taylor, 2014). It is often used in conjunction with an internal clock, so that the animal can adjust its goal direction taking into account the movement of the sun across the sky. In Chapter 2, we saw that in birds, the Wulst appears to have a special role in monitoring the sky. Consistent with this view, Budzynski et al. (2002) found that pigeons with lesions of the Wulst were impaired in the use of a sun compass, compared with control pigeons.

A STAR COMPASS

INDIGO BUNTINGS

Many migratory birds travel at night rather than by day. An example is the indigo bunting (*Passerina cyanea*), a bright blue denizen of the eastern United States during spring and summer, which molts to a dull brown in the fall and flies southward, mainly to Central America. Emlen (1975) showed that the bunting navigates by the stars. Early in life it learns to recognize constellations that move in a small circle as the night progresses, thus indicating the location of the North Pole. Emlen reared a group of buntings in a planetarium, under a fake night sky in which stars rotated around a stationary Betelgeuse (a bright star visible from the United States in the southern sky); a control group was reared in the same planetarium but under a normal sky. At the time of fall migration, all birds were tested under a normal sky. Those in the control group turned in their cages to face away from the North Pole, while those in the experimental group turned to face away from Betelgeuse. So buntings aren't born with a knowledge of constellations but do have an instinctive propensity to notice how stars move with respect to one another.

DUNG BEETLES

The only insects known to navigate by the stars are dung beetles. When some of these beetles, such as members of the Old World species *Scarabaeus satyrus*, find a supply of dung, they roll a ball of it into a burrow and lay an egg in it. Dacke et al. (2013) examined the cues used by these beetles as they navigated with their ball of dung away from the center of an experimental arena. Under a full night sky, they moved in a relatively straight line, but on an overcast night or when their view of the sky was blocked, they wandered aimlessly (see Figure 5.4). Their best guide was the moon, but they still did fairly well on moonless nights. To identify the celestial cue used by the beetles in the absence of the moon, the experimenters moved the wooden, 2-m diameter arena into a planetarium, where the contents of the sky could be manipulated. The key celestial feature was a diffuse streak of light representing the Milky Way. While dung beetles do not have the acuity needed to see an individual star, they can see this edge-on view of our own galaxy consisting of billions of stars, and use it for navigation.

A later study (Foster et al., 2021) showed that the behavior of the beetles can be disrupted in several ways by light pollution, which is becoming an increasingly severe problem for many nocturnal species. Light pollution makes the starry sky harder to see. If the pollution is direct, for example from a floodlight, beetles will congregate near it; if the pollution is indirect, in the form of skyglow, they become disoriented.

POLARIZED LIGHT

We have seen that on a sunny day, foraging bees can find nectar and pollen, return to their hive, and indicate the direction to the food—relative to the sun's position—through their dance. But what happens if the day is "partly cloudy" so that the sun is behind a cloud but areas of blue sky are visible? Bees can once again dance accurately. Something in the patches of blue sky is telling

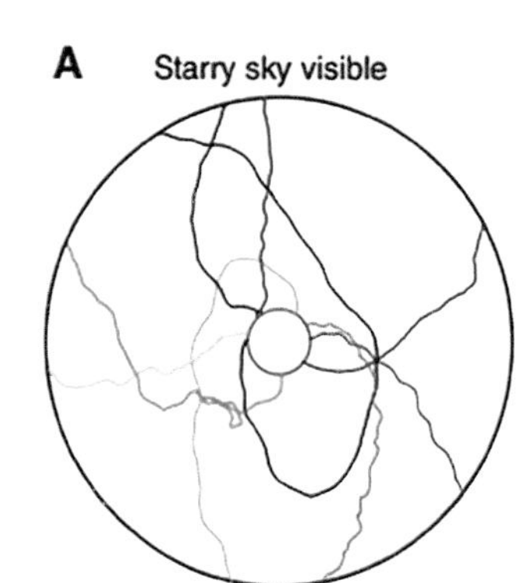

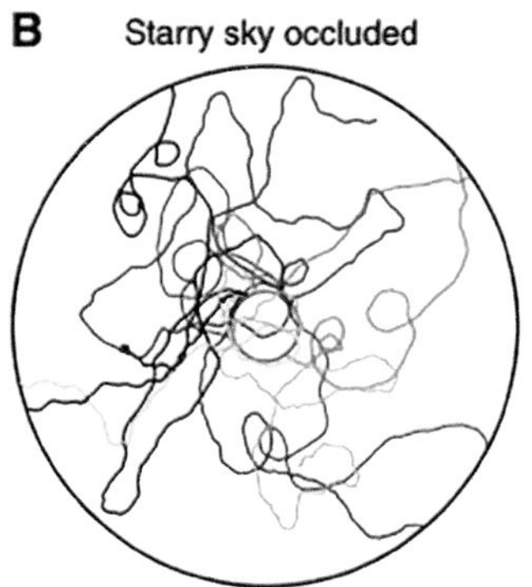

Figure 5.4 After forming a dung ball, a dung beetle normally heads straight home with it. To determine whether they navigate by viewing the sky, beetles were filmed after being released with their dung ball from the center of an arena. Those with a view of the starry sky traveled in relatively straight lines (individual tracks in A), but those wearing opaque caps that blocked their view of the sky (C) wandered aimlessly (B). Others wearing transparent caps were not impaired.

Source: **From Dacke, M., Baird, E., Byrne, M., Scholtz, C. H., & Warrant, E. J. (2013). Dung beetles use the Milky Way for orientation. *Current Biology*, *23*, 298–300. Used with permission of Elsevier Science & Technology Journals; permission conveyed through Copyright Clearance Center, Inc.**

the bee where the sun is. Von Frisch realized that this "something" is the polarization of skylight, a subject to which we now turn.

The sun gives off light in all directions. Because some of it travels directly to our eyes, we can see the sun's brilliant disk. (Don't look at it!) But much of the light arriving from the sun reaches our eyes indirectly, after bouncing off air molecules. This is scattered light. The sky is blue because blue light scatters more than other wavelengths. (Ultraviolet light scatters even more, but we can't see it.)

As a photon moves through space, it vibrates sideways. These vibrations are in a plane that is different for different photons. For example, if a photon is traveling directly toward you at eye level, it may be vibrating up and down, or to the left and right, or in some other orientation perpendicular to its overall direction of movement. Most beams of light are composed of photons that are vibrating in many different orientations.

Some beams of light, however, consist of photons that are all vibrating in the same orientation; in this case, we say that the light is *polarized*. For example, if a beam of sunlight strikes the surface of a lake at an angle, some of its photons will be reflected, while others will be refracted down into the water. Most of the reflected photons will be horizontally polarized, while most of the refracted rays will be vertically polarized.

An easy way to obtain polarized light is with a *polarizing filter* that allows photons vibrating in a particular orientation to pass, while absorbing others. Sunglasses often include a layer of vertically oriented polarizer, to reduce the glare of horizontally polarized reflections from the surface of water or other shiny horizontal surfaces.

Light from the blue sky is somewhat polarized, as a result of the way light scatters. The orientation of polarization (the *e-vector* of the light) varies from place to place in the sky and depends on the location of the sun. Roughly speaking, the e-vector of a point in the sky is perpendicular to a line connecting that point with the sun. This pattern of polarization indicates the location of the sun, even when the sun itself is invisible. Some animals use this pattern to determine where the sun is and, therefore, where they are headed.

POLARIZATION SENSITIVITY IN INSECTS

Bees. It has long been believed, based on careful research in an outdoor setting, that bees respond to the polarization of light, but it was hard to prove this conclusively in the natural environment, where polarization is just one of several potential cues to navigation. To settle the issue, Kraft et al. (2011) trained honeybees to make use

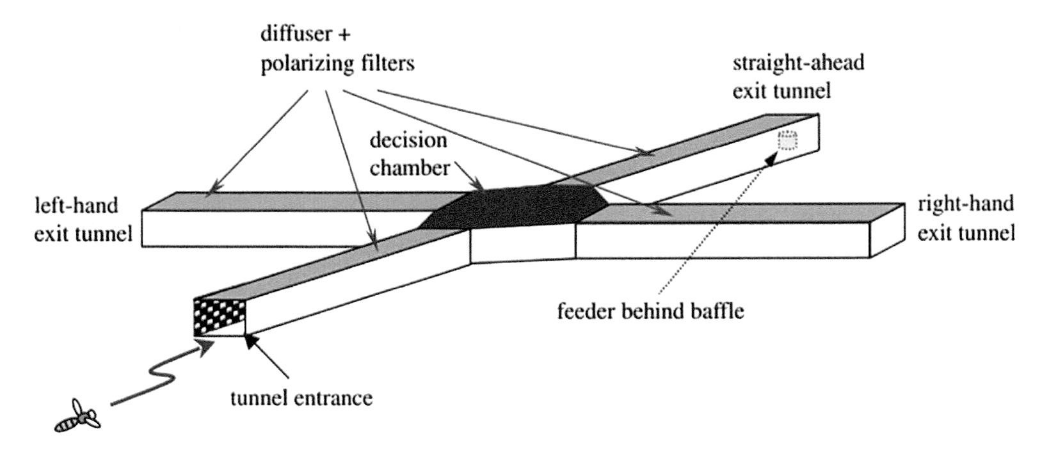

Figure 5.5 A maze used to study the ability of honeybees to use the polarization of light for navigation. After entering the maze and reaching the central decision chamber, bees traversed one of three exit tunnels. Polarized illumination indicated which exit tunnel led to a food reward. After learning the task, bees were tested without reward to ensure they were not using smell to identify the correct exit.

Source: **From Kraft, P., Evangelista, C., Dacke, M., Labhart, T., & Srinivasan, M. V. (2011). Honeybee navigation: Following routes using polarized-light cues. *Philosophical Transactions of the Royal Society of London B, 366*, 703–708. Used with permission of The Royal Society (U.K.); permission conveyed through Copyright Clearance Center, Inc.**

of polarization to find their way through a maze to obtain a reward. The maze was in the shape of a cross, with an entrance arm and three exit arms. Bees traversing the entrance arm reached a decision chamber at the center of the cross and were free to use any one of the exit arms to leave the maze. But only one of the exit arms contained a food reward, which was not visible from the decision chamber (see Figure 5.5).

Each of the four arms was lit from above by polarized light. In one experiment the entrance arm and one of the exit arms had light polarized parallel to the arm, while the other two arms had light that was polarized transversely; the bee's task was to choose the exit arm where the polarization (with respect to the arm) matched that in the entrance arm. Bees had little difficulty learning such tasks, showing that they were able to find their way based on the polarization of light.

But could they use this sensory ability to tell other bees, by means of the waggle dance, the way to a reward? To test this possibility, Evangelista et al. (2014) allowed bees in the laboratory to fly along a tunnel, at the end of which a food reward was waiting. The bees then returned to their hive and did a waggle dance. The independent variable was the angle of polarization of the tunnel's overhead illumination. Remarkably, the orientation of the bees' waggle dance was related to the e-vector of the light, as if the roof of the tunnel had been a patch of blue sky overhead as the bee flew toward a food source. The authors note that polarization would be an even more complete source of information in the natural environment, where the bee can see more than one patch of sky.

Ants. Given that ants and bees are on the same major branch (Order Hymenoptera) of the insect family tree, it is not surprising that their methods of navigation are similar. For example, ants also use the polarization of skylight in their foraging activities. In a series of experiments, Rüdiger Wehner (1994) showed how North African desert ants (genus *Cataglyphis*) can detect the polarization of light coming from different parts of the sky and use this information to determine their heading when the sun is invisible. Foragers wander about, looking for food; when they find some, they hurry back to the nest to recruit others. Remarkably, they do not retrace their steps on this return journey: Instead, they go directly home along a straight line, using polarized light to help them stay on course.

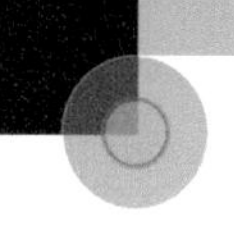

Ultraviolet (UV) receptors are essential to the response of bees, ants, and other insects to polarization. In one demonstration of this, a member of Wehner's research team kept pace with individual ants as they walked homeward, keeping a large colored filter over the ant so that only light from a particular part of the spectrum reached it. When the ant could see UV light, it stayed on course; but when it could see all wavelengths except UV, it wandered erratically (Wehner, 1976). Only in the ultraviolet region of the spectrum was the polarization of light useful to the ant, and therefore, the ant was using only its UV receptors for this purpose. Similar results in bees had been obtained earlier by von Frisch (1967). Insects' UV receptors thus serve double duty: They help with navigation and also with color vision.

The location of the sun and the orientation of polarized light are also used for navigational purposes by the monarch butterfly on its long-distance migrations (Reppert et al., 2010). Apparently, these sensory abilities and the central mechanisms needed to support them, such as an internal clock, are widespread among insects.

The polarization sensitivity of insects is achieved in a straightforward way, for the elongated pigment molecules in receptor cell microvilli are all locked in place, with their chromophores parallel to the long axis of the microvillus that houses them. Since photons are most efficiently absorbed by pigment molecules that are aligned with their vibrations, a receptor cell whose microvilli all point in the same direction will be selectively sensitive to that same plane of polarization.

Polarization sensitivity in most insect species that have been studied—including bees, ants, beetles, wasps, crickets, and dragonflies—is confined to a specialized area at the top of the eye, a location consistent with the importance of celestial observation for navigation (Wehner, 1994). In other parts of the eye, insects can lack polarization sensitivity either because central structures necessary for its analysis are lacking or because some receptor cells are twisted so that their microvilli do not all point in the same direction (Wehner, 1976).

POLARIZATION SENSITIVITY IN VERTEBRATES

The ability to sense the polarization of light is by no means limited to insects. Other arthropods have this ability, as do many molluscs, including the octopus. Humans, however, do not have it, and therefore, scientists assumed until recent decades that it is entirely lacking in vertebrates. The difference between vertebrate and invertebrate photoreceptors, described in Chapter 3, made this a plausible assumption. But it was quite wrong. Although absent in mammals, polarization sensitivity is found in some birds and in many fish. As is usually the case, when useful information is available in the environment, methods of capturing it will evolve.

This ability is of special value to fish, since polarization is a rich source of information in the complex underwater environment (Kamermans & Hawryshyn, 2011). Like other animals, fish can use the pattern of polarization of skylight (made more complex when it enters the water) to help with navigation. And polarized reflections from the iridescent scales of some fish may signal their presence to others, of the same or different species. In fact, there is currently intense interest in the question of whether and how fish communicate with one another by means of polarized light, but more research is needed before any general conclusions can be drawn (Marshall et al., 2019).

The way in which vertebrate visual receptors are able to respond selectively to polarized light has been a long-standing mystery. Rods and cones are activated when light is absorbed by the chromophore in their visual pigment molecules. As with invertebrates, the chromophore is most capable of absorbing photons with an e-vector parallel to its long axis. But visual pigment molecules in a rod or cone are not lined up parallel with one another: They are confined to disks that are stacked like dinner plates within the outer segment of a receptor, but they can drift and turn, independent of their neighbors, within the disk. Thus, a receptor cell is no better at absorbing, say, vertically polarized light than horizontally polarized light.

The special case of anchovies. But nature has found ways around this problem. One is found in—and may be unique to—anchovies. Electrical recordings from the optic nerve of the northern anchovy, *Engraulis mordax*, show that the overall response of the retina of this small fish differs in magnitude depending on the polarization of the stimulating light (Novales Flamarique & Hawryshyn, 1998). How does its retina achieve this differential sensitivity?

The answer is both simple and astonishing (Fineran & Nicol, 1978): The disks in some cone outer segments are turned sideways (see Figure 5.6). A photon, reaching the retina from the pupil and traveling along the long axis of the cone, will encounter these disks edge-on. It is more likely to be absorbed if its e-vector is parallel to the disk (and therefore parallel to the long axis of some of the chromophores within it) than if it is perpendicular to the disk (and therefore not parallel to any of its chromophores). So these receptors are differentially stimulated depending on the polarization angle of the incident light: They are *dichroic*.

In a large portion of the northern anchovy's retina, there are two types of cones, long and short. Both have sideways disks, but the disks in the long cones are oriented perpendicular to those in the short cones (Fineran & Nicol, 1978; Novales Flamarique, 2011). The long cones are most sensitive to incoming light that is horizontally polarized, while the short cones are most sensitive to vertically polarized light. Working together, then, these two types of cones can capture information about the angle of polarization of light.

The long and short cones contain different visual pigments: The long cones are believed to contain a green-sensitive pigment, while the short cones contain a blue or UV-sensitive pigment. Since light underwater is predominantly green and horizontally polarized, the ongoing response of the long cones will generally be greater than that of the short ones. Perhaps the short cones serve an alerting function, responding to unusual objects or events, such as a shiny blue predator (Novales Flamarque, 2011).

The role of double cones. In other fish, and polarization-sensitive vertebrates more generally, the basis of polarization sensitivity is not so clear, but the answer appears to involve remarkable structures called *double cones*. These consist of two

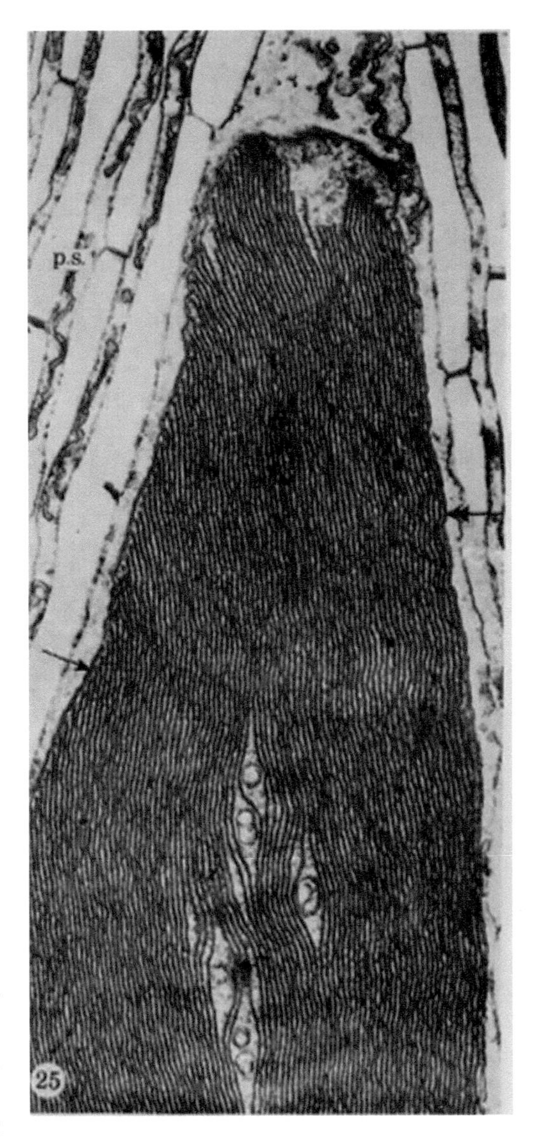

Figure 5.6 Outer segment of a cone photoreceptor in the anchovy retina. Its tapering profile points toward the back of the eye (upward in the figure). The thin lines inside the cone are discs, seen edge on, each containing thousands of visual pigment molecules. In most vertebrates, these discs are perpendicular to the long axis of the cone (see Figure 2.3), but in the anchovy, they are parallel to the long axis. This gives anchovies the ability to detect polarized light.

Source: **Fineran, B. A., & Nicol, J. A. C. (1978). Studies on the photoreceptors of *Anchoa mitchilli* and *A. hepsetus* (Engraulidae) with particular reference to the cones. *Philosophical Transactions of the Royal Society of London B*, *283*, 25–60. Used with permission of The Royal Society (U.K.); permission conveyed through Copyright Clearance Center, Inc.**

nonidentical cones, a chief cone and an accessory cone, which are pressed closely together and even share some intracellular elements. Double cones are found in many fish, amphibians, reptiles, and birds, but not in mammals; in other words, they are found, broadly speaking, in vertebrates that have polarization sensitivity.

One theory of double-cone functioning in fish such as the rainbow trout (Novales Flamarique et al., 1998) is that a diagonally oriented membrane between the two cones reflects a polarized component of incoming light, relaying it to single cones that are positioned nearby within the retinal mosaic. These recipient cones absorb the light and send a polarization-specific signal on to retinal networks. The recipient cones are thought to be UV-sensitive, since the trout's polarization sensitivity is greatest in the ultraviolet part of the spectrum (Hawryshyn, 2000).

In birds, the situation is more complicated. Polarized light influences navigational behavior by interacting with the birds' magnetic compass (Muheim et al., 2016). This is a subject that we will discuss in Chapter 10.

ALAN AND NAVIGATION

As we saw in Chapter 2, widespread use of artificial light at night (ALAN) can have harmful effects on reproduction in fish. But it interferes with ecological balance in other ways as well. One of these is its effects on animal navigation. Here, we will consider two examples.

MOTHS

It is proverbial that moths are drawn to a flame, and in fact, they are to any bright light, especially lights rich in short-wavelength (blue) light. A mercury or LED streetlight will draw moths out of surrounding trees and shrubbery at night, revealing their presence to insectivorous bats.

Many moth species, upon hearing the echolocating cry of a bat approaching in the dark (a subject discussed in Chapter 7), will navigate out of the way, by dropping directly downward, going into a downward spiral, or zigzagging erratically. Is this escape behavior affected by ALAN? To find out, Svensson and Rydell (1998) observed the responses of individual winter moths (genus *Operophtera*) to bursts of ultrasound simulating a bat's cry. The behavior of moths close to a mercury vapor lamp was compared with others in a wooded area more than 50 m from the lamp.

Almost all the moths tested in the woodland reacted as expected to the ultrasound, diving out of the path of a presumed bat. But of those tested within 1 m of the lamp, fewer than 40% did so. Apparently, the bright lamp was such a powerful attractant that it made the moths oblivious to the danger signaled by the auditory stimulus.

This imbalance in the contest between bat and moth may sound like good news for bats in the short run, but it is bad for moths and the night-flowering plants they pollinate. The ecological web that binds many species together in a given habitat is delicate, and can be disrupted by light pollution in ways that may not be immediately obvious.

BIRDS

It has long been observed that bright lights at night can attract and disorient birds. In a unique study, Van Doren et al. (2017) document these reactions to the intense, upward beams of blue light of the Tribute in Light, an annual display in New York City commemorating the 2001 terrorist attack on the World Trade Center. The two beams shine from dusk to dawn on the night of September 11 but are periodically turned off for about 20 minutes to mitigate their effect on birds migrating through the area.

Over a seven-year period, the investigators used radar measurements to compare the number of birds within 500 m of the memorial during periods of illumination and periods of darkness on the same nights. On average, there were 3.4 times as many birds in the area when the beams were on compared to when they were off (see Figure 5.7).

The birds also flew more slowly during periods of illumination; visual observations from the ground confirmed that many were circling or flying erratically. In addition, sound recordings revealed that birds were vocalizing more in the artificial light than in the dark. Such flight calls can be maladaptive if they induce conspecifics to join the caller in moving toward light from tall buildings (Winger et al., 2019).

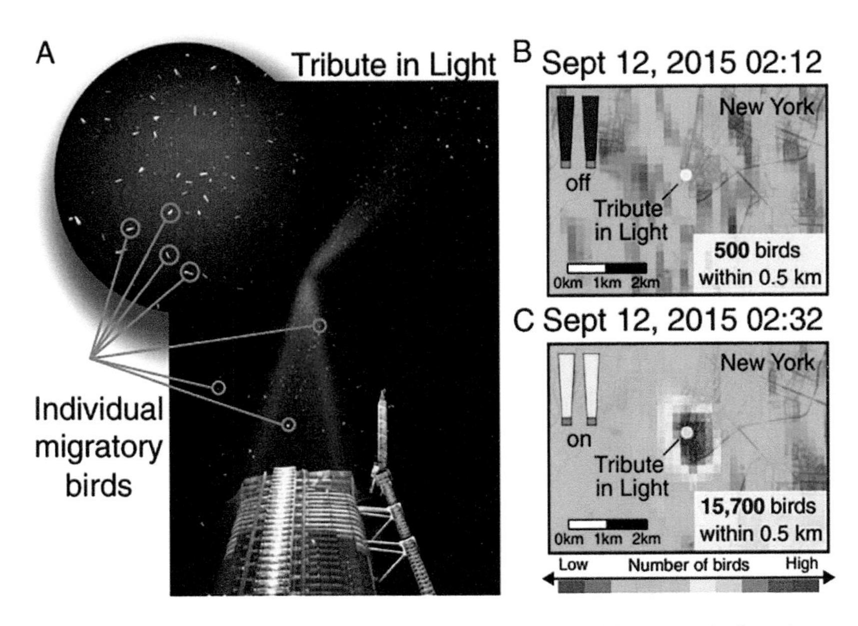

Figure 5.7 The effect of the Tribute in Light display on bird navigation. (A) Photographs from the ground show many birds in and near the two beams of blue light, which appear to converge because of linear perspective. (The building at the bottom of the picture is not part of the memorial.) Color-coded maps show that the number of birds near the display was much higher when the lights were on (C) than when they were off (B).

Source: **From Van Doren, B. M., Horton, K. G., Dokter, A. M., Klinck, H., Elbin, S. B., & Farnsworth, A. (2017). High-intensity urban light installation dramatically alters nocturnal bird migration. *PNAS*, *114*, 11175–11180. Used with permission of PNAS.**

SUMMARY

One of the most important functions of vision is to determine where objects are located. This requires that light coming from different directions activate different visual receptor cells. In primates, for example, light from the animal's left activates visual receptors in the right half of each eye's retina, causing perception of a stimulus on the left.

Perceiving an object's location in space involves not just its location in the visual field but also its distance from the perceiver. This depth perception is based on several types of information. The most important of these is motion parallax, in which the animal's own movements (such as the peering movements of a praying mantis) result in a relative shifting of objects across the field of view that reveals their locations in depth. Pictorial cues, such as texture gradients, contribute to depth perception in mammals, birds, and other animals with advanced nervous systems. And animals with front-directed eyes have stereopsis, a type of depth perception based on binocular disparity.

Visual perception of large-scale spatial relationships helps animals to navigate, whether across a field or during a long migration. Landmarks, such as a specific tree, can be used on short journeys. Seasonal migrations involve celestial cues such as the position of the sun or the stars, and internal compasses based on these cues often take into account the time of day.

Even when the sun is not visible, the polarization of light from the sky can indicate its position. Most visual receptors in invertebrates can detect this polarization because of the highly structured arrangement of their visual pigment molecules. Vertebrate photoreceptors do not have this advantage, but in some species, other ways of detecting polarization have evolved.

Artificial light at night disrupts navigation in many nocturnal species, by attracting and disorienting animals and blotting out their view of the sky.

6 Touch

For human beings, touch (broadly defined as perception of mechanical or thermal stimulation of the body surface) is the most crucial sensory modality—the one it is hardest to live without. Individuals with no sense of touch would be greatly challenged. How could they perceive whether an object was heavy or light, hard or soft, rough or smooth? How could they eat and drink? Sense dangerous heat or cold? Become socialized, starting in infancy? And reproductive behavior would be less strongly motivated without mutual touch.

The same is true throughout the Animal Kingdom. No other sense is so essential. Moles that live underground are nearly blind, and molluscs have little hearing; yet both groups thrive without these modalities. But even the lowliest animal on the most obscure branch of the tree of life recoils from the touch of the biologist's probe, showing that it has, and therefore needs, a sense of touch.

We will begin this chapter with a description of the receptor cells for touch in human skin. Most striking of these are mechanoreceptors: myelinated sensory endings with accessory structures—capsules—that accentuate specific components of stimulus force. Others, without capsules, are specialized at the molecular level to respond to warmth, cold, or noxious stimulation. Tactile information is represented centrally in somatosensory cortex, where maps of the body surface emphasize mobile body parts such as the hands and lips.

The cutaneous receptors of other mammals are similar to ours, but receptors increasingly different from our own, such as those associated with scales, appear in other vertebrate classes. Fish have a unique lateral line system that responds to pressure waves in a water environment; arthropods and molluscs have still other types of cutaneous receptors. A nearly universal principle, however, is the presence in animals of both slowly adapting and rapidly adapting mechanoreceptors.

Later in the chapter, we discuss emotional aspects of the sense of touch. Gentle, affiliative touch is common among mammals and, to some extent, other vertebrates; there is evidence that it promotes emotional and physical health, especially during development. Pain, in contrast, is inherently aversive; its avoidance in representatives of all three phyla reviewed here has been used as an argument for widespread animal sentience.

THE HUMAN SENSE OF TOUCH

RECEPTOR CELLS IN THE SKIN

In humans (and many other vertebrates), it is not skin cells that are sensitive to touch, but specialized receptors in or just beneath the skin. These receptor cells are connected to nerve fibers that carry touch messages to the animal's nerve centers—the spinal cord and brain. Receptors are of many different kinds. Some appear as simply the tapered peripheral ends of nerve fibers. Traditionally called "free" nerve endings because they are not embedded in anatomically

DOI: 10.1201/9781003362319-6

recognizable structures, they are nevertheless specialized, at the molecular level, for responding to specific types of stimulation.

For example, some sensory nerve endings have specialized receptors in their membrane that are activated by painful heat, opening a channel for the entry of ions into the cell and triggering a message that is sent along the fiber. These *TRPV1 receptors* respond not just to heat but also to chemicals called vanilloids, such as capsaicin, the "hot" ingredient in chili peppers (Caterina et al., 1997, 2000). Other free nerve endings may look indistinguishable from those that contain TRPV1 but have instead other members of the transient receptor potential (TRP) family of receptor molecules that make them sensitive to other types of stimulation, such as cold (McKemy et al., 2002). These studies were carried out by David Julius's research team at the University of California, San Francisco.

Most free nerve endings respond best to stimulation that is strong enough to be painful. Pain is such an important topic that we will deal with it in a section of its own later in the chapter.

There are, however, sensory nerve endings in the skin whose endings are not free but enclosed in tiny capsules. These encapsulated nerve endings all respond to gentle mechanical stimulation and so are called *mechanoreceptors.* They are receptors for feeling the breeze on your skin, the pen you are writing with, and the texture of tree bark. Different types of capsules provide different mechanical advantages, making the enclosed nerve fiber especially sensitive to a particular type of force or event.

MECHANORECEPTORS

Mechanoreceptors fall into two broad categories. *Slowly adapting mechanoreceptors* respond continuously to a steadily applied force, as when you press on an avocado to determine, by how firmly it presses back, whether it is ripe. Other mechanoreceptors are *rapidly adapting,* responding only to changes in force; they let you know whether a fly captured in your hand is still alive. We will first describe the main types of mechanoreceptors found in humans and other mammals and then see how closely they correspond to those in other vertebrates.

A SLOWLY ADAPTING MECHANORECEPTOR

The *Merkel receptor,* named for anatomist Friedrich Merkel (1845–1919), is an example of a slowly adapting mechanoreceptor. It consists of two parts: a Merkel cell and a nerve ending. The Merkel cell is a rounded cell located on the border between the epidermis and the dermis. On its upper surface are a number of fingerlike prongs extending into the epidermis (see Figure 6.1). When pressure is applied to the surface of the skin, these prongs act as levers that induce strain in the cell, eliciting a graded electrical response. The nerve ending itself is a flat disk, which is in close contact with the Merkel cell and is activated by it. The receptor is considered slowly adapting because when steady pressure is applied to the skin, the response of the fiber declines very gradually.

Merkel receptors enable us to perceive the shape, size, and solidity of objects in the environment, as when we try to find a light switch in

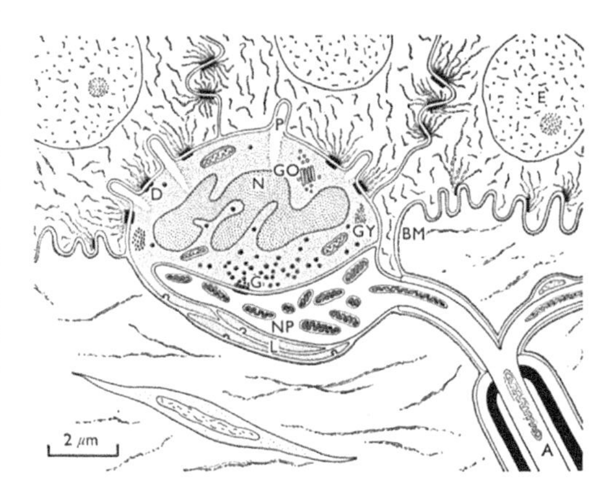

Figure 6.1 A Merkel receptor. This slowly adapting mechanoreceptor is located between the epidermis (above) and the dermis (below). Fingerlike prongs (P) extend from the Merkel cell into the epidermis and are deflected when pressure is applied to the skin. This activates the cell, causing signals to be synaptically conveyed to the flattened nerve ending (NP) and from there to a sensory axon (A). The same axon typically receives signals from more than one Merkel receptor, as indicated here.

Source: From Iggo, A., & Muir, A. R. (1969). The structure and function of a slowly adapting touch corpuscle in hairy skin. _Journal of Physiology, 200_, 763–796. Used with permission of John Wiley & Sons—Books; permission conveyed through Copyright Clearance Center, Inc.

the dark. They are the receptors that enable us to read braille or to discover a tick on a dog's ear.

How does mechanical force applied to the membrane of a Merkel receptor, or to other types of mechanoreceptors, actually trigger an electrical response? The answer was discovered by Ardem Patapoutian's research group at Scripps Research in La Jolla, California (Coste et al., 2010; Ranade et al., 2014). They found that receptor molecules in the membrane of mechanoreceptors, especially a molecule called *Piezo2,* are channels that open when force is applied, allowing positive ions to enter the cell and initiate neural activity that is responsible for touch sensations. For their work on somatosensory receptors, Ardem Patapoutian and David Julius were jointly awarded the Nobel Prize in Physiology or Medicine for 2021.

A RAPIDLY ADAPTING MECHANORECEPTOR

An example of a rapidly adapting mechanoreceptor is the *Pacinian corpuscle,* discovered by Filippo Pacini (1812–1883). These oval structures are on the order of 1 mm in size, making them the only encapsulated endings that can be seen with the naked eye. They are located deep in the dermis and in subcutaneous tissue. They are extremely sensitive to vibration.

A Pacinian corpuscle consists of a series of thin layers of connective tissue surrounding a nerve fiber. It somewhat resembles a tiny onion, but with thinner layers and many more of them. The nerve fiber runs down the center of the receptor, exiting as an axon that travels to the spinal cord. When you press a key on your computer, the force it exerts on your fingertip passes into the skin and reaches Pacinian corpuscles. This force is propagated through all the layers of the corpuscle and presses on the nerve fiber, causing it to fire. But if the downward pressure on the corpuscle continues, fluid between its layers will ooze to the sides, reducing the squeeze exerted on the nerve fiber and causing it to stop firing; that's why this type of mechanoreceptor is rapidly adapting.

Mendelson and Loewenstein (1964) demonstrated this by experimenting on individual Pacinian corpuscles in a laboratory dish. They recorded from a corpuscle's axon while applying pressure to its surface with a tiny probe. When they dissected most of the receptor's layers away, it changed from being rapidly adapting to being slowly adapting.

The fact that Pacinian corpuscles are rapidly adapting makes them extremely responsive to vibration, which is pressure that stops and starts in rapid alternation. The axon fires, usually with just one action potential, to each burst of pressure, then quickly returns to its resting state and is ready to respond to the next burst. Some Pacinian corpuscles respond to vibration frequencies as high as 500 Hz.

What has driven the evolution of sensitivity to such high frequencies? Part of the answer is that these frequencies play an important role in the perception of fine textures, such as the surface of a pebble or a leaf. As you actively explore these surfaces, small irregularities in the surface catch momentarily on fingerprint ridges and then break free, applying a series of brief forces to the receptors. Bensmaïa and Hollins (2003, 2005) recorded the resulting vibrations of the skin and showed that they sometimes contain frequencies in excess of 300 Hz.

It is now recognized that a complex surface, such as the outside of a scallop shell, is perceived through the activity of both slowly and rapidly adapting mechanoreceptors. Merkel receptors respond to individual bumps, ridges, and other macroscopic features, enabling the brain to build up a spatial representation of the surface. In contrast, information about the microstructure of the surface is captured by Pacinians and other rapidly adapting mechanoreceptors, based on the vibrations that occur when the hand moves across them. This duplex nature of tactile texture perception was first recognized by the psychologist David Katz (1925/1989) and has since been confirmed by multiple lines of evidence (Hollins, 2010; Goodman & Bensmaia, 2018).

There are two additional types of encapsulated endings: *Meissner corpuscles* are rapidly adapting mechanoreceptors that respond to low frequencies of vibration. *Ruffini cylinders* are slowly adapting and respond to lateral stretching of the skin, such as occurs on the back of your hand when you make a fist. Ruffinis contribute mainly to proprioception (awareness of body position) rather than to tactile sensation. The interested reader is referred to Gescheider et al. (2009) for additional information.

RECEPTOR ACTIVITY AND PERCEPTION

Ochoa and Torebjörk (1983) showed that there is a close link between receptor activity and perception. These researchers recorded the electrical activity of individual nerve fibers in the forearms of stoical volunteers—fibers bringing sensory messages from the hand to the brain. As expected, fibers carrying signals from Merkel receptors became active when gentle pressure was applied to the hand; a different set of fibers, coming from Pacinian corpuscles, responded when high-frequency vibration was applied.

Next, the experimenters used electricity to artificially activate individual fibers—a procedure called microstimulation—causing them to deliver spurious messages to the brain. When a single Merkel fiber was activated, volunteers reported a sensation of light pressure on the hand, as if a small object were resting on it; when a single Pacinian fiber was activated, a sensation of vibration was reported; and stimulation of a Meissner fiber produced a fluttering or tapping sensation. This remarkable study is one of several showing that activity in a single mechanoreceptor is enough to trigger a sensation and that the nature of the sensation depends on which fiber is firing.

In hairy skin, there are fewer capsules because adequate mechanical advantages are provided by hairs themselves. Nerve fibers wind around the base of the hair, below the surface of the skin. When the hair is pushed out of its normal position, this stretches and squeezes the nerve fiber, making it fire, but the hair quickly adjusts to the external stimulation by bending, so that the follicle is restored to its resting orientation and activation of the nerve fiber stops. Hair follicle receptors are thus rapidly adapting mechanoreceptors.

CENTRAL PATHWAYS

Axons from mechanoreceptors, bundled into nerves, travel to the spinal cord, and most continue an uninterrupted journey to the brainstem. Here, they synapse on cells in brainstem nuclei, the *nucleus cuneatus* (the recipient of fibers from the upper limbs) and *nucleus gracilis* (receiving fibers from the lower limbs). Second-order neurons relay tactile signals to the thalamus, and from there to the parietal lobe of the cerebrum, where they end on cells in *primary somatosensory cortex*.

This cortical area contains a map—actually, several maps—of the body surface. These somatosensory maps are distorted, devoting more cortical area to some parts of the body than others. Most notably, the fingers, lips, and tongue are disproportionately represented. This cortical magnification of some parts of the body surface is analogous to the outsize foveal representation in visual cortex that was discussed in Chapter 2.

TOUCH AND HAPTICS

The foregoing description of the receptors for touch is based mainly on research on the human hand. The hand is of special interest because of its role in exploring and capturing information about the environment, an active process called *haptic perception*. The hand is well adapted for this task because of its mobility: It can carry out a number of exploratory procedures, such as grasping, rubbing, pressing, and lifting, which allow a person to answer specific questions about the size, surface texture, hardness, weight, and other properties of an object (Lederman & Klatzky, 1987) (see Figure 6.2).

A remarkable feature of the hand is its high tactile acuity, the ability to resolve fine details. This helps us explore and identify small objects, such as the different keys on a key ring when we are trying to find the right one on a dark night. Extensive research has shown that tactile acuity is primarily mediated by Merkel receptors (Johnson, 2002).

One measure of tactile acuity is *two-point threshold*, the smallest separation between two pointed stimuli—like the points of a compass, but not sharp enough to break the skin—that are perceived as distinct. Measurements on different body sites show that two-point threshold is very large (about 20 cm) on the back but gradually decreases as measurements are made farther out along the arm, reaching a minimum of about 1 mm on the fingertips. The only area of the

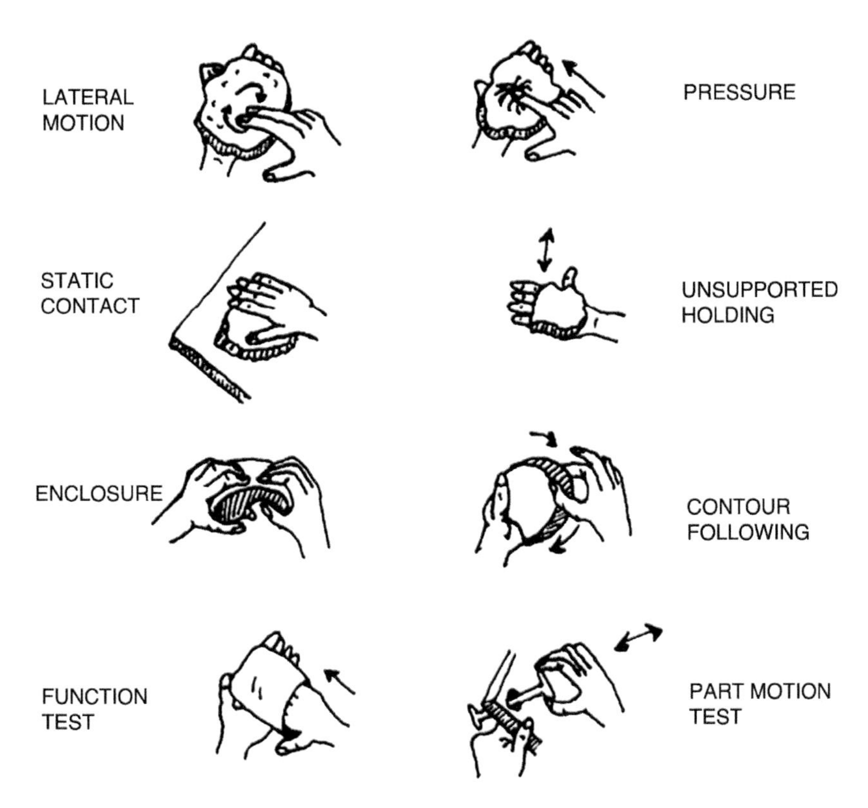

Figure 6.2 Exploratory procedures (EPs) used during haptic examination of objects. Different EPs are useful for determining different object properties. For example, lateral motion provides information about the surface texture of an object, while unsupported holding reveals its weight.

Source: **From Lederman, S. J., & Klatzky, R. L. (1987). Hand movements: A window into haptic object recognition. *Cognitive Psychology*, *19*, 342–368. Used with permission of Elsevier Science & Technology Journals; permission conveyed through Copyright Clearance Center, Inc.**

body equal in tactile acuity to the fingertips is the mouth, where we use the lips and tongue to examine food (e.g., for a fishbone) before swallowing it. In other words, tactile acuity is greatest on highly mobile parts of the body with which we investigate our surroundings. This difference in tactile acuity is the result of two factors: the fact that Merkel receptors are more numerous, and therefore more closely spaced, in fingers and lips than elsewhere (Johansson & Vallbo, 1979), and the fact that more cortical area is devoted to the skin of these highly mobile structures.

TOUCH IN OTHER VERTEBRATES

To some extent, the above account of receptors in the skin of the humans applies to other vertebrates as well. For example, reliance on both slowly and rapidly adapting mechanoreceptors appears to be universal among vertebrates. However, the structure and distribution of these receptors vary across animal groups.

MAMMALS

In most mammals, almost all of the body surface is covered with fur, so hair follicle receptors play a larger mechanoreceptive role than they do in people. In addition, most mammals walk on all fours and have forelimbs less adapted than ours to manipulating objects. Their face therefore plays a major role in exploring the environment. One manifestation of its importance is the widespread presence of vibrissae—long, movable whiskers—in rodents, carnivores, and many other mammals.

RAT VIBRISSAE

Carvell and Simons (1995) carried out a psychophysical experiment to see whether rats can use their vibrissae to make precise judgments of surface texture. A blindfolded rat was placed on a starting platform and had to jump to one of two other platforms to receive a food reward. Each of these latter platforms featured a grooved surface, which the rat, while still on the starting platform, could touch with its whiskers. The two surfaces presented on a trial were slightly different, and to obtain a reward, the rat had to jump to the platform with the smoother surface. To carry out the task, the rat swept its whiskers back and forth across the two surfaces in turn and then made a decision.

There were two experimental conditions. In one condition, both surfaces were coarse, with ridges and grooves 1 mm or more in width; they differed from each other only slightly. In the other condition, fine textures were used instead: Both surfaces had ridges and grooves of 0.5 mm or less. Rats easily mastered both conditions.

Next, the experimenters imposed a greater challenge, trimming all but one of a rat's whiskers and then testing it on the same discriminations as before. The rats continued to perform well in the fine texture condition, indicating that they were discriminating based on the slightly different vibrations created as their one whisker swept across the surfaces. However, the rats were no longer able to discriminate between two coarse surfaces, suggesting that this discrimination depended on spatial information obtained when some vibrissae touch a groove while others touch a ridge. These findings of Carvell and Simons are consistent with the duplex nature of tactile texture perception described earlier in humans, and with the fact that some vibrissae (unlike most hairs elsewhere on the rats' bodies) are served by slowly adapting as well as by rapidly adapting mechanoreceptors (Furuta et al., 2020).

Despite the greater role played by hair follicle mechanoreceptors in nonprimates, Merkel receptors and Pacinian corpuscles occur in both hairy and glabrous skin and appear to be present in all mammals. But their distribution varies among groups in idiosyncratic ways. In cats, for example, Merkel receptors are located inside small raised bumps in hairy skin called touch domes (Iggo & Muir, 1969). Each dome contains one axon that receives signals from dozens of individual Merkel receptors. Since the domes do not have hairs on them, the pressure exerted on a touch dome by a small object is not spread by hairs to neighboring regions, which would degrade tactile acuity.

STAR-NOSED MOLE

A much more complex tactile structure is found in the star-nosed mole, *Condylura cristata*, a hamster size mammal that lives in marshes and other wet areas (see Figure 6.3). Nearly blind, it relies heavily on touch as it forages in tunnels for worms, insect larvae, and other small invertebrates (Catania, 2012).

The animal gets its name from the array of 22 fleshy appendages that surround the nose, like the rays of a star. The surface of each ray is covered with bumps called *Eimer's organs*. Each Eimer's organ is a sensory structure containing mechanoreceptors of three types: slowly adapting Merkel receptors; rapidly adapting layered capsules resembling tiny Pacinian corpuscles; and rapidly adapting free nerve endings (Marasco & Catania, 2007). An Eimer's organ is thus a small machine for extracting all the properties—including hardness, roughness, and movement—of whatever touches it.

Figure 6.3 A star-nosed mole, *Condylura cristata*. Note its star and its powerful claws, adapted for digging.

Source: **Used with permission of Shutterstock. Photo Contributor, *Agnieszka Bacal*.**

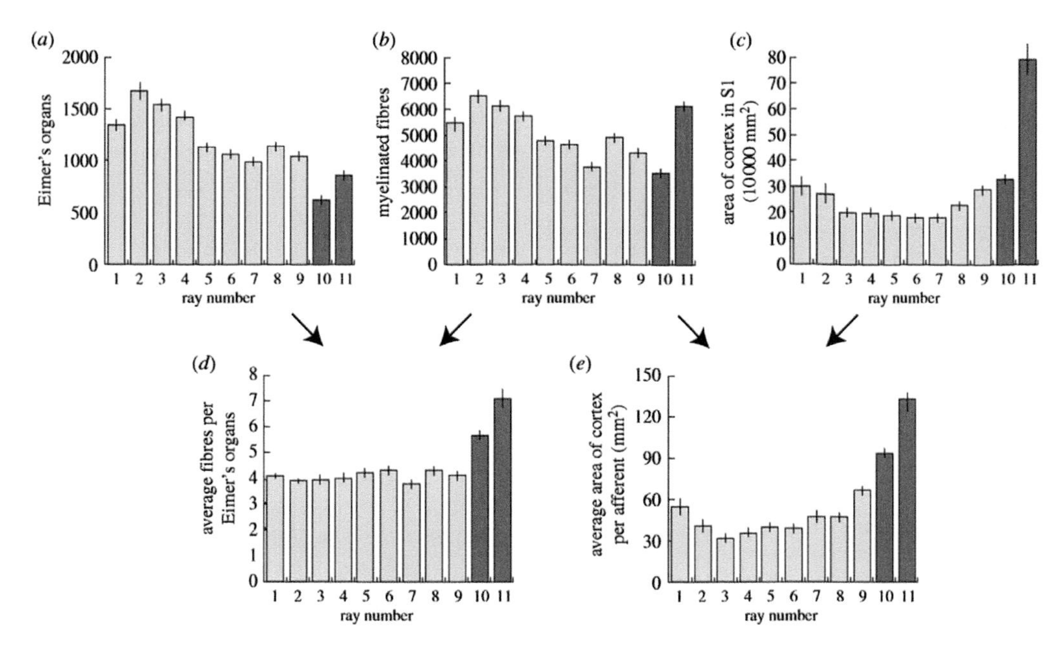

Figure 6.4 In the star-nosed mole, the area of somatosensory cortex devoted to ray 11 is much greater than for other rays (c). This is partly due to the fact that ray 11 gives rise to a large number of sensory fibers (b) but is also due to the fact that more cortex is devoted to signals from each of these afferents (e). Mechanisms intrinsic to the cortex thus play an important role in cortical magnification.

Source: **From Catania, K. C. (2011). The sense of touch in the star-nosed mole: From mechanoreceptors to the brain. *Philosophical Transactions of the Royal Society B, 366*, 3016–3025. Used with permission of The Royal Society (U.K.); permission conveyed through Copyright Clearance Center, Inc.**

The lower rays of the star, directly in front of the mole's mouth, are the most heavily innervated. When the mole senses a potential food item with its other rays, it quickly pivots to examine the item more thoroughly with these lower rays and reaches between them to seize the prey with its tweezer-like teeth (Catania & Remple, 2005). It is so efficient at this foraging behavior that it is able to consume several morsels in a second.

The star-nosed mole's somatosensory cortex has sharply defined anatomical regions corresponding to the individual rays of the star. The amount of cortex devoted to the lower rays, especially ray 11, is much greater than that devoted to the upper rays. This cortical magnification is analogous to the distorted representation of the retina on primate visual cortex, suggesting that the representation of ray 11 is a sort of *tactile fovea* (Catania & Kaas, 1997).

Is cortical magnification simply a reflection of a higher innervation density in some parts of the sensory surface, i.e., some arms of the star, or does it also reflect mapping processes intrinsic to the cortex? Figure 6.4, from Catania (2011), answers this question. Cortical magnification depends on both factors.

ECHIDNA

A sensory structure remarkably similar to Eimer's organ—the *push rod*—has evolved independently in the echidna, or spiny anteater, which is native to Australia and New Guinea (see Figure 6.5). The name Echidna derives from a monster in Greek mythology. The real echidna is no monster, but like its fellow monotreme, the platypus, it is an oddity: an egg-laying mammal. Its keratinized snout would make mechanoreception difficult if it did not contain soft spots housing the push rods. Each push rod contains Merkel receptors, Pacinian corpuscles, and some unencapsulated endings, a combination that enables both slowly and rapidly adapting responsiveness

Figure 6.5 *Tachyglossus aculeatus*, the short-beaked echidna. This Australian monotreme feeds on ants and termites that it captures with its agile, sticky tongue.

Source: **Used with permission of Shutterstock. Photo Contributor, Paul Looyen.**

(Augee et al., 2006) as the echidna rummages for ants and termites. Echidnas are also sensitive to low levels of electricity, a subject that we will discuss in Chapter 10.

BIRDS, REPTILES, AND AMPHIBIANS

RAPIDLY ADAPTING RECEPTORS

Other vertebrates have mechanoreceptors similar to ours (as well as some that are clearly different), and it is sometimes a matter of definition whether these are varieties of the same receptor types or should be considered distinct. For example, birds have a *Herbst corpuscle* that is similar to our Pacinian corpuscle. In some birds, large numbers of Herbst corpuscles are located in tiny pits in the bird's bony beak (du Toit et al., 2020). An example is *Bostrychia hagedash*, the hadeda ibis, a native of Sub-Saharan Africa named for its distinctive call. The hadeda forages in water or soft soil with its beak, using Herbst corpuscles to pick up vibrations caused by the movement of insects, worms, or other small animals.

Vibration receptors similar to our Pacinian corpuscles but differing in size, lamellar thickness, and other structural properties have also been found in some reptiles and amphibians: for example, in the mouth of the Nile crocodile (Putterill & Soley, 2003) and the skin, especially of the fingers, of the "edible frog," *Rana esculenta* (Düring & Seiler, 1974). Computer simulations and neural recording indicate that despite structural differences, vibration frequency tuning is similar across species, usually peaking at 40–50 Hz (Quindlen-Hotek et al., 2020); our own Pacinian corpuscles are an exception, peaking at around 250 Hz (Gescheider et al., 2009).

SLOWLY ADAPTING RECEPTORS

Merkel receptors in some birds resemble their mammalian equivalent; in ducks and other wildfowl, however, they are found inside unique structures called *Grandry corpuscles*, consisting of several Merkel cells arrayed in parallel, with flattened nerve terminals sandwiched between them (Halata et al., 2003). Merkel cells are also found in reptiles (Landmann & Halata, 1980) and amphibians (Whitear, 1989).

SCALES

An important feature of the integument of many vertebrates is the presence of scales. They are absent in amphibians, and nearly so in mammals (the pangolin being the exception), but widespread among fish, reptiles, and on the feet and ankles of birds. Scales are stiff plates that grow out of the skin. They offer protection from injury and from dehydration and help some animals glide smoothly along the ground or through water. Scales are not all the same: They are of different types and have evolved independently several times. Because scales change the flexibility of the skin, cutaneous mechanoreceptors in scaled animals can differ in form and context from the versions found in soft skin. For example, the Stokes' sea snake (*Hydrophis stokesii*), a large, venomous snake found in South Asian and North Australian waters, has specialized sensory structures on the scales on its head. These scale sense organs, or *sensilla*, are bumps consisting of dermal papillae pushing upward on a thin region of the scale's superficial layer (Crowe-Riddell et al., 2019). The sensilla contain nerve endings similar to mammals' Meissner corpuscles. Deeper in the skin, lamellar corpuscles resembling small Pacinian corpuscles occur.

FISH

The presence of encapsulated nerve endings in fish was once controversial, but a recent study (Cohen et al., 2020) demonstrates their presence in the adhesive disk of the sharksucker remora, *Echeneis naucrates*. This meter-length fish attaches itself to larger fish or other marine organisms, feeding on their table scraps or on small parasites that are also attached to the host. The remora's dorsal fin is modified as a device—the *adhesive disk*—for attaching the remora to its host. The fleshy lip of the disk forms a water-tight bond with the host; ridges inside the disk then change their alignment so as to produce suction, which locks remora to host (see Figure 6.6). The lip is amply supplied with push rods reminiscent of those in the echidna: Each contains a number of Merkel receptors and, below them, several Pacinian corpuscles.

There is evidence that Merkel receptors are also present in the pectoral fins—that is, the forelimbs—of some fish (Hardy et al., 2016). Electrical recordings from sensory nerve fibers serving the pectoral fins of the round goby (*Neogobius melanostomus*), a bottom-dwelling fish, show that these fibers encode information about surfaces that the fin comes into contact with (Hardy & Hale, 2020) as the goby moves along. Each fiber responds to pressure applied to a specific region of skin, the fiber's receptive field. These receptive fields range from 2 to 5 mm across, indicating a spatial precision of tactile encoding approaching that of the human hand. A ridged surface such as a clam shell elicits a periodic response from the fiber as individual ridges and grooves move sequentially across its receptive field. In this way, the goby acquires information about the texture of the shell.

Figure 6.6 A remora, showing the adhesive disk with which it attaches itself to a host.

Source: **Reproduced by permission of Shutterstock. Photo Contributor, Shane Gross.**

This surprising discovery shows the great age of encapsulated endings, since fish diverged from other vertebrates more than 350 million years ago. However, these ancient structures, so important to our own sense of touch, are rare in fish. Their scarcity is explained by the fact that the primary function of mechanoreception in fish is the detection of currents and vibration of water—a task for which a different set of structures, the *lateral line system*, has evolved.

THE LATERAL LINE SYSTEM

A lateral line is a row of receptor structures that runs from front to back along each side of a fish. Actually, the lateral line system usually consists of more than just these two lines: It includes additional rows of receptors that extend onto the animal's face or other parts of its torso. Each of these "lines" is a partially closed channel or tube, with evenly spaced openings through which water can enter or exit. And midway between adjacent openings, on the inner surface of the channel, is a sensory structure called a *neuromast*. A neuromast is a clump of receptor cells, each bearing a number of cilia. All of the cilia on a neuromast are embedded in a gelatinous cap called a *cupula*. When a cupula is bent by water pressure, these hairlike filaments also bend, and receptor cells are activated.

When a fish is resting in still water, the water pressure is the same on the two sides of each cupula, so no bending occurs. Only when the water pressure on the two sides of a neuromast is different does the cupula bend. If it bends forward along the lateral line, some of the receptor cells in the neuromast will be activated; if it bends backward, other receptors in the same neuromast will be activated. In either case, the activated receptors transmit signals to nerve fibers that convey this sensory information to the central nervous system.

What the fish most needs to detect is movement of another animal in its vicinity. All but the very slowest movements will create pressure waves that move away from the source at the speed of sound (1480–1500 m/s in water). If, say, a hungry trout is approaching a minnow from the rear, pressure waves created by its advance will race ahead of the trout, reaching the minnow. A ridge of high pressure will enter the back end of the minnow's lateral lines first, bending cupulas forward and thereby warning the minnow in time (perhaps!) for it to escape.

A lateral line system is only useful in water; pressure waves in air created by, say, a snake slithering toward a mouse are too gentle to be of benefit to the mouse. To detect a predator (or prey) by touch, terrestrial vertebrates rely instead on Pacinian corpuscles to detect faint vibrations traveling through the ground. And so, vertebrates lost their lateral lines when they emerged from the water.

TOUCH IN ARTHROPODS

Arthropods, like vertebrates, have a variety of types of mechanoreceptors. However, these are distinctive to arthropods and different from those of vertebrates. There are, for example, no Pacinian corpuscles in insects. But the major division of mechanoreceptors into rapidly adapting and slowly adapting types applies equally to both phyla.

INSECTS

The most common type of mechanoreceptor in insects is the *sensillum*, a structure that includes a nerve ending. There are two main types. In a *trichoform sensillum*, this nerve ending is attached to the base of a hairlike structure called a *seta*. Movements of the seta evoke a rapidly adapting response from the nerve fiber.

An example of the use of trichoform sensilla is given by the tobacco hawkmoth, *Manduca sexta*. This large moth feeds on nectar, which is sucked up from flowers through a long, thin proboscis. It is drawn to a flower by its smell but must then locate its nectar-bearing center. Vision is of little help, since the moth is crepuscular, that is, active at twilight. To locate the opening in the center of the flower, it relies instead on touch, specifically the many sensilla that cover its proboscis. Deora et al. (2021) studied the moth's behavior using a high-speed digital camera and artificial flowers made with a 3D printer. They found that the moth hovered above a flower and explored its surface with the tip of its proboscis. With its sensilla, it was able to detect the edge and surface curvature of the flower; it systematically swept its proboscis inward from the edge, again and again, until it located the hole in the center, within which "nectar" (sucrose solution) was available.

Another type of insect mechanoreceptor is the *campaniform sensillum*. Like the trichoform sensillum, it contains a flexible nerve ending, but in this case located inside a bulge in the insect's cuticle (exoskeleton). These domes are located near joints in the exoskeleton, which are the places where it bends when, for example, a cockroach walks across the counter or squeezes into the breadbox. Unlike the trichoform sensillum, the campaniform sensillum is slowly adapting.

In addition to sensilla, insects have a considerable diversity of other types of mechanoreceptors, and many of these differ from one type of insect to another. Social insects routinely communicate by means of vibration and have receptors, on their legs and elsewhere, that are extremely sensitive to it. In the waggle dance, for example, a honeybee returning from a foraging run uses vibration and other cues to inform her nestmates as to the availability of nectar. They crowd around and touch her as she dances, thus learning about the location and value of the find.

It is worth noting that the sense of touch, broadly defined, includes proprioception—knowing where the parts of your body are. For example, your elbow feels different when extended than when bent. Many active touch ("haptic") experiences, such as using a screwdriver to tighten a screw, involve both mechanoreception (sensing pressure of the screwdriver on your skin) and proprioception (sensing the effort involved in turning the screwdriver). Proprioception is an equally prominent part of touch in arthropods, with many nerve fibers reporting the flexion or extension of limbs, the flapping of wings, etc.

SCORPIONS

A series of experiments by Philip Brownell and Roger Farley show that nocturnal scorpions (*Paruroctonus masaensis*) in the Colorado Desert make good use of their vibration sense. In response to an insect moving nearby, the scorpion turns toward the prey and runs forward in the dark to seize it with its pedipalps, large appendages ending in pincers.

To record and measure the movements of the scorpions in darkness, the researchers illuminated them with UV light, causing them to fluoresce yellow-green. They found that the animals can turn accurately to face a disturbance (which the experimenters created by pushing a thin wooden rod into the sand), while at the same time taking its distance into account. That is, the scorpions made a tight turn to face a nearby stimulus, but a looser turn to face a more distant one. They judge distances accurately up to about 15 cm, although they can detect a disturbance much farther away (Brownell & Farley, 1979b).

Several lines of evidence establish that the scorpion is responding to vibration traveling through the sand, rather than to sounds conveyed through air. First, it hunts down not only insects walking on the sand but also burrowing ones such as desert cockroaches, which live below ground. Second, a moving insect suspended in air near the scorpion elicits no reaction. And third, scorpions tested on a platform with an air gap in the middle do not respond to a disturbance on the other side of the gap.

Brownell and Farley (1979a) discovered that the scorpion detects vibrations with two different types of sensilla in their tarsi (feet): hair sensilla and slit sensilla. In some other arthropods, these sensilla respond to sound, as will be described in the following chapter. However, for the desert scorpion, they belong to the sense of touch. Vibration traveling through sand is a complex process that involves two different waves differing in speed; the scorpion's hair sensilla respond to the faster wave, and the slit sensilla to the slower but more directionally informative wave.

SPIDERS

Spiders are extremely sensitive to vibration. Those with webs detect it when a victim is caught in their web and thrashes about, trying to escape. And jumping spiders, when on the hunt for a webmaking spider in its home, conceal their approach by shaking the owner's web in a way that mimics the wind.

Some spiders, like many other animals, can form associations between stimuli in different sensory modalities, such as touch and smell. Such configural learning has been studied in the whip spider, *Phrynus marginemaculatus* (Flanigan et al., 2021). These nocturnal hunters with an elaborate social life are not true spiders; they occupy a separate branch of Class Arachnida. They have large, convoluted mushroom bodies, which perhaps account for their learning abilities. Flanigan and colleagues tested the spiders' ability to recognize their shelter, a piece of PVC pipe provided by the experimenters, when they returned home after a night of hunting. When an animal was first placed in its shelter, one or more sensory cues were provided as a basis for later recognition. Spiders in one group had a shelter scented with a floral or fruity odor; those in a second group had a shelter bordered with either rough or smooth sandpaper. Finally, spiders in a third group were placed in a shelter with a particular combination of odor *and* texture, such as floral smell and smooth texture.

During a later test, the whip spiders were given a choice of two different shelters. Those in the first group had no trouble choosing the shelter with the familiar odor, and those in the second group easily identified the shelter with the familiar texture. The most interesting results concerned those whip spiders whose home shelter had a particular combination of odor and texture. When it was available, they still chose a shelter with this same combination. But when tested with shelters having just one cue—odor or texture—they performed only at a chance level. They could not recognize "home" based on odor or texture alone. They had learned to approach not just a certain odor or a certain texture, but a combination of the two.

Such configural learning may be an important part of the spider's everyday life, for example, learning to seize a roach, but not to seize other objects just because they are brown or moving along the ground. The findings of Flanigan et al. may indicate that whip spiders can bind the different features of an object together into a unified percept, an ability that would pose for sensory scientists an invertebrate version of the *binding problem* described in Chapter 2.

CRUSTACEANS

Mechanoreception in crustaceans has been extensively studied, because the large size of many of them facilitates experimentation. As in insects, hairlike filaments are the primary receptors. An example are the pit receptors on the carapace of the red swamp crayfish, *Procambarus clarkii* (Mellon, 1963). The carapace is a part of the exoskeleton covering the top of the head and thorax of the animal. Scattered across the carapace are pit receptors where sensilla extend outward, and each sensillum is innervated by two nerve fibers, which extend a short distance into the hair, one along the anterior portion of its inner surface and the other along the posterior portion. When the hair is bent forward, the posterior fiber is stretched and responds, and when the hair is bent backward, the other responds. These hairs respond vigorously if another object touches them, but in everyday life, they are probably used more to monitor currents in the water.

The most highly refined sensory structures in crayfish and other crustaceans are the antennas. The crayfish has two pairs of antennae, extending forward out of the head. The pair closer to the midline are shorter than the other pair and are usually called *antennules*. Antennules themselves bifurcate midway along their length, with a lateral and a medial branch. The lateral branches of the antennules have been the subject of considerable study because they contain receptors for both touch (feathered sensilla) and olfaction (aesthetasc sensilla).

There is a curious interaction between the two modalities. It is useful to realize that when odorants reach the crayfish, they normally do so not by passive diffusion, the way a drop of food coloring diffuses in a bowl of water, but by being carried along on eddies in the water, such as those caused by another animal swimming by. These water currents bend and thereby stimulate the feathered sensilla, which send tactile messages to the central nervous system, triggering a reflexive downward flicking of the antennule's lateral branch (Mellon & Hamid, 2012). And this flicking, in turn, exposes the aesthetasc sensilla to a fresh burst of odorant, just as sniffing does in a mammal. Tactile stimulation thus leads to more effective olfactory stimulation.

TOUCH IN MOLLUSCS

Touch, or more accurately *haptics* (active touch, as when objects are manipulated), has been extensively studied in cephalopods, especially the octopus, whose arm suckers are remarkably prehensile. In attacking, say, an oyster, an octopus attaches suckers to the two halves of its shell and pulls vigorously. If this doesn't work, the octopus uses its radula, a rasping tongue lined with small teeth, to drill holes in the shell, through which toxins may be injected. Then, it pulls some more, eventually opening the oyster, whose flesh is quickly eaten.

These coordinated, precise actions show that the octopus is using information about the shape, weight, shell texture, and other properties of the oyster. Vision is no doubt important as the octopus approaches the oyster, but once the attack is underway, its view of the oyster is blocked by its own arms, making touch the critical sense. And some octopuses hunt in murky water, at night, or at considerable depth, where vision makes little contribution and the animal must rely to a great extent on touch.

What and where are the octopus's mechanoreceptors? Anatomical study shows that both the suckers and the lip (a fleshy ridge surrounding the beak) are richly supplied with receptors of several types (Graziadei, 1964; Emery, 1975a). Most of these receptors have cilia, which in some cases extend upward to the surface of the epithelium but in other cases are enclosed within the cell and located well below the surface. The sensory modality to which these receptors belong is uncertain. Those with cilia in contact with the surface of the epithelium are thought to be chemoreceptors, whereas those well below the surface are more likely to be mechanoreceptors. Additional research is needed to resolve the question.

An additional part of the mechanosensory apparatus of cephalopods, including the octopus, is the *lateral line analog system*. This has developed independently of its namesake in fish and aquatic amphibians, a remarkable example of convergent evolution. It consists of lines of ciliated mechanoreceptors that are activated when water moves past them. The lines primarily run fore and aft along the head but sometimes extend onto the arms as well. An individual receptor "fires" in response to bending of its cilia in a particular direction. When vibrations in the water travel along the line, alternately pushing the cilia forward and backward, the local field response has twice the frequency of the vibration itself (Budelmann & Bleckmann, 1988). This result shows that cells responding to forward movement and others responding to backward movement are interspersed within the line.

The advantage of the lines is that they provide the cephalopod with detailed information about disturbances in the water, such as those created by the swimming movements of an approaching predator. The value of the lateral line analog system was tested in the brief squid, *Lolliguncula brevis*, by York and Bartol (2014). Individual squids were placed in a tank with two summer flounders, *Paralichthys dentatus*, aggressive predators that lurk on the bottom but will swim into action to overtake their prey. The experimenters pharmacologically ablated the lateral line analog of some squid but left other individuals intact. The intact squids survived more 10-minute sessions with the flounders than did the ablated ones, indicating that the lateral line analog system helps squid evade predators.

The lateral line system of fish and aquatic amphibians is more refined than that of cephalopods: The tuning of their system to fore-and-aft movements is based on the same principle of ciliary orientation as in molluscs but is sharpened by the fact that the receptors are in a canal. This "intracoastal waterway" protects receptors from the buffeting of steady currents, so that they are stimulated only by moment-to-moment changes in water pressure along the animal's longitudinal axis. And it is these changes that carry the most valuable information about events in a fish's environment.

SOMATOSENSATION AND EMOTION

BENEFICIAL TOUCH

So far we have considered the sense of touch as a means by which information about the environment is obtained. But in many animals, it is closely tied in with affect, in a way that vision and hearing are not. Among humans, gentle touch, such as a caress, is generally felt as pleasurable, and the same is clearly true among other mammals with which we interact. Our dogs and cats like to be patted and rubbed, and there is anecdotal evidence that other mammals who have been domesticated—farm animals, horses, and the occasional elephant—do as well. Our own order, the primates, are especially given to grooming one another, an activity that is welcomed by the recipient, probably as much for the tactile contact as for the cleaning and insect removal that it entails (see Figure 6.7).

Nor is this liking for physical contact limited to mammals. Parakeets and other pet birds will approach their human caregiver to be patted, and fish in an aquarium will sometimes do the same.

Figure 6.7 Crab-eating monkeys (*Macaca fascicularis*) grooming one another.

Source: **Used with permission of Shutterstock. Photo Contributor, Acon Cheng.**

Among invertebrates, however, such behavior is rare. There is no evidence of which I am aware that insects seek body contact with conspecifics, outside of mating and specific utilitarian contexts, such as the contact worker bees make with one another during the waggle dance to learn the location of a food source. And contact with other species is usually in the interest of devouring or parasitizing them, as in the praying mantis's contact with the locust.

Octopuses, although by most accounts the most intelligent of invertebrates, appear to have little interest in affiliative touch (Godfrey-Smith, 2016). Even mating, which involves the transfer of a sperm packet, is often an arm's-length affair. Most octopuses lead solitary lives and do not nurture their young once the eggs have hatched. Even when adult octopuses live in close quarters, as in "Octopolis," a site off the east coast of Australia described by Godfrey-Smith, males typically interact by fighting, presumably to establish and maintain dominance. A more subdued form of social interaction is that, when one octopus roams past the dens of others, they may extend an arm to probe the passerby. It is often unclear whether this is a hostile poke or a simple greeting.

At least among mammals, physical contact with infants is important for both their physical and behavioral development. The most compelling demonstrations of this are the classic experiments of Harry Harlow, who raised infant rhesus monkeys in various conditions of social isolation. As expected, monkeys reared by their mothers did well, as did those raised with siblings. In contrast, those raised in isolation showed profound emotional and behavioral disturbances, such as excessive fear, rocking, and self-injurious behavior (Harlow et al., 1965). These disturbances were somewhat alleviated if the isolated infants were provided with a cloth-covered mannequin, which they clung to for a measure of comfort. They greatly preferred this cloth "surrogate mother" (in Harlow's terminology) to another made of wire mesh, even if the infants were fed by a bottle attached to the wire mannequin (Harlow, 1958).

Harlow's experiments, carried out in an era when research with animals was not as carefully regulated as it is today, raise ethical concerns. But his findings influenced society (Vicedo, 2009) by increasing awareness of an infant's need for physical contact with its mother or other caregivers.

More recent research indicates that gentle massage of human infants, especially premies in an neonatal intensive care unit (NICU), improves their health in a variety of ways (Pados & McGlothen-Bell, 2019). The primary mechanism appears to be stress reduction mediated by activity in the vagus nerve, a large nerve that responds to cutaneous pressure and influences the activity of several internal organs (Field et al., 2011). The same mechanisms may play a role in the health benefits of massage in adults (Field, 2014).

NOCICEPTION AND PAIN

We start our consideration of pain by distinguishing it from *nociception,* the registration by the nervous system of a damaging, or potentially damaging, stimulus. It is made possible by *nociceptors,* specialized nerve endings in skin or other body tissues that respond selectively to intense levels of stimulation. These high-threshold fibers typically respond to strong heat or cold (thermal nociceptors), to stretching, pinching, or other deformation of tissue (mechanical nociceptors), or to irritating chemicals. Some, called bimodal or polymodal nociceptors, respond to more than one modality of noxious stimulation.

Unlike low-threshold mechanoreceptors such as Pacinian corpuscles, nociceptors have no encapsulated endings; their specializations lie at the molecular level. In mammals, these neurons fall into two anatomically distinct categories, based on the thickness of their axons, which extend from the innervated tissue to the central nervous system. Some of these axons are a few microns in diameter and are covered with myelin, a cholesterol-rich coating that speeds nerve impulses on their way. These *A delta (Aδ) fibers* carry signals that reach the central nervous system in a fraction of a second. The other, much more numerous type of nociceptors are *C fibers*: These are less than a micron in diameter and conduct action potentials very slowly because they have no myelin. Both types are thinner and slower than the axons of low-threshold mechanoreceptors.

Nociception is probably universal among animals with nervous systems. Extensive research on nociceptors has been carried out in a range of mammals, from mice to people, and the results are generally consistent with the description given above. Research on non-mammals has been less extensive but has documented the existence of nociceptors in birds (Gentle et al., 2001), frogs (Hamamoto & Simone, 2003), fish (Sneddon et al., 2003), crayfish (Puri & Faulkes, 2015), fruit flies (Hwang et al., 2007), sea slugs (Illich & Walters, 1997), and other animals (Smith & Lewin, 2009).

PAIN IN HUMANS

In contrast to nociception, *pain* is a subjective sensory experience that has a negative affect. By definition, pain is aversive. Generally speaking, pain occurs when the brain processes signals from nociceptors. For example, when you stub your toe, signals from Aδ nociceptors cause sharp, sudden pain, whereas signals from C nociceptors cause a deeper, aching pain that wells up gradually and may continue long afterwards. In some clinical conditions, pain can even be of central origin, occurring in the absence of nociception.

When a noxious stimulus is delivered to human volunteers undergoing functional magnetic resonance imaging (fMRI), researchers can see what parts of the brain are activated. They find that many parts of the brain, including cortical structures, such as the insula, and subcortical regions, such as the periaqueductal gray (PAG), are involved in the overall neural response. But which structures are essential to the experience of pain, rather than just for nociception?

Some scholars (e.g., Rose et al., 2014) maintain that pain depends on cerebral cortex and, in fact, on a highly developed portion of it called *neocortex*. A word about terminology is needed here. All vertebrates have a cerebrum, with a roof called the *pallium*. A portion of the pallium, characterized by a layered architecture and efferent connections to subcortical motor structures, has gradually become more elaborate over the course of vertebrate evolution: In reptiles and mammals, it is called the cerebral cortex. Like the pallium of the oldest vertebrates (Suryanarayana et al., 2017), the reptilian cortex has three layers. In mammals, however, a six-layered region called neocortex appears alongside three-layered regions and gradually becomes the dominant portion. (The pallium of birds is unique, but its circuitry resembles that of neocortex [Stacho et al., 2020].) If neocortex is a prerequisite for pain, it would follow that only mammals are capable of a true pain experience.

Those who emphasize the importance of neocortex for pain point to the fact that in humans, information about bodily states, including painful ones, is represented in the *insula*, a cortical area buried within the lateral fissure. A strong case has been made (Craig, 2002) that activity here is key to feelings such as pain, hunger, and sexual arousal, and perhaps even to our sense of self. However, Damasio et al. (2013) studied a patient whose insulae on both sides of the brain had been severely damaged but who nevertheless reported normal bodily feelings, including pain. While research continues, a reasonable interim conclusion is that pain in humans results from the combined activity of a set of brain regions, subcortical as well as cortical (Damasio & Carvalho, 2013). Cortex may be more involved in modulating, analyzing, and evaluating pain than in generating the basic unpleasant sensory experience.

PAIN IN FISH

But even if it were established that pain *in humans* requires cortical activity, this would not prove that the same is true of other animals. So let us ask directly whether there is evidence that animals without a cortex do in fact feel pain. Recent behavioral research suggests that they do. For example, studies in the rainbow trout, *Oncorhynchus mykiss*, show that it responds to noxious stimuli in a variety of ways (Sneddon et al., 2003). The noxious stimulus was an injection of acetic acid into the lips; control fish received only a saline injection. Both groups of fish were injected while anesthetized; after they

recovered from the anesthesia, their behavior was observed. Those in the acid group rocked from side to side, rubbed their lips on the sides and floor of the tank, and increased their rate of respiration.

Perhaps all of these responses could be regarded as reflexive. But in a follow-up study (Ashley et al., 2009), trout treated in this way were tested to determine whether more complex aspects of their behavior were affected by the acid injection. To learn how the fish would respond to indications of a predator in the vicinity, the experimenters added alarm pheromone (extracted from macerated trout skin) to the water in their tank. Control fish responded normally to the pheromone, rushing about and hiding in a covered part of the tank; fish who had received acid did not do this, suggesting that they were distracted by pain from the threat of a predator.

Another experiment from Sneddon's (2015) lab provides evidence of pain in zebrafish (*Danio rerio*). The fish were allowed to move back and forth between two chambers in their aquarium: one chamber "enriched" with gravel, plants, and other interesting objects; the other bare and brightly lit. Normally, the fish chose to spend most of their time in the enriched chamber. Adding lidocaine, a local anesthetic, to the water in the bare chamber did not increase its appeal. But if, in addition, a noxious lip injection was given to the fish, they afterwards spent most of their time in the bare chamber with lidocaine. The results suggest that the fish changed their behavior to reduce the aftereffect of the injection, indicating that it was aversive.

PAIN IN INVERTEBRATES

OCTOPUS

A similar study on Bock's pygmy octopus, *Octopus bocki*, has been reported by Crook (2021). Animals were given an acetic acid injection into an arm while sedated, and upon awakening were confined to one chamber of a three-chamber tank. When tested hours later, after the injection had worn off, they avoided that chamber. The results indicate that the noxious stimulus had caused pain which made their period of confinement aversive. Control octopuses injected with saline instead of acid showed no such avoidance (see Figure 6.8).

The procedures used by Sneddon (2015) and Crook (2021) are examples of the *conditioned place aversion (CPA) paradigm*, widely used, mainly in rodents, to determine the subjective effects of drugs or other stimuli. Rats will avoid a room where they had an unpleasant experience or seek out a room where they had a pleasant experience—in which case the paradigm is called *conditioned place preference (CPP)*. These paradigms make use of Pavlovian conditioning, in that a conditioned stimulus (the room) takes on the affective qualities of the unconditioned stimulus (the drug) with which it is paired.

Pavlovian conditioning paradigms such as CPP also "work" in humans (e.g., Childs & de Wit, 2009), whether or not we realize that conditioning is occurring. But the procedure does require that the unconditioned stimulus be pleasant or unpleasant. The point of the Sneddon (2015) and Crook (2021) studies is that the acetic acid must have produced a strong negative feeling, such as pain.

HERMIT CRAB

Studies carried out on the hermit crab, *Pagurus bernhardus*, provide a nuanced understanding of pain in an arthropod. These decapod crustaceans take possession of and move into snail shells that they find on the beach (see Figure 6.9). If a crab encounters another shell that careful inspection shows to be a better fit or of better quality than its current one, it will move to the new shell (Jackson & Elwood, 1989); and two crabs will sometimes fight over a prestigious shell (Dowds & Elwood, 1983).

Elwood and Appel (2009) collected about 100 crabs, took away their mollusc-shell "homes," and provided them with new shells that had been fitted with wires so that mild-to-moderate electric shocks could be delivered to the occupant. The crabs were then randomly assigned to shock and no-shock groups, and those in the

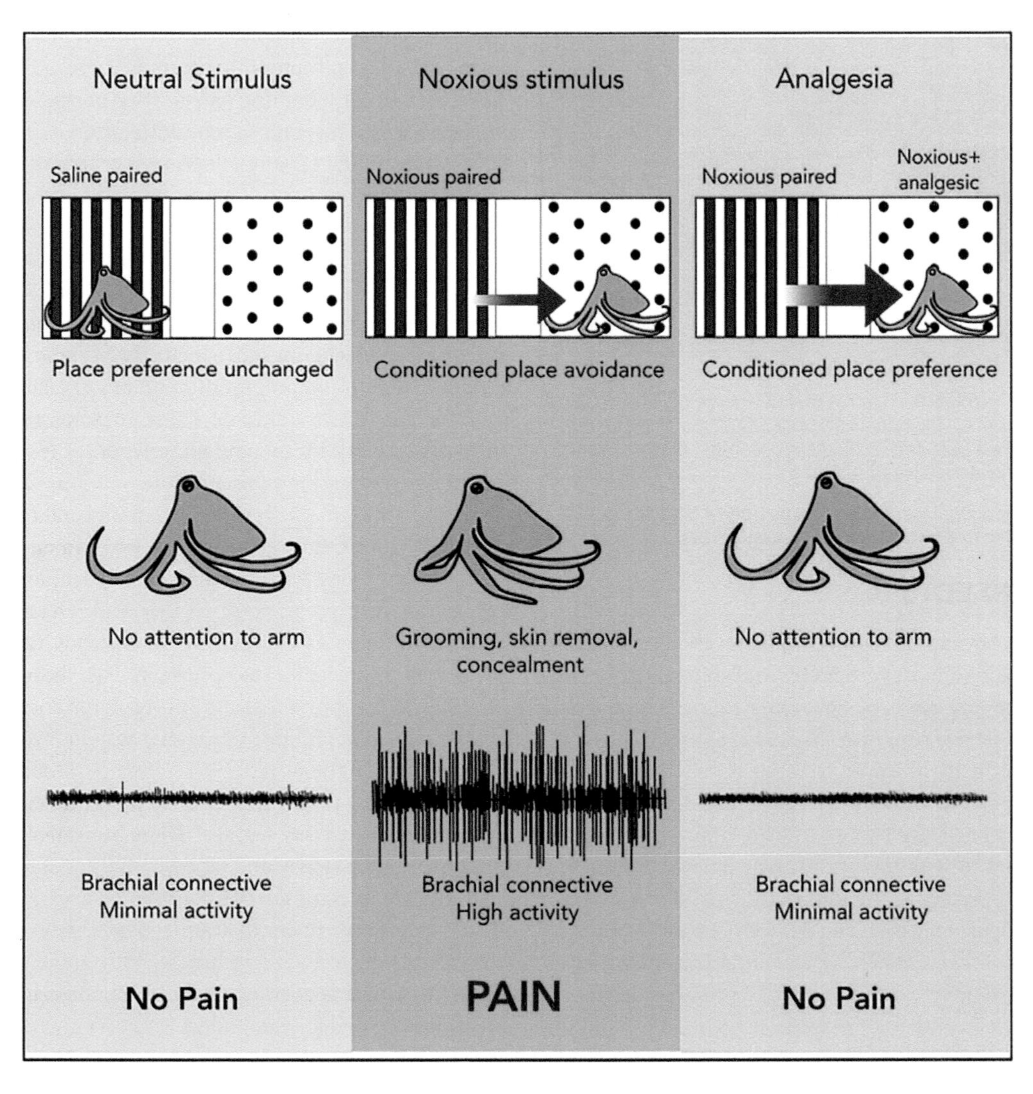

Figure 6.8 Results of experiments to determine whether an octopus feels pain. Animals were first given a choice between striped and dotted chambers. They were then given an arm injection of either saline (left column) or acetic acid (center column) and confined to their preferred chamber for 20 minutes. A later retest showed that individuals who had received acid now preferred the other chamber. The change was especially marked in animals who were injected with acid in one chamber and then given lidocaine, a local anesthetic, in the other chamber (right column). The results imply that an octopus can feel pain, a conclusion supported by behavioral observations and neural recordings (bottom).

Source: **Crook, R. J. (2021). Behavioral and neurophysiological evidence suggests affective pain experience in octopus. *iScience*, *24*, 102229. © 2021 Robyn J. Crook. Used with permission of Elsevier and the author.**

shock group received a series of brief shocks extending over 3 minutes. After a short delay, the experimenters provided all crabs with an extra shell that they were free to move to. Only 55% of those in the no-shock group opted to move, but 82% of those in the shocked group did so. This indicates that its association with shock was treated as a shortcoming of the old cell, though apparently this was not the only factor considered. In other words, the shock was aversive.

Figure 6.9 A hermit crab (*Pagurus bernhardus*) with its shell. Elwood and Appel (2009) found that if the shell delivered electric shocks, most crabs chose to move to a different one.

***Source:* Used with permission of Shutterstock. Photo Contributor, JorgeOrtiz_1976.**

INSECTS

A strong case can be made for the occurrence of pain in most vertebrates and some invertebrates, on the basis of behaviors that are not simply reflexive responses to noxious stimuli. It is more difficult, however, to make a case for the occurrence of pain in insects, where the evidence is inconclusive (Adamo, 2016). Injured insects tend not to withdraw from their normal activities or to protect an injured body part: For example, a locust may continue to feed normally even while it is being eaten by a mantis (Eisemann et al., 1984).

SUMMARY

A sense of touch is universal among animals. It can be passive, as when raindrops fall against your skin, or active, as when you rub cloth between your fingers to see how thick it is. Active touch is usually called haptics.

In either case, tactile stimuli activate mechanoreceptors in the skin that send signals along nerve fibers to the spinal cord or the brain. These receptors consist of nerve endings enclosed in tiny capsules that filter or amplify specific components of the stimulus. They fall into two categories: slowly adapting mechanoreceptors such as Merkel receptors that respond to steady force, and rapidly adapting ones such as Pacinian corpuscles that respond to intermittent force. In hairy skin, many rapidly adapting mechanoreceptors are nerve fibers wound around the base of hairs.

In humans, each type of mechanoreceptor gives rise to a specific sensory experience, such as localized pressure or vibration. Haptic examination of an object involves specific hand movements to determine properties of an object such as its shape and roughness. These exploratory procedures work by causing stimulation of different types of mechanoreceptors.

Other mammals have mechanoreceptors similar to ours; other vertebrate classes have different types, along with the occasional Pacinian or Merkel. The presence of rapidly and slowly adapting categories is a general rule, surviving variations in receptor morphology. Fish have specialized receptors that are incorporated into scales, and a lateral line system that responds to disturbances in the water.

Arthropods have mechanoreceptors called sensilla on their body surface. There are different types, some slowly and others rapidly adapting. The most common are hairlike structures containing nerve fibers that are activated when the hair is deflected. Arachnids, both spiders and scorpions, make effective use of vibration in locating prey and for camouflage.

Less is known about mechanoreceptors in molluscs. In cephalopods, they are abundant, alongside chemoreceptors, in the arm suckers and near the mouth.

Touch can be either pleasant or unpleasant. At least in mammals, affiliative touch by conspecifics is reinforcing and contributes to the normal development of young animals. Pain, in contrast, is defined by the fact that it is aversive. Animals ranging from fish to hermit crabs to octopuses react to noxious stimuli not just in a reflexive way but also with lingering avoidance; this suggests that at least some members of all three phyla discussed in this book feel pain.

Hearing

Hearing is one of our most cherished senses, the sensory system normally used for language (although vision provides a viable alternative in the case of sign language) and for music. Both arthropods and vertebrates use hearing extensively, and in birds, the auditory system is especially well developed. Molluscs were long thought to lack an auditory sense, but recent research indicates that cephalopods, at least, can hear low-frequency sounds.

We will begin this chapter with a description of the human auditory system: how our ears respond to sound and how the brain processes the resulting neural signals. Next, we will describe the evolution of the vertebrate ear, from fish to mammals, and explore the role that hearing plays in the lives of different classes of vertebrates. Finally, we will discuss hearing in invertebrates.

HUMAN HEARING

Sounds are created when objects move or collide, causing a disturbance of air molecules. For example, when a tuning fork is struck on the side of a table, the tines of the fork vibrate, moving quickly and repetitively in and out. Consider just one tine of the fork: As it moves rightward, it compresses the air molecules to its immediate right. As these are shoved closer to one another, they collide more often and the air pressure in that small region increases. Some of these colliding molecules are knocked to the right, where they press against previously uninvolved molecules, and the region of high pressure moves slightly rightward. At the same time, the tine moves back to the left, creating a region of low density and therefore of low pressure. As the tine vibrates, successive peaks and troughs of pressure, collectively constituting a sound wave, move away from the fork.

STRUCTURE OF THE EAR

When a sound wave reaches your ear, it is funneled by the fleshy structure on the side of your head, the *pinna*, into the ear canal, at the end of which is the *eardrum*. The wave's peaks of pressure push the eardrum inward, and its troughs pull the eardrum outward, so it vibrates in lockstep with the sound wave.

Sounds differ in frequency—how many cycles of the vibration occur in a second. The standard unit of frequency is the Hertz (abbreviated Hz), named after Heinrich Hertz, a nineteenth-century German physicist. A sound wave created by striking a 50-Hz tuning fork will cause the eardrum of a listener on the other side of the room to move back and forth 50 times per second.

Farther into the skull is the *inner ear*, a complex, fluid-filled structure that contains the auditory receptors. Sound must reach them for hearing to occur. But a gap exists between the eardrum and the *oval window*, the membrane that serves as the entrance to the inner ear. The air-filled compartment separating the eardrum from the oval window is called the *middle ear* (see Figure 7.1)

DOI: 10.1201/9781003362319-7

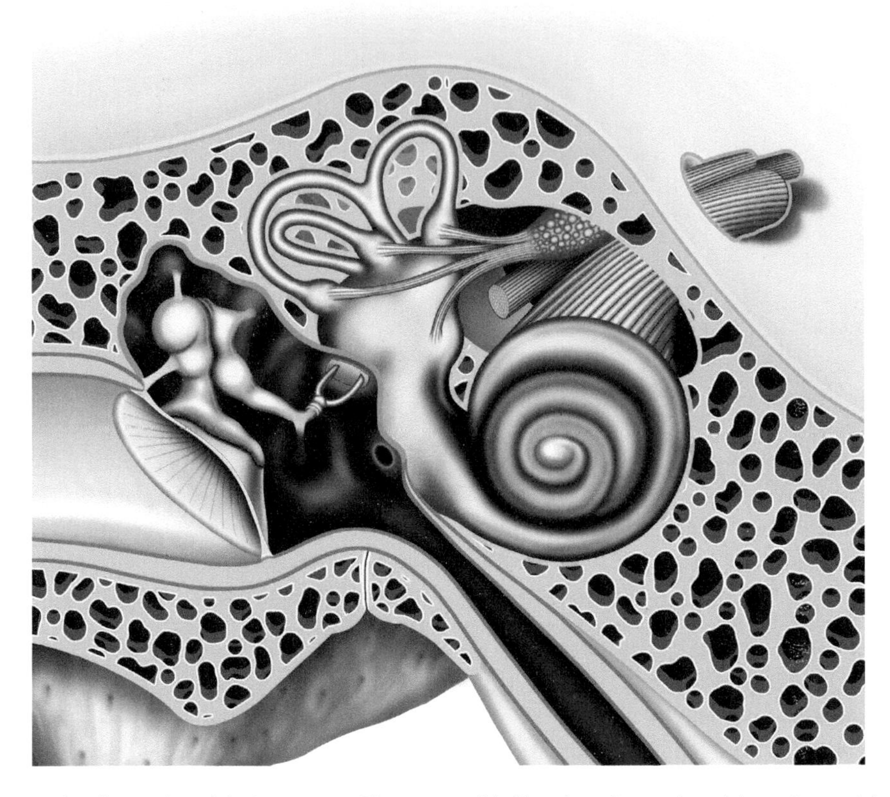

Figure 7.1 Interior portion of the human ear. The ear canal (left) ends at the eardrum (shown in purple), which separates the outer ear from the middle ear. The middle ear is an air-filled chamber containing three ossicles: (left to right) the hammer, anvil, and stirrup. The middle ear is connected to the throat by the Eustachian tube (lower right). The inner ear, consisting of the organs of equilibrium and the cochlea, is on the right. These sensory structures give rise to the branches of the auditory nerve. Note that the footplate of the stirrup is attached to the inner ear's oval window.

Source: **Used with permission of Shutterstock. Illustration Contributor, ilusmedical.**

If the middle ear contained nothing but air, only a small fraction of the sound energy passing through the eardrum could reach the oval window and cause it to vibrate also. But a bridge of bone, consisting of three tiny *ossicles*—the smallest bones in the human body—extends from the eardrum to the oval window. These ossicles are connected end to end, with the *malleus* (hammer) attached to the eardrum, followed by the *incus* (anvil), followed by the *stapes* (stirrup), which is attached to the oval window. The ossicles convey sound pressure to the inner ear and, in fact, amplify it.

Amplification is important, because the pressure of sound waves striking the eardrum is much less than the pressure needed to produce comparable movement of the oval window, which has fluid, the inner ear's lymph, behind it. Two mechanisms are responsible for this amplification. The first is concentration of the sound: The ossicles funnel the sound energy striking the whole of the eardrum onto the much smaller oval window. The second mechanism is leverage: The ossicular chain is flexibly attached to the walls of the middle ear by ligaments that allow it to pivot about a fulcrum that is closer to the oval window than to the eardrum.

Passing through the oval window, the sound wave enters the cochlea, a coiled structure resembling (and named for) a snail shell. A flexible structure called the *basilar membrane* extends from the base to the apex of the cochlea, dividing it into an upper portion and a lower portion. Arranged in rows on the top of the basilar membrane are the *hair cells*, the receptor cells for hearing.

HOW THE COCHLEA WORKS

Once inside the cochlea, the stimulus is transformed from an ordinary sound wave into a traveling wave that moves along the basilar membrane, following the coils of the spiral from the base to the apex. It moves like a rope does if you grab one end and shake it up and down. If you shake it quickly, the result will be a high-frequency wave that has closely spaced peaks, but if you shake it slowly, the traveling wave will have a lower frequency and a longer wavelength. Elucidating this remarkable mechanical transformation was the life's work of Hungarian scientist Georg von Békésy (1899–1972), who won the 1961 Nobel Prize in Physiology or Medicine for his discoveries (Békésy, 1960).

As the traveling wave moves along the basilar membrane, it gradually gets bigger, reaches a peak, and then gets smaller. Von Békésy discovered that the spot on the basilar membrane where the traveling wave reaches its maximum size depends on the frequency of the sound wave. High-frequency waves quickly rise to their maximum amplitude near the base of the cochlea, whereas low-frequency waves continue increasing in amplitude until they are approaching the apex. In fact, there is an orderly *tonotopic map* of sound frequency on the basilar membrane.

Different parts of the basilar membrane respond most vigorously to different frequencies because the mechanical characteristics of the membrane change gradually from the base to the apex of the cochlea. Imagine the membrane as consisting of narrow strips running from side to side across it, like planks in a suspension footbridge. Near the base, these strips are stiff but only loosely connected to one another, so they can be pushed up or down independently; the result is that this region of the membrane can conform quite well to a high-frequency (and therefore short-wavelength) traveling wave. Near the apex, however, the strips are flexible but tightly woven together, so they are well suited to conform to low-frequency, long-wavelength traveling waves.

Hair cells, positioned on the upper surface of the basilar membrane, extend their "hairs" (cilia) upward into the lymph. As the basilar membrane in a given location rises and falls, the hairs bend first one way and then the other. Aiding in this process is a flabby overlying structure called the *tectorial membrane*. It is poorly named, for it is more a flap of tissue than a membrane. It runs the length of the cochlea but is attached to the wall on only one side. So when the traveling wave causes the basilar membrane to move upward, it pushes against the tectorial membrane, and the two scrape across one another. This increases the bending of the hairs. The process is reversed when the basilar membrane moves down again.

Intracellular recordings from the hair cells show that they are depolarized by movement of the hairs in one direction, and hyperpolarized by movement in the opposite direction. The hair cells make synaptic contact with nerve fibers, transmitting to them information about the up-and-down movements of each small portion of the basilar membrane. Fibers of the auditory nerve fan out in the cochlea, going to different parts of the basilar membrane. An auditory nerve fiber tends to fire when the part of the basilar membrane it serves moves up and to stop firing when that portion of the membrane moves down.

PITCH PERCEPTION

What does all this tell us about our ability to hear? A person with good hearing can hear frequencies as low as 20 Hz and as high as 20,000 Hz. As the frequency of sound rises, the sensation it produces also changes, primarily along the psychological dimension of *pitch*. A low-frequency sound, such as is produced by distant thunder, is perceived as a deep rumble; a high-frequency sound, such as the squeak of a mouse, is heard as high in pitch. In between are the pitches of human voices, musical instruments, bird song, rain, and everything else that we hear.

PLACE CODING

A classic problem in auditory science is how the responses of the cochlea to different frequencies of sound enable us to experience these different pitches. One of the first to address the question of pitch coding was the German physician and physicist Hermann von Helmholtz (1821–1894).

He knew that people with long exposure to loud sounds of a specific frequency (a frequent condition among workers in early factories) suffered hearing loss that was most severe near that frequency. At autopsy, Helmholtz discovered that in such cases, the basilar membrane was most damaged—with hair cells torn loose and washed away—at a specific location. The data, though only partial, suggested a *tonotopic map*, with high frequencies represented near the base of the cochlea and low frequencies at the apex.

Helmholtz therefore proposed the idea of *place coding* of pitch: that the pitch you hear is determined by the place on the basilar membrane that responds most vigorously to a sound. Helmholtz did not know how the basilar membrane actually works; that was left for von Békésy and others to discover in the following century. But their subsequent work confirmed his early insight.

However, place coding is not the whole story on pitch perception. Another process, called *temporal coding*, is also at work. To understand it, we must take a deeper look at the structure of sound waves.

FOURIER ANALYSIS

A sound wave can be represented on a computer screen by a graph in which time is on the X-axis and air pressure is plotted on the Y-axis. As a sound wave passes through a fixed point in space, the air pressure at that point rises, falls, and rises again. The rate at which the cycle repeats is the frequency of the wave. Importantly, sound waves have different shapes. For example, a flute has a simple, smoothly curving wave, closely resembling a sine wave. But other sounds have more complicated waveforms. A piano note rises steeply to a peak and then the pressure declines slowly, with several minor ups and downs along the way before the cycle comes to an end and begins again. The same note played on a violin has still another waveform.

The world is full of sound waves—millions of them, having unique waveforms. The French mathematician J. J. Fourier (1768–1830) wanted to find a single way of describing them all mathematically. He was interested in all types of waves, but we will describe his ideas by using sound waves as an example.

Fourier discovered that any waveform can be thought of a set of sine waves of different frequencies added together. Sine waves, in other words, are the basic building blocks, or *Fourier components*, from which all waves are built. Any wave other than a sine wave has at least two Fourier components and is called a *complex wave*. The shape of the complex wave uniquely determines how many Fourier components it has and what their frequencies and amplitudes are.

Fourier developed mathematical procedures for analyzing a wave into its sinusoidal components. Any sound, even something like the sound of a branch breaking off a tree and falling to the ground, can be broken down in this way. But the method is most useful for analyzing sounds that are periodic, with repetition of the waveform from one cycle to the next. Sounds made by musical instruments fall into this category, as do many sounds made by animals, such as vocalizations.

In periodic sounds, the Fourier component that has the lowest frequency is called the *fundamental*, and its frequency is the same as that of the overall sound wave. The frequencies of other Fourier components, called *harmonics*, are multiples of the fundamental frequency. For example, if a sound wave has two components—100 Hz and 300 Hz—the 100-Hz component is the fundamental and the 300-Hz component is the third harmonic.

Figure 7.2 is a recording of human speech illustrating its frequency structure. Whenever the person is speaking, many frequencies are present in the sound. During vowel sounds, the vocal cords are vibrating, and *formants*—bands of sound energy around certain frequencies—can be seen. These are caused by resonances in different parts of the mouth and throat, which change shape during speech. Formants enable the listener to distinguish one vowel from another.

Sensitivity to formants is an aspect of audition that is not unique to humans. Dogs (Root-Gutteridge et al., 2019), songbirds (Ohms et al., 2012) and other animals are able to distinguish among the vowels of human speech on the basis of their formants. And some animals, including mynah birds (Klatt & Stefanski, 1974), elephants (Stoeger et al., 2012), and killer whales

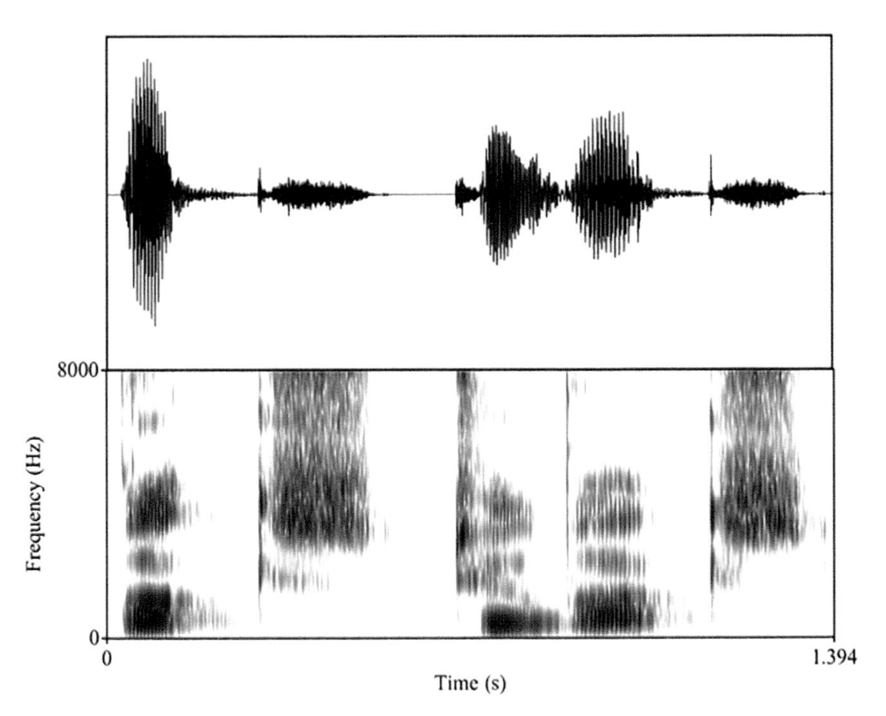

Figure 7.2 A recording of a person saying "books" and then "two books," displayed in two ways. The upper panel shows how the overall amplitude of the sound changes over time. The lower panel, called a sound spectrogram, gives more information about its frequency structure. Frequency is plotted on the vertical axis, and dark areas indicate the occurrence of sound at certain frequencies. During vowels, energy is concentrated in frequency bands called formants; during consonants, the distribution of sound energy across frequencies is more continuous.

Source: **From Abramson, A. S., & Whalen, D. H. (2017). Voice onset time (VOT) at 50: Theoretical and practical issues in measuring voicing distinctions. *Journal of Phonetics*, *63*, 75–86. Used with permission of Elsevier Science & Technology Journals; permission conveyed through Copyright Clearance Center, Inc.**

(Abramson et al., 2018), can imitate human speech sounds and sometimes the vocalizations of other species, by producing similar formant structures.

TEMPORAL CODING

Fourier analysis has been of great value to sensory scientists, for example by making possible the discovery of formants. But does it help us understand how the ear works? The answer to this question began with Georg Ohm (1789–1854), the German physicist who discovered Ohm's Law relating voltage to electric current. He was also interested in music, and fascinated by the perceptual richness—the *timbre*—of musical sounds. By listening carefully to, say, a piano note, Ohm realized that he could hear several tones within it, and he determined that these were harmonics of the note. His conclusion that the ear is doing a Fourier analysis on sounds, somehow separating their sinusoidal components, is called Ohm's *Acoustic* Law, to distinguish it from his electrical one.

The cochlea performs this Fourier analysis by spatially separating the components of a complex wave. High-frequency harmonics (like high-frequency tones presented in isolation) produce maximal movement of the basilar membrane at the base of the cochlea, near the oval window, while low-frequency components produce maximal displacement near the apex. So different groups of auditory nerve fibers respond to different components.

But it turns out that the *way* they respond is also important. Recall that auditory nerve fibers

tend to fire when the part of the basilar membrane that they serve moves up, but not when it moves down. This means that the nerve impulses in these fibers are *phase locked*, occurring only during the rising phase of the sound wave's cycle. So the brain receives a series of action potentials that are evenly spaced in time, like the steady beating of a drum. From its tempo, the brain can compute the frequency of the sound wave and cause you to hear the appropriate pitch. This is called *temporal coding* of pitch.

But there's a problem with this idea. Nerve fibers can fire just so often, because they need to pause briefly between action potentials. Generally speaking, neural firing rates top out at about 800 impulses per second. Yet we hear pitches corresponding to much higher frequencies. This paradox is partly resolved by the fact that each small region of the basilar membrane sends many nerve fibers to the brain. No single fiber can respond to each cycle of a high-frequency wave, but if neighboring fibers fire in alternation, the overall beat of the signal will still be accurate, and can contribute to pitch—as long as all action potentials are phase locked.

However, at frequencies above 5,000 Hz, phase locking does not occur, and therefore, temporal coding does not contribute to pitch. So below 5,000 Hz our pitch sensations are determined by a combination of two processes (place coding and temporal coding), but above that frequency, by only one process (place coding).

Importantly, sounds above 5,000 Hz don't have the same musical quality as lower frequency sounds. If a familiar tune is transposed several octaves higher, so that the notes all have frequencies above this value, we can still hear the notes, and can tell which are higher in pitch than others, but we cannot recognize the tune. This means that music perception is based primarily on temporal coding, and so is the perceived richness of human voices, birdsong, and other sounds in the natural world that are important to us.

LOCALIZING SOUNDS

We saw in earlier chapters that much of the brain is devoted to the processing of visual information. The same is true of auditory information, which travels to brainstem structures and then to the cerebral cortex. The most important of the brainstem structures for our discussion are the left and right *superior olives*. Their primary function is to compute the spatial location of the sound's source. They do this by precisely comparing the intensity and timing of the neural signals from the two ears.

Some information about the location of a sound does not require two ears. For example, if a familiar sound, such as the bark of the family dog, is quite loud, we can reasonably assume that Fido is close by. And if the siren of a police car is rising in pitch, as a result of a Doppler shift, we can tell that it is approaching. But people who are deaf in one ear cannot tell much about the direction from which a sound is coming, except by turning their head.

Comparing the intensity of sound at the two ears helps to indicate its source because the sound is more intense at the nearer ear. This is particularly true at high frequencies, where the head casts a *sound shadow* that reduces intensity at the farther ear. At low frequencies, the long-wavelength sound wave can bend around the head, so it is almost the same intensity at the two ears.

However, comparison of the arrival times of a sound at the two ears is another powerful, and remarkably precise, location cue. A sound wave coming directly from the left must travel about a quarter of a meter farther to reach the right ear than the left ear. Since sound waves travel through air at about 340 m/s, this additional distance will require less than a millisecond. In fact, we can tell which side a sound is coming from even if the source is only slightly off the midline, meaning there will be an even smaller difference in arrival times, one measured in microseconds. The way in which interaural time differences are computed will be more fully described later in the chapter.

AUDITORY CORTEX

After analysis in the superior olive and other brainstem nuclei, auditory signals make their way to the temporal lobe of the cortex, where they terminate in the *primary auditory area* (A1). Along the way, groups of fibers carrying information about different frequencies of sound retain their spatial relationships with one another. Arriving in area A1, they terminate in such a way that a tonotopic map is formed: a neural recreation of the mechanical one on the basilar membrane. Cells at the front end of A1 respond to high frequencies, and cells farther and farther back respond to gradually decreasing frequencies.

But cells in primary auditory cortex are responsive to numerous properties of a sound in addition to its frequency, such as its intensity, complexity, and the location and movement of its source. From A1, auditory signals are relayed to other cortical areas—some close at hand, and others farther away. There is sorting by type of information, so that signals useful for identifying a sound—such as whether it is the crowing of a rooster or the croaking of a frog—tend to stay in ventral regions of cortex, especially the temporal lobe, while signals useful for localizing a sound travel to more dorsal regions, especially in the parietal lobe. There are, in other words, *sound identity and sound location pathways* in auditory cortex analogous to the visual pathways described in Chapter 2 (Arnott et al., 2004; Clarke et al., 2000). There are also specialized areas, such as Broca's area and Wernicke's area, that are involved in perceiving and producing speech. Collectively, activity in these areas is the physiological correlate of our auditory perceptions—the beauty of music, the meaning of language, and the dangers and opportunities signaled by a host of environmental sounds.

HEARING IN FISH

THE ORIGIN OF EARS

Some sensory structures have changed more than others over the course of vertebrate evolution. Our eyes, for example, are similar in their basic structure to those of fish. But the same is not true of ears, which have undergone major, fundamental changes since the emergence of Phylum Chordata in the Cambrian Period half a billion years ago.

Early fish didn't have ears. However, as will be recalled from Chapter 6, all fish have channels called *lateral lines* running down their sides, which are open at intervals to the surrounding water. These channels are lined with *neuromasts*, clusters of receptor cells with hairlike extensions. A gelatinous *cupula* sits atop the hairs and bends them as it is pushed back and forth by currents passing up or down the channel.

Early in fish evolution, a part of the lateral line system on the sides of the head began to develop into a primitive ear. It became sealed off from the surrounding ocean and was filled with a type of lymph. This internal ear has several component structures, one of which, the *saccule*, is primarily concerned with hearing; the other components contribute to balance, helping the fish to remain upright. The saccule contains a sensory epithelium with hair cells, the auditory receptors. Loosely attached to the tips of the hairs is an *otolith*, a hard structure made of calcium carbonate, that resembles a flat pebble.

In terrestrial animals, sound waves arriving through the air and striking the body surface are to a large extent reflected or squelched. But fish live in a medium that is comparable in density to their own bodies. Vibrations therefore sweep unimpeded through the animal, so that everything vibrates in sync. Everything, that is, except the heavy otoliths. These move more sluggishly than the rest of the fish and therefore pull the hair-cell filaments to and fro, causing the receptors to respond.

Another factor in fish hearing is the *air bladder*, a sealed, air-filled chamber that helps many fish remain buoyant. Water is not compressible but air is, so the air bladder undergoes transient changes in shape and size as vibrations pass through it. If the air bladder is next to the inner ear, as it is in some fish, its movements can influence the saccule's hair cells. In other fish, the air bladder is separated from the inner ear; in them, the *Weberian ossicles*, a chain of small bones that evolved from vertebrae, serve to connect the two structures. Air bladders and Weberian ossicles are unique to fish; other vertebrates don't have them.

AUDITORY SENSITIVITY

How well do fish actually hear? To answer this question for goldfish (*Carassius auratus*), Fay

(1969) used animal psychophysics to measure their threshold for hearing at different frequencies. The experiment made use of the fact that their respiration involves sucking in water through their mouths and then pushing it out through their oxygen-absorbing gills. The goldfish moves its mouth slightly every time it sucks in water, about once per second.

Fay recorded these mouth movements and found that, when he delivered a brief, mild shock near the fish's tail, its rate of respiration slowed for a few seconds. If an audible sound consistently preceded the shock, the fish learned to slow its rate of respiration even before the shock was delivered, an example of *classical conditioning*. So the experimenter was able to use this conditioned response to determine whether or not the fish heard sounds of 12 different frequencies. Figure 7.3 shows the results in the form of an audibility curve: a graph of the fish's detection threshold as a function of frequency. Fay discovered that goldfish can hear low frequencies of sound (they are most sensitive to 350 Hz), but that above 1,000 Hz they become markedly less sensitive and can't hear anything above 3,000 Hz.

PITCH PERCEPTION

In a subsequent experiment, Fay (1970) used similar methods to determine whether goldfish can distinguish sounds of different frequencies from one another. In each test, the two sounds were carefully adjusted in intensity, based on the audibility curve, so they would be equally loud for the goldfish; the question was, then, whether they were perceived as differing in pitch. Fay found that the fish could tell two simple tones apart if they differed in frequency by at least 5%. For example, a 210 Hz sound could just be distinguished from one with a frequency of 200 Hz. This is only about one-tenth as keen as a human's frequency discrimination but is still a remarkable ability. Fish, after all, have no cochlea and no basilar membrane, so a place mechanism of pitch perception cannot operate.

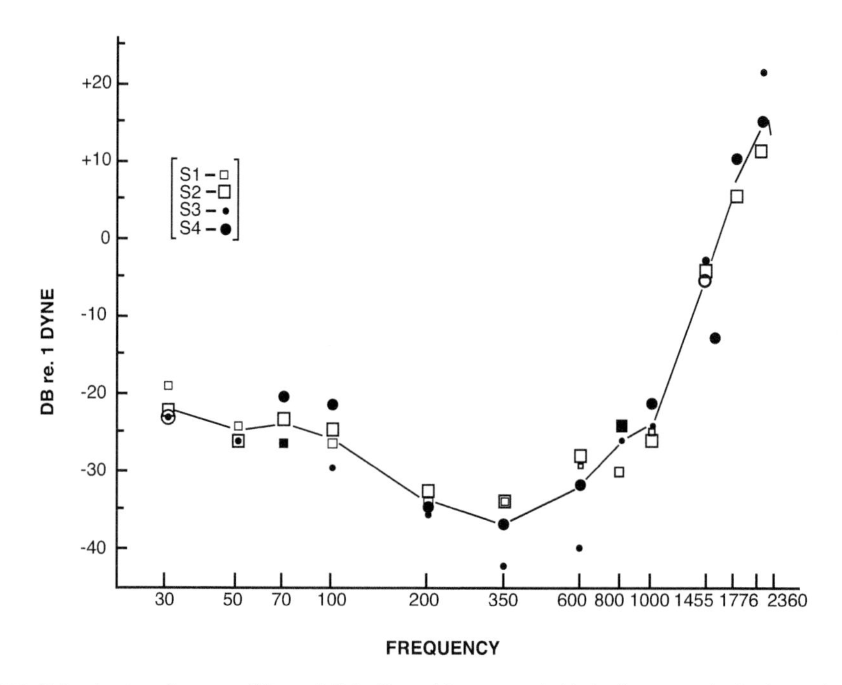

Figure 7.3 Behavioral audiogram of the goldfish. Sound frequency in Hz is shown on the horizontal axis, and detection threshold (the lowest sound intensity at which the fish's respiration rate measurably dropped) is plotted on the vertical axis. Different symbols represent the data from four individual fish.

Source: **From Fay, R. (1969). Behavioral audiogram for the goldfish. *Journal of Auditory Research*, 9, 112–121.**

So how can they tell frequencies apart? The likely answer is, by temporal coding. Goldfish auditory nerve fibers do in fact respond to sounds in a phase-locked way. This is not surprising given that the otolith tugs on the hair cells' cilia first in one direction and then the other as each cycle of a sound wave sweeps through the saccule.

To look for direct evidence that the goldfish's pitch perception is based on temporal coding, Fay (1972) tested to see whether the fish experience an auditory illusion called *periodicity pitch*. This occurs in humans when a high-frequency tone, called the carrier, is amplitude modulated, that is, repeatedly increased and decreased in intensity at a lower rate, the modulation frequency. Such a sound has only one Fourier component (at the carrier frequency) that waxes and wanes in strength. But in listening to the sound, people hear two pitches, corresponding to the frequency of the carrier and—this is the illusion—the modulation frequency. The reason for the illusion is that auditory nerve fibers fire in sync with the modulation, and this neural periodicity is interpreted by the brain as pitch.

To determine whether fish experience periodicity pitch, Fay first conditioned them to respond to a sinusoidal tone with a frequency of 40 Hz. Then, he presented an amplitude-modulated test stimulus: a 1,000 Hz carrier modulated at a frequency of 40 Hz. The goldfish responded to this as well, suggesting that they heard a nonexistent 40 Hz tone in the amplitude modulated (AM) signal. As a control, Fay presented test stimuli with other modulation frequencies, and the fish did not respond strongly to these. This evidence of periodicity pitch in goldfish indicates that their pitch perception depends on temporal coding.

LATERAL LINES AND HEARING

It is sometimes wondered whether the lateral line system, from a section of which the internal ear developed, might contribute to a fish's hearing. It too has hair cells with cilia that are pushed back and forth, in their case by the cupulas hovering over the neuromasts. But the neuromasts are directly buffeted by water containing chemicals and debris; this should make the lateral line response to sound less precise than that of the saccule.

To assess the possible role of the lateral line system in hearing, Higgs and Radford (2013) first measured the audiogram of goldfish using a physiological method, recording auditory brainstem potentials with electrodes positioned just under the skin. The fish were then placed for 3 hours in a tank containing a solution of streptomycin, an antibiotic that destroys neuromasts. Finally, the fish were returned to the testing tank, and their audibility functions determined a second time.

The results showed that streptomycin produced a fourfold increase in threshold at the lowest frequencies tested (100 and 200 Hz) but had negligible effect at higher frequencies. Higgs and Radford interpret their findings to mean that low frequencies are both heard by the ear and felt by the lateral line system but that only the auditory system can detect higher frequencies. An analogous phenomenon occurs in humans: we can both hear and feel music from a pipe organ, but vibrations from a flute can be detected only by the ear.

Our discussion of hearing in fish has so far been limited to the familiar and well-studied goldfish. But there are thousands of species of fish, and their auditory systems and abilities vary widely.

For example, Mann et al. (1997) discovered that the American shad (*Alosa sapidissima*) can hear much higher frequencies than the goldfish—frequencies that we call *ultrasound* because they are higher than 20,000 Hz, the upper limit of human hearing. The researchers used classical conditioning to measure the detection threshold of the fish at each of a series of frequencies spanning a wide range. The shad's audibility curve resembled that of the goldfish at frequencies below 2000 Hz, with threshold being lowest at about 400 Hz. And like the goldfish, the shad was not sensitive to frequencies between 2,000 and 10,000 Hz. But at still higher frequencies, its threshold dropped again, showing an ability to hear sounds up to 180,000 Hz to which the goldfish is deaf.

This ultrasonic sensitivity enables the shad to escape from dolphins and other cetaceans who hunt using echolocation, a topic discussed later in the chapter. Mann et al. suggest that the shad's ability to hear high frequencies may depend on narrow, air-filled tubes that extend from its air bladder to the inner ear, but exactly how these work is not understood.

HEARING IN AMPHIBIANS

Unlike fish, whose internal ears receive auditory stimuli that pass without difficulty through the head, amphibians have ears with specialized structures that pick up sound waves in air and pass them along to the fluid-filled inner ear. We will focus on frogs of the ranid family, like the bullfrog, since their hearing has been studied more than that of other amphibians. These frogs have an eardrum, as we do, but it is nearly flush with the skin, rather than lying at the end of a protective ear canal; and there is no pinna.

On the other side of the tympanic membrane is the middle ear. A single ossicle, the *columella*—precursor to our stapes—bridges the gap to the frog's oval window. As in other terrestrial vertebrates, the middle ear serves to amplify sounds so they are able to effectively stimulate the inner ear. Its evolution, from fish to mammals, is summarized in Figure 7.4.

The frog's inner ear, however, is quite different from our own, in that it contains no basilar membrane. The hair cells are instead gathered into two regions of epithelium, the *amphibian papilla* and the *basilar papilla*, that are attached to immovable structures deep within the inner ear. The elongated amphibian papilla is the larger of the two, containing something like a thousand hair cells; the more compact basilar papilla has fewer than 100. A tectorial membrane hovers over each. Sound waves pass through the lymph that fills the inner ear, jostling the tectorial membranes which in turn activate hair cells (Van Dijk et al., 2011). Low frequencies activate primarily the amphibian papilla, while the basilar papilla is more sensitive to high (>1000 Hz) frequencies.

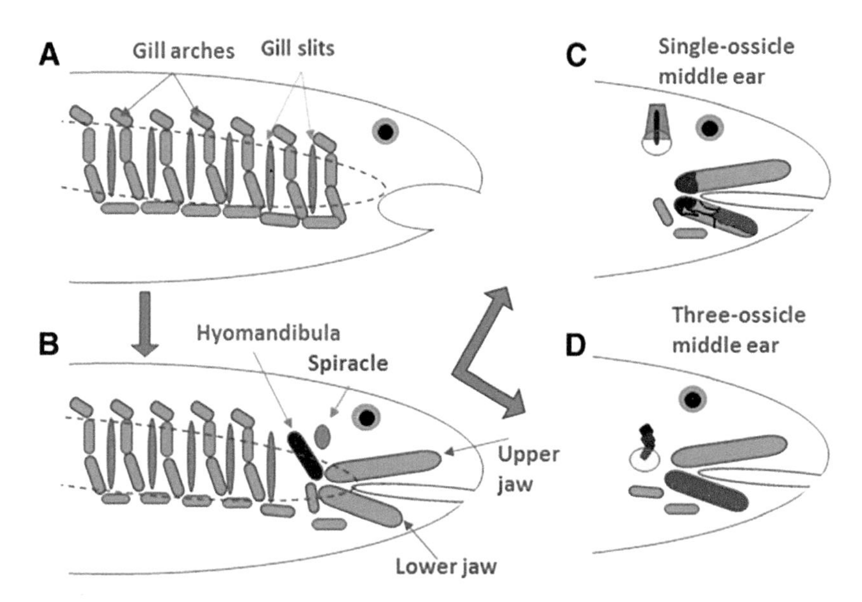

Figure 7.4 Evolutionary origin of middle ear ossicles. In early, jawless fish (A), the six pairs of gill slits are supported by bony or cartilaginous gill arches. (B) Over time, components of the two most anterior gill arches move forward to become the jaws, while the first gill slit morphs into the spiracle, an opening to the surface of the fish's head. (C) In amphibians, another gill arch component, the hyomandibula, migrates into the spiracle to become the columella. (D) In mammals, two components of the jaw have migrated into the middle ear to become the malleus and incus.

Source: **From Manley, G. A. (2017). Comparative auditory neuroscience: Understanding the evolution and function of ears. *Journal of the Association for Research in Otolaryngology, 18*, 1–24. Used with permission of Springer Nature BV; permission conveyed through Copyright Clearance Center, Inc.**

Hearing serves a variety of functions for anurans, but the most important of these is the role it plays in courtship and mating. During breeding season, male frogs gather into groups called *leks* and issue loud vocalizations that advertise their presence to females. Sexually receptive females move toward the sound and, arriving at the lek, choose among potential suitors. The strength of this phonotactic response varies in a seasonal way that correlates with the level of estradiol and other gonadal hormones in the female frog's bloodstream; and hormone injection affects the probability of response (Arch & Narins, 2009).

Do hormones change the female frog's sensitivity to advertisement calls, or only her motivation to respond to them? To address this question, Miranda and Wilcynski (2009) recorded multi-unit neural responses from the *torus semicircularis* of green tree frogs (*Hyla cinerea*). This midbrain structure, homologous to the inferior colliculus of mammals, is a major center of auditory processing in the frog's brain. The investigators found that it responds both to advertisement calls and to the individual frequencies of sound that make them up.

Next, the researchers implanted some frogs with testosterone, a hormone that might be expected to have effects opposite to those of the estradiol used in earlier studies. The testosterone had no effect on the auditory sensitivity of male frogs but substantially decreased the sensitivity of females, both to advertisement calls and to their component frequencies. Interestingly, the change was especially marked at high frequencies to which the basilar papilla responds, raising the possibility that this small inner-ear structure plays a special role in sexual communication.

The major conclusion of the study is that hormonally driven changes in an animal's behavior can be mediated, at least in part, by changes in sensory processing.

HEARING IN REPTILES

Fish and amphibians lay their eggs in water, which bathes and supports the developing embryos. The development of an *amnion*—a set of protective membranes surrounding the embryo—enabled the emergence of vertebrates that can remain on land throughout their lives. *Amniotes* include the reptiles, birds, and mammals.

Most reptiles (but not snakes) have a middle ear resembling that of amphibians, with a columella, with cartilaginous extensions, that bridges the gap from eardrum to oval window. There is a marked difference, however, between the inner ears of amphibians and amniotes (Manley, 2017). In amniotes, the auditory papilla with its hair-cell receptors rests not on a solid backing, but on a moveable *basilar membrane*. This flexible structure, anchored only around its margins, enabled the evolution of complex movement patterns important for tonotopic mapping. The amniote inner ear has only one auditory papilla, not the two that are found in some amphibians.

Knowledge about the hearing ability of reptiles is based largely on electrophysiological recordings from auditory nerves or other neural structures; psychophysical data are limited because of the difficulty of training reptiles. However, where both psychophysical and neurophysiological results have been obtained in the same animal, they are in good agreement (Martin et al., 2012). In this study, the animal was a female loggerhead turtle (*Caretta caretta*), 31 years of age, living in an aquarium in Florida, and tested underwater. The flash of an LED light indicated to the turtle that a trial was beginning. Her task was to press a response paddle when she heard a sound during the trial, and to refrain from pressing it when there was no sound. Correct choices were rewarded with bites of squid and fish. The intensity of the test sound was adjusted from one block of trials to the next, being raised after blocks in which she performed at a chance level and lowered after blocks in which most responses were correct. In this *tracking procedure*, sound intensity gradually approached and then hovered about the turtle's threshold. The data showed that the animal's hearing was best at low frequencies but tapered off above 1,000 Hz. Auditory evoked potentials, recorded from the top of the turtle's head, were consistent with these psychophysical findings.

Hearing limited to frequencies below 5,000 Hz is the general rule among reptiles, although there are exceptions (Manley & Kraus, 2010).

This is sufficient to hear the vocalizations of conspecifics, which are usually in the interest of courtship or dominance.

HEARING IN BIRDS

The remarkable vocal and auditory abilities of birds are in striking contrast to the simpler abilities of reptiles. This is primarily due to their enlarged brains, with a greater capacity for neural processing. But developments in the inner ear are also partly responsible. The most important of these is the elongation of the basilar membrane in birds. With a greater length of this structure, more precise tonotopic mapping—in terms of millimeters of basilar membrane per octave—is possible (Manley, 2017). In general, larger birds have longer basilar membranes and hear best at lower frequencies. For example, many songbirds hear best at about 4 kHz, while large birds such as the emu are most sensitive at 1–2 kHz (Gleich et al., 2005).

Hearing in birds and the sounds they make for others to hear are a subject that has fascinated people throughout history. Naturalists can recognize the calls or songs of many dozens of bird species, and scientists have carried out a vast array of experiments to understand why and how birds use sound. Some sing to attract a mate; some to threaten intruders or warn of predators; some to mimic sounds they hear, for reasons that are sometimes obscure. And some birds listen quietly to detect or localize the sounds made by other animals. We will consider audition and associated behaviors in two very different birds: cowbirds and owls.

COWBIRDS

Cowbirds are best known for being brood parasites: They lay their eggs in the nests of other species, leaving the hatching and rearing of their offspring to the unsuspecting hosts. Emancipated from the duties of parenting, adult cowbirds have a lot of time for social interaction, in which singing plays a crucial role. Most of the singing is done by males, who sing to females to court them (see Figure 7.5). If a female is intensely courted at close range by a singing male, she may make a copulation solicitation display (similar to a lordosis response in mammals), making it possible for mating to occur.

Figure 7.5 A male brown-headed cowbird, *Molothrus ater.*

***Source:* Used with permission of Shutterstock. Photo Contributor, Steve Byland.**

Males also sing to one another, in singing contests. One male will sing to another, generally stringing several short songs together. The listener then responds, *countersinging* the same songs. These contests are one of the ways in which a dominance hierarchy is established. For example, if a dominant male lands on a perch, subordinate males are expected to move aside. Dominant males often thwart the attempts of subordinates to sing at close range to females. Thus a male's social status has a powerful effect on his mating success, defined as the number of copulations achieved in an aviary or in the wild (White et al., 2010).

In the laboratory, the recorded songs of different males can be played to a female in a sound-attenuating chamber, and ranked in terms of their *potency*, i.e., their ability to elicit the copulation solicitation display. Song potency is another key factor in a male's mating success: It is necessary for mating, although not always sufficient because of constraints imposed by the males' dominance hierarchy (West et al., 1981).

Females' ability to discriminate among the songs of different males is a subtle auditory skill that is updated in response to ongoing experience, as shown by West et al. (2006). The researchers housed two groups of females in separate same-sex aviaries. Prior to breeding season, the prerecorded songs of males were played repeatedly in both aviaries. In addition, one of the aviaries received a 12-day visit from other males, whose songs were also recorded by the experimenters. During breeding season, the females were individually tested both with songs from the original set and with songs from the visiting males. Females in the visited group responded with more copulation solicitation displays to the songs of the visitors, while females in the unvisited group preferred the original songs. A female's song preferences are thus dynamic and modifiable, even in adulthood.

The neurobiological basis of birdsong was not well understood until the discovery by Fernando Nottebohm and his colleagues (1976) of a *song control system* in the brains of songbirds. This is a set of neural structures, the most important of which is the HVC, high in the pallium, which contains the memories of songs and their motor programs. (HVC was originally an abbreviation but is now the official name of this structure.) Both male and female songbirds have an HVC, but it is larger in males.

To evaluate the role of the HVC in the ability of female cowbirds to perceive the quality of males' songs, Maguire et al. (2013) lesioned this structure in a number of females. The birds were anesthetized, and ibotenic acid, a neurotoxin, was injected into the HVC. Sham-operated control birds received injections of an inactive liquid. After they recovered from the surgery, the birds were individually tested with recordings of male songs to assess their ability to make potency judgments. The control females responded normally, but the lesioned females did not discriminate among the songs, making copulation solicitation displays to all of them (although not to the songs of other species) (see Figure 7.6).

Returned to their respective aviaries, the lesioned and control females had different effects on the behavior of the flocks. Re-introduction of the control birds produced no measurable impact on the dynamics of the aviary. Re-introduction of the lesioned birds, however, produced major disruptions. They responded with solicitation displays both to their own mate and to other males who sang to them. This resulted not only in a weakening of their own pair-bond but that of other birds in the flock. Singing by males to

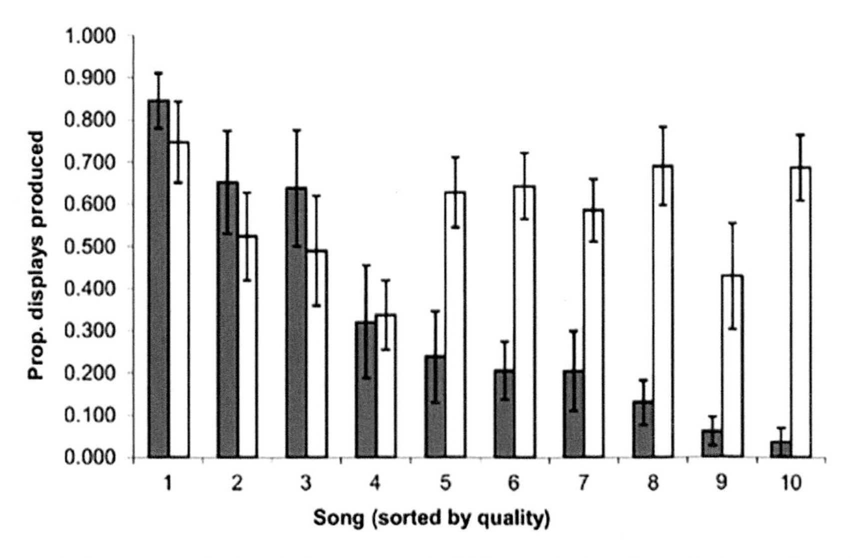

Figure 7.6 Auditory processing in a brain structure, the HVC, enables female cowbirds to evaluate the songs of different male cowbirds. Recorded songs were played to individual females whose HVC had been chemically lesioned (unfilled bars) or left intact (filled bars). The response measure was the proportion of trials on which a song elicited a copulation solicitation display. Data were tallied separately for songs differing in "quality" (effectiveness) as established in earlier research, song 1 being the most effective and song 10 the least effective. Intact females responded more to high-quality than to low-quality songs, but lesioned females were oblivious to the quality of the songs.

Source: **From Maguire, S. E., Schmidt, M. F., & White, D. J. (2013). Social brains in context: Lesions targeted to the song control system in female cowbirds affect their social network. *PLoS One*, 8, #5, e63239. © 2013 Maguire et al.**

other males, a way of maintaining the stability of the dominance hierarchy, was also affected. This study shows that nuanced perception of male song by the females plays a key role in the social functioning of cowbirds. The same is probably true of many other songbirds.

BARN OWL

Nocturnal owls have good vision in dim light. In the early evening, these predators are on the lookout for rodents, snakes, and other small animals. But when light levels are even lower—what most people would call total darkness, although in nature it is never quite that—they are still able to catch a scurrying mouse (see Figure 7.7).

A classic experiment with a barn owl (*Tyto alba*) showed that in this situation, the owl relies on hearing, rather than another sensory modality such as olfaction, to locate its prey (Konishi, 1973). The experimenter attached a string to a mouse's tail and a piece of crumpled paper to the other end of the string. The mouse was placed in a light-free room where the owl waited. As the mouse walked noiselessly across foam rubber, the paper, which it was dragging along, rustled. The owl flew down from its perch and attacked the paper: It was attacking the source of the sound, ignoring any odor or warmth radiating from the mouse.

The barn owl's remarkable ability to catch prey in darkness depends mainly on its keen

Figure 7.7 A barn owl (*Tyto alba*) watches and listens for its dinner to scurry by.

Source: **Used with permission of Shutterstock. Photo Contributor, Albert Beukhof.**

sensitivity to sound, which Konishi (1973) measured by training the animal to take off from its perch when it heard a sound. At the frequencies it hears best (4–6 kHz), the owl is about three times as sensitive as a person is to his/her best frequencies (2–4 kHz).

But sensitivity alone cannot explain the owl's auditory localization, which requires precise comparison of the sound received by the two ears. Like humans, barn owls compare the times of arrival of a sound, and its intensities, at the two ears. These comparisons are not matters of conscious deliberation but are made automatically in the brainstem.

The pathways that auditory signals travel from the ears to the avian brain are complex. Once hair cells are activated by sound, they synaptically transmit signals to auditory nerve fibers. These extend from the cochlea to the brainstem, where they synapse on neurons in the cochlear nucleus on the same side of the brain. Some of its neurons, in turn, send axons to the *nucleus laminaris* (homologous to the superior olive of mammals), a large structure on the dorsal surface of the brainstem in the pons-medulla region. Although there are technically a left and a right nucleus laminaris, they are close to the midline and function as a single structure.

How this brainstem structure works was discovered by Carr and Konishi (1990). When there is a rustling sound, hair cells in both ears are activated and send signals that eventually reach the nucleus laminaris. The left and right ears' signals traverse a left-right array of laminaris neurons, but in opposite directions. A cell in the array will respond vigorously if the two signals reach it at exactly the same time, but not if one arrives before the other. So if a neuron is in (say) the left half of the nucleus, the signal from the right ear will have to traverse more of the array to reach it than does the signal from the left ear. How can this cell ever give a vigorous response?

The answer is that when a sound is presented on one side, it will reach the closer ear sooner than the farther one, and the resulting signal will get a head start that can exactly make up for the longer distance it has to travel across the array to reach a specific cell. In other words, different coincidence-detecting cells in the nucleus laminaris respond to sounds at different locations in space. Cells on the left side of the nucleus will respond to sounds that are on the owl's right, and vice versa. So based on which cells are firing, the owl can tell where a sound is coming from.

There is a complication. The barn owl's left ear is slightly higher on its head than the right one. Thus, a sound could arrive slightly sooner at the left ear if it comes from the owl's left, or from overhead. To resolve this ambiguity, the owl computes the difference in intensity of a sound at its two ears, in addition to the difference in arrival time. Differences in arrival time are strongly influenced by the *azimuth* (left/right position) of a sound source, but only moderately by its *elevation*, whereas the reverse is true for differences in intensity. So by combining these two independently computed differences, the owl can localize the sound with high accuracy (Moiseff, 1989; Takahashi et al., 1984; Viete et al., 1997).

There is another way in which the hearing of the barn owl (and presumably many other birds) is remarkable: It declines very little with age. Humans are familiar with *presbyacusis*, the gradual elevation of auditory thresholds, especially at high frequencies, that tends to occur as we grow older: High-frequency impairments of 60 decibels or more are common by age 70. The same process has been shown to occur in other mammals as well. The main contributor to presbycusis is a loss of hair cells, particularly at the base of the cochlea, due to the "wear and tear" of a lifetime of auditory stimulation.

To obtain comparable data on barn owls, Krumm et al. (2017) tested the hearing of both young (less than two years old) and old (ages 13–17) individuals. Birds were trained to sit on a perch facing a loudspeaker. Only when they heard a tone were they to fly to a different perch to receive a reward. The lowest intensity to which they consistently responded was taken as threshold. Across a range of frequencies, the old birds were only slightly less sensitive than the young ones; even at 12 kHz, the upper end of the owl's audible range, the difference between the groups was a modest 6 dB. The maintenance of hearing across the lifespan in barn owls is believed due to an ongoing regeneration of hair cells, a process that has been documented in several bird species.

HEARING IN MAMMALS

It will be recalled that young people can hear frequencies of sound between 20 Hz and 20 kHz, a range of 3 log units (almost 10 octaves). Their sensitivity, represented in an audibility curve of threshold versus frequency, is U-shaped, with best sensitivity at about 3 kHz. Many other mammals have a similarly broad range, but their audibility curve is often shifted to the left or right along the frequency axis compared to that of humans. Dogs, for example, can hear frequencies between about 60 Hz and 40 kHz. Thus, a dog whistle, audible to dogs but not humans, is designed to emit frequencies between 20 and 40 kHz.

We anthropocentrically refer to frequencies below 20 Hz as *infrasound*, and frequencies above 20 kHz as *ultrasound*. Generally speaking, small animals hear best at high frequencies and large animals at low frequencies. In part, this is due to the fact that middle ear ossicles and other peripheral components of the auditory system are themselves smaller in small animals, and thus resonate at higher frequencies. But high-frequency hearing also helps small animals localize sounds. Unless a sound has a short wavelength compared to an animal's head, it will bend around the head and not cast a crisp sound shadow, making interaural intensity differences negligible.

In this section, we will consider the hearing of three mammals: elephants, who are sensitive to infrasound; bats, who are sensitive to ultrasound; and whales, whose auditory systems are adapted to the aquatic environment to which they have returned.

ELEPHANTS

The basic hearing ability of an elephant was measured by Heffner and Heffner (1982). Their subject was a seven-year-old female Indian elephant (*Elephas maximus*) in a zoo. The tethered

animal faced an array of three response panels in a wall within her enclosure. She pressed a central panel with her trunk to initiate a trial, listened for a tone, and pressed the left panel if she heard one, the right one if she did not. Correct responses were rewarded with a sip of flavored sugar-water delivered into a trough below the response panel.

Before looking at the results of this experiment, it is worth noting that guidelines for behavioral research with animals have evolved considerably in the last 40 years. Especially in the case of large mammals, respect for their autonomy is now an important requirement. An additional factor in the case of the Indian elephant is that since 1986, it has been classified as endangered by the International Union for Conservation of Nature (IUCN). It is now the norm for researchers to study elephants in ways that do not constrain their actions.

This elephant's audiogram had the expected U shape, with audible frequencies extending from 16 Hz to 12 kHz, with threshold being lowest at about 1 kHz. All three of these values are below the corresponding values for humans (approximately 20 Hz, 20 kHz, and 3 kHz, respectively). She also had good auditory localization, in perceiving whether a tone was coming from a speaker to the left or right of the response panels.

A more naturalistic study, carried out on free-ranging African elephants (*Loxodonta africana*) in a national park in Namibia, examined the way in which they respond to the calls of other individuals (Langbauer et al., 1991). From a tower, the experimenters made video and audio recordings of elephants at a water hole below. Male elephants generally arrived alone, or in small all-male groups. Females, in contrast, consistently traveled in groups and were accompanied by their offspring (see Figure 7.8).

Once the animals had arrived at the water hole and had a drink, the researchers played back the prerecorded calls of other individuals—mostly females—using speakers up to 2 km from the water hole. These calls had fundamental frequencies in the 15–35 Hz range and were occasionally heard as faint rumbles by the researchers. They were about 30 seconds in duration. The initial response to a call was the

Figure 7.8 A female elephant and her calf head toward a water hole.

Source: **Used with permission of Shutterstock. Photo Contributor, Michael Sheehan.**

same for male and female elephants: raising their head, spreading their ears, and freezing as they listened. After that, however, males and females responded differently. Females vocalized, perhaps responding to the distant caller, and then set out in groups (with their calves) in the direction of the call, but not walking far. The males, in contrast, seldom vocalized; instead they quietly set out directly toward the call, often walking more than a kilometer in a straight line. The researchers speculate that the males refrained from vocalizing in order not to draw the attention of other males to the possible presence of a receptive female.

Whatever the reasons for their behavior, the elephants' reactions clearly showed that they heard the calls, even at a distance of 2 km. And because the calls were, for technical reasons, played back at less than their original intensity, Langbauer et al. concluded that in the wild, calls can be heard at distances of some 4 km. This is a remarkable feat on the part of elephants, due both to their auditory sensitivity and to the fact that low frequencies of sound travel farther than high frequencies.

Studies using similar methods focused on the ability of African elephants to extract social information from the playback calls. McComb et al. (2000) recorded the responses of family groups (i.e., clusters of related adult females and their offspring) to the calls of other females with whom they had different degrees of familiarity. When the calls were from members of other families with whom the subjects had frequent, friendly interactions, they called in response and moved toward the loudspeaker; when the calls were from other families they had encountered but did not often interact with, the subjects paused and listened but generally did not respond; and when the calls were from unfamiliar females, the subjects became agitated and huddled close together. In other words, female elephants are affiliative, seeking to approach others they know while avoiding those who are unfamiliar, and that when they hear a call, they are able to perceive the familiarity of the caller.

A female elephant whose call had been recorded died during the study. Twenty-three months after her death, this call was played back to her family group. They called in response and moved toward the loudspeaker. So elephants really do have long memories, including auditory memories.

A more recent study (Stoeger & Baotic, 2017) focused on the responses of male elephants, in a national park in South Africa, to the calls of females. Two sets of calls were recorded: Some were from females living in the same park, with whom the males were therefore familiar, and other calls were from females living elsewhere. Choosing a time when a male elephant was alone and eating calmly, the experimenters played two calls, 10 minutes apart. One was from a familiar, and the other from an unfamiliar female; the order of the two calls was counterbalanced across subjects. The male elephants showed a clear ability to discriminate between the calls. They spent twice as much time listening attentively—as shown by pausing and lifting their ears—to the unfamiliar calls as to the familiar ones. And they spent more than eight times as long facing the speaker and approaching it when hearing an unfamiliar call. Mating opportunities with unfamiliar females appear to have a high priority for male elephants, a preference that may have adaptive value by reducing inbreeding.

BATS

We turn now from elephants, a group of mammals that use infrasound, to another group—bats—that use *ultrasound*. There are well over 1,000 species of bats, which differ in many respects, including the ways they use their hearing. Some eat insects, others eat fruit, and still others (a small minority) drink the blood of other animals. We will focus on one species, the big brown bat (*Eptesicus fustus*), which relies heavily on audition both for social communication and for navigation and hunting.

The big brown bat does not weigh a lot (about 20 grams—less than an ounce) but earns its name from its third-of-a-meter wingspread. It is widely distributed across North and parts of South America. By day, it roosts in enclosed spaces such as caves, hollow trees, attics, and barns; it becomes active at night, hunting for nocturnal flying insects.

Big brown bats can hear frequencies ranging from about 5 kHz up to more than 100 kHz. Their audiogram is roughly U-shaped but is shifted far to the right along the frequency axis compared to the human audiogram (Simmons et al., 2016). In other words, most of the bat's hearing is in the ultrasonic range; so it is unsurprising that the same is true of its vocalizations.

Some bats vocalize by clicking their tongues (Yovel et al., 2011), but a majority, including big brown bats, create sound by the vibration of vocal cords in the larynx. An individual vocalization lasts less than 100 ms and consists of multiple harmonically related frequencies, such as 25 kHz, 50 kHz, and 75 kHz. These are actually not pure tones, but frequency bands, called *formants,* like those in human speech. During a vocalization, the frequencies of the formants may remain relatively constant, but more often rise or fall in parallel, a change called *frequency modulation*.

SOCIAL VOCALIZATIONS

Bats emit two types of vocalizations, which serve different functions. We will first consider *social vocalizations* and then the calls that are part of *echolocation,* a method of hunting and navigation. Social calls are generally longer than echolocation calls and sometimes extend into lower frequencies that are audible to humans. However, even echolocation calls convey some social information, such as the caller's sex, age, and family membership (Kazial & Masters, 2004; Masters et al., 1995). Research on big brown bats

suggests that both types of calls develop from the infant's isolation calls, along separate developmental pathways that diverge in the second week of life (Monroy et al., 2011).

Social vocalizations communicate various messages, such as that the caller is seeking a mate, claiming an insect, or threatening aggression. Gadziola et al. (2012) found that the vocalizations of the big brown bat can be sorted on the basis of their physical properties into a large number of categories, specific to particular behavioral contexts. For example, when bats are resting together on a perch and gently jostling one another, they emit what the researchers call appeasement vocalizations, such as a long call in which the frequencies of the harmonics gradually decrease, and then at the end rise slightly up again. When the bats are agitated, as when an intruder bat approaches and tries to get onto the perch, they may emit the bat's equivalent of a growl, consisting of noise spanning a wide frequency range.

The overall point is that the bat's social vocalizations are complex, reflective of their complex social interactions.

ECHOLOCATION

Big brown bats are also among the many bat species that employ echolocation. As it flies about, avoiding obstacles and catching insects, an echolocating bat is guided by the echoes of its own calls. These are more stereotyped than the social vocalizations, consisting of short but intense (over 120 dB: Hulgard et al., 2016) bursts of sound. Within each call, the frequencies of all the formants drop rapidly and then briefly level off. Usually one of the formants is more intense than the others, and its central frequency is called the *dominant frequency* of the call.

Let's consider in more detail how echolocation works. Say a bat is searching for insects on a dark night. Several times a second it emits a call lasting perhaps 10 ms. Say further that a moth is hovering in the air 5 m in front of the flying bat. Given the speed of sound of 340 m/s, the bat's call will take about 15 ms to reach the moth and reflect off its body. In another 15 ms, the echo will arrive back at the bat's ears. The bat hears both its own cry and the echo, and from the time between them, its brain computes the distance to the moth.

As the bat approaches the moth, it emits calls more often, and thus gets more frequent updates on the moth's location. At the same time, the round trip of a call and its echo gets shorter. For example, when the bat is 50 cm from the moth, the round trip will take only 3 ms. This poses a potential problem for the bat: If call duration remained fixed at 10 ms, the call's faint echo would begin arriving back at the bat's ears while the call itself was still leaving the bat's mouth, and would therefore be inaudible. The bat avoids this problem by making its calls shorter and shorter as it closes in on the moth. Call intensity also declines, since less sound is needed to produce an audible echo from a near object.

An individual call's dominant frequency drops very rapidly and then levels off. During this constant-frequency portion, the bat can detect small differences between the frequency of the echo and the frequency of the outgoing call. These differences indicate relative movement between the bat and the moth. If the bat is drawing closer to the moth, it will encounter successive cycles of the reflected wave at a higher rate than they were emitted. But if the moth is getting away, the echo will have a lower frequency than the outgoing call. These informative changes in frequency are called *Doppler shifts*. Even the fluttering of the moth's wings produces tiny, alternating Doppler shifts that may help the bat identify its prey.

Once a bat has detected the presence of a moth, it determines its direction by comparing the echoes arriving at its left and right ears. It then aims its outgoing calls in that direction, by turning its head to the left or right. Calls spread outward very broadly, but with a central axis in which the sound energy is more intense.

The importance of this sonar beam axis has been demonstrated in an elegant study by Surlykke et al. (2009), in which bats flew through a circular hole in a net to seize a tethered insect. Video cameras using dim red light invisible to the bats recorded their movements, and an array of microphones simultaneously captured the direction of their calls. As they flew toward the hole in the net, they aimed their sonar beam alternately at its right and left edges, ensuring safe passage between them, but even before passing through the hole, they shifted their sonar axis to the insect, perceived through the net.

The researchers note that the bat's ability to direct the beam toward one thing at a time is analogous to the way we direct our gaze toward objects of interest so that their images will fall, sequentially, on the fovea. As it swerves around obstacles, avoids insects it can tell are unpalatable, and closes in on its prey, the bat is constantly aiming its calls in different directions.

A remarkable feature of bats' echolocation is that they are able to hear both their own very intense calls and the faint echoes of those calls that arrive a few milliseconds later. Humans would be unable to do this, because each loud call would produce a temporary increase in auditory threshold that would render the subsequent echo inaudible. In some bats, small muscles attached to the middle ear ossicles contract just before the bat emits a call, reducing ossicle movement and thus attenuating the sound reaching the inner ear (Jen & Suga, 1976); the muscles relax a few milliseconds later, allowing the echo of the call to be heard.

Even with this protective mechanism in play, however, bats' auditory receptors are frequently exposed to intense sound, yet continue to function. The resilience of the bat's auditory system was demonstrated by experiments (Hom et al., 2016; Simmons et al., 2016) in which big brown bats were tested after being exposed to an hour of 110 dB broadband noise. Their auditory thresholds and ability to navigate an obstacle course were unaffected. The biological mechanisms underlying bats' ability to echolocate accurately despite their routine exposure to very high sound intensities are not well understood.

JAMMING

A bat's echolocation can sometimes be interfered with by the calls of other bats, a phenomenon called *jamming*. These may be bats of the same species or of different species; and the jamming may be an inadvertent result of bats hunting in close proximity, or caused by special jamming calls (Jones et al., 2018).

For example, the Brazilian free-tailed bat, *Tadarida braziliensis*, emits special jamming calls when a nearby bat of the same species is closing in on an insect. The jamming call's dominant frequency rises and falls sinusoidally at a high rate. Confused by the temporal overlay of the jamming signal and its own echoes, the hunter now usually fails to catch the moth, increasing the chances that its jamming competitor can do so (Corcoran & Conner, 2014).

Big brown bats are also affected by the jamming calls of other bats but have devised countermeasures so that their hunting efficiency is not compromised. They change their echolocation calls so that they are more distinguishable, in their spectral or temporal properties, from the jamming call (Bates et al., 2008; Jones et al., 2018).

Surprisingly, big brown bats are more vulnerable to jamming by a tiger moth, *Bertholdia trigona*, which begins emitting rapid clicks as a bat approaches. Corcoran et al. (2009) showed, by videotaping individual bats as they attempted to seize a loosely tethered moth, that these clicks interfere with the bat's hunting ability. With control moths of a non-clicking species, the bats were able to catch the moth on about 90% of trials, but with clicking *B. trigona*, the bats made contact only about 20% of the time. To confirm that this difference was due to clicking, rather than to some other difference between moth species, the experimenters lesioned the clicking mechanism in some *B. trigona* individuals. These silent moths were invariably captured. By detailed analysis of the bats' flight paths, the investigators (Corcoran et al., 2011) concluded that clicks in close temporal proximity to echoes interfere with the bat's ranging mechanism, so that it perceives a blurred acoustic image of the moth.

AUDITORY CORTEX

The auditory cortex of the big brown bat contains *combination cells* that respond best to the combination of a call and its echo (Dear et al., 1993). For each cell, there is a specific best delay, corresponding to a specific distance to the reflecting object.

In some other echolocating bats, there are combination cells that respond selectively to stimuli located at particular azimuths (left/right direction) and elevations. In other words, these cells are tuned to detect an echo coming from a specific location in three-dimensional space. In still other bats, there are combination cells that detect differences in frequency between a call and its echo—Doppler shifts indicating relative movement of the target and the bat. The existence of these different types of combination cells shows that the bat's cortex is highly specialized to support echolocation.

CETACEANS

Bats are not the only mammals that use echolocation. Many whales, including dolphins, have highly developed biosonar that allows them to perceive animals and objects in their environment even in darkness or when the water is turbid. And like bats, cetaceans have other types of vocalizations that serve social functions. But the ways in which whales both produce and hear sound are different from those of other mammals, as we will see in this section.

THE WHALE EAR

The earliest cetaceans, descended from terrestrial mammals, entered the water about 50 million years ago. Their ears, which had evolved for hearing in air, were not ideal for an underwater environment, for the following reason.

Air has a low acoustic impedance, while an animal's body has a high impedance. For terrestrial mammals, this mismatch means that sound waves are, for the most part, reflected by the head; only via the delicate machinery of the middle ear can sound waves reach the inner ear and its receptors.

Water, on the other hand, has an impedance comparable to that of body tissues, so when a terrestrial mammal is underwater, sounds flow into the head and spread widely throughout the skull, a process called *bone conduction*. When these vibrations reach the cochlea, they agitate the basilar membrane in a cruder way than sounds conveyed by the ossicles.

In whales, this deleterious effect of bone conduction has been greatly reduced by evolutionary changes in ear anatomy, most of which occurred in the first ten million years of their aquatic existence. Whales still have a middle ear and an inner ear, but these structures are no longer in direct contact with the skull. Instead, they are part of the *tympanoperiotic complex*, a structure protected from the skull's vibrations by sound-attenuating air sinuses (Nummela et al., 2007).

Another evolutionary change is that in whales, the ear canal is no longer operative. Instead, a privileged route to the auditory system is through fatty tissue within and around the whale's lower jaw. This *acoustic funnel* (Yamato & Pyenson, 2015), a novel structure that evolved from the malleus in combination with another bony element, the goniale, is thought to capture sound from the jaw fat and channel it to the other ossicles.

Some 34 million years ago, cetaceans split into two groups, toothed whales and baleen whales, and evolution of the auditory system continued independently in each group. *Toothed whales (odontocetes)* include sperm whales and beaked whales, as well as dolphins and porpoises. They are predators, and many use high-frequency echolocation to home in on prey. *Baleen whales (mysticetes)*, a group that includes the blue whale and the humpback, are named for a rigid, comblike structure attached to their upper jaw. They feed by taking in water, then closing their mouth and filtering the water back out through the baleen, retaining only plankton. In contrast to odontocetes, their hearing is skewed toward low frequencies.

WHALE VOCALIZATIONS

Toothed whales. Odontocetes do not vocalize with their larynx, as terrestrial mammals do. Instead, they make sound with a functionally analogous structure, the *phonic lips*. These are not in the throat, but in the nasal passage, which leads to a nostril, the blowhole, on the top of the head. The phonic lips can pull apart to allow air to flow upwards, or move quickly together, stopping the flow of air and producing an echolocation click.

Click rate can vary, depending on the species and situational factors, from about 10 Hz to 300 Hz, but the clicks themselves can contain ultrasonic frequencies approaching 200 kHz (Madsen et al., 2023). Once produced by the phonic lips, sound passes forward to the melon, a large fatty structure on the whale's forehead. This acts like a lens to focus the sound and project it forward into the environment.

A toothed whale's phonic lips can operate in different modes or *registers*, which are mediated by changes in the lips' tension, length, and other physical properties. Named for analogous registers in the human voice, these are the *vocal fry, chest, and falsetto registers.* Echolocation clicks are generated in vocal fry register. The sounds emitted in chest and falsetto registers are nearly continuous rather than pulsatile; these social vocalizations are rich in low frequencies and are often described by human listeners as grunts and whistles, respectively (Madsen et al., 2023).

Tønnesen et al. (2020) studied the use of echolocation by sperm whales, the largest of all tooth-bearing predators, as they swam in the Azores. The investigators attached a sound-recording tag to the whale's nose with a suction cup, to record both its very intense clicks and their echoes. The time elapsing between emitted clicks and their echoes revealed that the whale received information about potential prey at distances well in excess of 100 m. When the whale targeted an animal for capture, the clicks sped up, producing a buzz. The investigators found that buzzing began while the sperm whale was still more than 20 m from its prey. Sperm whales thus appear to monitor a field of echoes before choosing one animal as their prey.

Baleen whales. The way in which mysticetes vocalize was long unknown, for they do not have phonic lips; and no vocal folds, the structures that produce sound in terrestrial mammals, had been found in the larynx. But in 2007, anatomists Joy Reidenberg and Jeffrey Laitman discovered, in the larynx of six mysticete species including the humpback whale (*Megaptera novaeangliae*), a previously overlooked U-shaped structure which they called the *U-fold.* Although different in shape from vocal folds, the U-fold's innervation and its connection to other parts of the larynx indicate that it is homologous to vocal folds. Later work by Damien et al. (2019) supports the view that mysticete vocalization depends on sound-producing vibrations of the U-fold.

The sounds made by baleen whales have been the subject of much speculation. Humpback whales, in particular, vocalize for hours at a time. Their songs have been described as haunting, at times resembling the sound of a violin or a human voice. Since singers are almost always males, their songs have long been thought to play a role in mating.

The role of singing was explored by Joshua Smith and his colleagues (2008) at the University of Queensland, who recorded the behavior of humpback whales off the east coast of Australia during the breeding season. An array of hydrophones recorded songs and allowed the singers to be tracked, while visual observations from a nearby hilltop documented the whales' movements and associations with one another. DNA testing of small skin biopsies from whales' backs enabled the investigators to determine the sex of most individuals.

Smith and colleagues discovered that singers often joined mother–calf pairs (Figure 7.9) and continued singing while accompanying them for as long as two hours. This association may serve to provide the male with mating opportunities, while the song gives evidence of his fitness.

Lone singers were often joined for shorter durations by non-singing individuals, most of whom (although perhaps not all) were males. The reason for this fraternization is unclear; it

Figure 7.9 Two humpback whales, probably a mother–calf pair, in the Pacific Ocean.

***Source:* Used with permission of Shutterstock. Photo Contributor, Imagine Earth Photography.**

may have to do with male coalition building. But the fact that the singer usually falls silent when joined suggests an alternative explanation: that the song may indicate the presence of females nearby, and by tagging along, other males hope to effortlessly increase their own mating opportunities (Smith et al., 2008).

Mercado (2018) has proposed that singing in humpback whales may also serve as a form of biosonar. This is a surprising idea, given that humpback whale songs consist largely of sound frequencies within the human audible range, whereas *toothed* whales echolocate using ultrasound as they hunt for fish and octopus. But this frequency difference is plausible given that low-frequency echoes can reveal, better than high ones can, large and faraway objects—useful information to 100-ton mysticetes who swim long distances searching for conspecifics while avoiding undersea obstacles.

Still, the idea that humpback whales use sonar is at present only a theory. Testing it will require measuring the echoes that a whale receives and determining whether these affect its behavior.

WHALES AND MILITARY SONAR

Sounds of human origin are making the oceans noisier, with harmful effects on a wide range of animals (Duarte et al., 2021) including many species of whales and dolphins. One type of anthropogenic underwater sound, military sonar, has a particularly strong effect on cetaceans, sometimes causing them to beach themselves and die.

It now appears that military sonar is similar in its effect to the sounds made by echolocating killer whales, fearsome predators of other cetaceans (Miller et al., 2022). When they hear either sound, potential prey stop their own noisy foraging and flee in haste, sometimes stranding themselves.

The initial evidence for this link between anthropogenic sound and whale strandings was correlational, the two often occurring in close spatial and temporal proximity (D'Amico et al., 2009). But the case was strengthened by experimental research with Blainville's beaked whales (*Mesoplodon densirostris*) carried out by Tyack et al. (2011).

Members of this odontocete species forage by diving deep into the ocean, where they use echolocation both to navigate and to locate prey. So Tyack and colleagues conducted the research in a deep water basin between islands in the Bahamas, where an array of hydrophones had been mounted on the seafloor by the U. S. Navy. These hydrophones, in conjunction with a tag attached to individual whales by a suction cup that was remotely released after 18 hours, were used to monitor the animals' locations and the sonar clicks they were emitting.

In the absence of experimental auditory stimulation, whales foraged on the sea floor using echolocation. In contrast, when recordings of naval sonar were played, the whales abruptly stopped clicking and headed toward the surface, just as they did when the sound of a killer whale was played. Rising into shallow water improves visibility, enabling whales to flee an area more rapidly.

Even people can learn to echolocate, to a degree, and many blind individuals are quite skilled at it: Faint echoes from the person's own footsteps or other self-generated sounds can reveal, to the practiced ear, proximity to an obstacle or the approach of another person. A century ago, this ability was not well understood and was attributed to gentle air currents and other nonauditory sources of information, but psychophysical experiments (for a summary, see Hollins, 1989) revealed its true basis.

HEARING IN ARTHROPODS

We humans draw a sharp line between hearing and touch: Hearing is what we do with our ears, and touch is something we do with our skin. But in invertebrates, with sensory systems quite different from our own, the distinction is not always clearcut. Sound is, after all, only a vibration. If, say, a spider reacts to sound, is it hearing or feeling the vibration?

There is no one universal criterion by which hearing and tactile perception of vibration can be distinguished. Extrapolating from our own experience, we can say that vibration sensitivity is more likely to be auditory than tactile if it has at least one of the following properties: (1) It is mediated by receptors located internally, rather than on the body surface and/or (2) the vibration is transmitted through air or water, rather than through solid objects. As we will see, hearing in arthropods has the second of these characteristics, while hearing in molluscs has both.

INSECTS

MOTHS

We earlier referred to the fact that the echolocation cries of bats prompt moths to take evasive action and, in some cases, to jam the bats' calls—a testament to their hearing ability and quick responsiveness. Moths' ears, like those of most hearing insects, are specialized *chordotonal organs* (Göpfert & Hennig, 2016). Most chordotonal organs are associated with junctions between sections of the insect's exoskeleton and serve a proprioceptive function such as monitoring the movement of wings. In moths, a chordotonal organ in each antenna, called *Johnston's organ*, detects sounds that cause slight movement of the antenna's last segment, the delicate flagellum.

CRICKETS

In other insects, the ears are in other parts of the body. In this section, we will focus on crickets, who hear with chordotonal organs on their front legs. Each ear includes a tympanum, an eardrum that responds to pressure differences between its outer and inner surfaces. And behind the tympanum are the auditory receptors, each housed in a structure called a *scolopidium*. The scolopidium protects the receptor's dendrite but allows it to sway in response to the tympanum's movement. The receptor's axon conveys signals to the brain, where the source of a sound is identified, localized, and (if it is behaviorally relevant) responded to.

Phonotaxis

Several types of auditory stimuli are particularly important to crickets. One is the calling song of male crickets. Females move toward the song, a response called positive phonotaxis. Males produce this song by *stridulation*, that is, by scraping the edge of one wing across a serrated structure on the other wing. The sound has a particular frequency, generally audible to humans, and consists of a series of chirps. For flying crickets, hearing and responding to the echolocation calls of bats are another imperative—in this case, the phonotaxis being negative.

Moiseff et al. (1978) studied these responses in the Pacific field cricket, *Teleogryllus oceanicus*. Female crickets were suspended in an air current to induce flying movements. Then, electronically generated pulses of sound mimicking either the low-frequency calling song of a male cricket or the high-frequency echolocating call of a bat were presented on one side. The female crickets responded with sideways movement of their abdomen and hind legs, as if to turn toward the former sound and away from the latter. Positive phonotaxis was greatest when the sound making up the pulses had a frequency near 5 kHz, and negative phonotaxis was most easily elicited by frequencies between 40 and 70 kHz.

This study proves that crickets can hear sounds over a wide frequency range, that within that range they can distinguish high from low frequencies, and that this discrimination makes possible frequency-specific responses crucial to the animal's survival and reproduction.

In a later study (Hedwig & Poulet, 2004), steering responses were recorded with greater temporal resolution. Whether flying or walking on a trackball, female Mediterranean field crickets (*Gryllus bimaculatus*) turned toward the simulated call of a male cricket. By presenting sound pulses alternately from the left and right, the researchers discovered that each pulse triggers a small positive phonotaxic response. However, to elicit robust responses the pulses must be delivered at a certain rate, which is species dependent. This and the sound frequency making up the pulses enable females to distinguish the calls of males of their own species from those of other species.

Pulse Rate Discrimination

A brain circuit underlying the cricket's selective response to a particular pulse rate has been discovered by Schöneich et al. (2015). The first step in the auditory pathway consists of sensory fibers that carry signals from each ear to the prothoracic ganglion on the same side of the

body. Next, a single neuron ascends from the prothoracic ganglion to the protocerebrum, the most anterior structure in the cricket's brain; it faithfully reports the occurrence of each pulse of sound.

Having reached the protocerebrum, the ascending axon branches. One branch proceeds directly to a key interneuron called a *coincidence detector,* while the other branch terminates on other neurons that delay the response to each pulse by about 40 ms before likewise delivering it to the coincidence detector. If the delayed response to the first pulse of sound in a chirp reaches the coincidence detector just as the direct response to the second pulse arrives, the two will combine and the coincidence detector will give a strong response. This will happen only if the delay interposed by the circuit matches the interval between pulses in the chirp. A final neuron in the circuit, called a *feature detector,* has a high threshold so that it fires only when it receives a vigorous response from the coincidence detector. Schöneich et al. found that the circuit is tuned to enable detection by females of the calls of conspecific males.

Sound Localization

But how do crickets localize the sound, in order to turn in the correct direction? They can do so with remarkable precision, an ability called *directional hyperacuity.* For example, Mediterranean field crickets can tell whether the sound is to the left or right even when it is only 1° from straight ahead (Schöneich & Hedwig, 2010). Differences in arrival time or intensity of the sound at the two ears—the cues used by vertebrates—are not great enough to explain this degree of accuracy. The key mechanism instead involves a set of interconnected hollow tubes, called *tracheae.* One branch of the tracheal system extends into the ear, behind the tympanum, while another opens to the surface of the thorax at a spiracle; in addition the left and right tracheal systems interface at the midline.

The amplitude of movement of the cricket's eardrum, like our own, depends on the pressure difference between the drum's outer and inner surfaces. In our ears, the pressure on the inner surface is relatively constant: It changes only when the Eustachian tube connecting the middle ear with the throat briefly opens. But in the cricket, the spiracle is always open to sound waves, so that the pressure on the inner surface of the tympanum is constantly being updated.

If a sound is coming from (say) the left side, its roundabout path to the inner surface of the left tympanum is slightly longer than its direct path to the outer surface, and its arrival is therefore slightly delayed. When it is delayed by half a cycle, so that a trough arrives at the inner surface of the tympanum just as a crest arrives at its outer surface, the pressure difference and the resulting movement of the tympanum will be greatest. The same will not be true of the right tympanum, so there will be a substantial difference between the response amplitudes in the two ears. If the dimensions of the tracheal system match the frequency of the male calling song, the result will be a precise and robust intensity code for the direction of this sound.

Enhancing the cricket's ability to localize sounds on the left or right are the interactions of two mirror-symmetric interneurons, one in each prothoracic ganglion. These *omega neurons* are named for the pattern, resembling the Greek letter, formed by their axon and dendrite. Each omega neuron is activated by sound reaching the ipsilateral ear, but in addition the two cells inhibit one another (Selverston et al., 1985). This *lateral inhibition* increases with the activity level of the cell doing the inhibiting; the result is that a sound coming from the left will cause the left omega neuron to fire vigorously, while the right omega neuron is suppressed (Pollack, 1988, 2015). When this difference is communicated to the brain, it ensures a directionally appropriate response by the cricket.

SPIDERS

Most research on audition in arthropods has been carried out on insects. Some recent work establishes, however, that spiders also can detect sounds. They do not have eardrums, as crickets and many other insects do, but rely instead on sensory hairs on their legs, which respond to both sounds and tactile stimuli. (Some insects also hear in this way.) In both cases—hearing and touch—the stimulus moves the hair, and it is to this movement that the receptor responds.

In this context, hearing means responsiveness to vibrations that are transmitted through the air rather than through the ground.

Shamble et al. (2016) measured the hearing ability of *Phidippus audax,* the bold jumping spider, which is widely distributed in North America. In creeping toward and then jumping on its prey, a jumping spider relies mainly on vision. But when the researchers made a sound, the spider froze, a startle response that makes a spider less likely to attract the attention of movement-sensitive predators. To make sure that the spider was detecting airborne sound, rather than vibrations traveling through the substrate, the investigators carried out tests while the spider stood on a heavy metal block that transmitted negligible vibration.

By using different frequencies of sound, they found that the spiders were most sensitive at about 80 Hz. This low-frequency sensitivity would enable the spider to hear, for example, a predatory wasp that flapped its wings at about this rate. Shamble and colleagues also recorded from neurons in the spider's brain, targeting an area near the *arcuate body,* a main sensory region. Consistent with the behavioral results, these brain cells responded to sounds at frequencies up to about 300 Hz, with peak responsiveness at 80 Hz.

Auditory sensitivity to a wider range of frequencies was demonstrated in a different species, the ogre-faced web-casting spider *Deinopis spinosa,* which is found in both North and South America (Stafstrom et al., 2020). (An Australian member of this genus is shown in Figure 7.10.) It captures prey with a net that it holds in its four front feet. If the prey is in front of the spider, it lunges forward to ensnare it, guided by vision. But if the spider hears an out-of-view insect flying overhead, it does a backflip to catch it with the net. This backflip response is triggered only by sound frequencies below 1 kHz.

However, the researchers found that single cells in the spider's brain respond not only to these low frequencies of sound but also to higher frequencies (1–5 kHz) that do not trigger any observable behavior. Stafstrom and colleagues hypothesize that this high-frequency hearing may alert the spider to the presence of birds, perhaps ensuring that it refrains from movement (Figure 7.11).

Figure 7.10 An Australian net-casting spider (*Deinopis subrufa*) waits with its net.

Source: **Used with permission of Shutterstock. Photo Contributor, Robyn Butler.**

How can the ogre-faced spider hear sounds that are inaudible to the jumping spider? Stafstrom et al. conclude that sensory hairs serve as low-frequency auditory receptors in both types of spider, but that ogre-faced spiders hear high frequencies with a different sensory structure, the *metatarsal organ.* This series of parallel grooves is located on the metatarsus, a segment of the spider's leg to which the terminal segment, or tarsus, is connected. When the delicate tarsus moves back and forth in response to vibration, it exerts stress on the metatarsal organ and activates its receptors. The auditory role of the metatarsal organ was confirmed by experiments in which mechanical dampening of the tarsus's movement reduced the leg's neural response to high-frequency sound.

In summary, hearing is a sense that is widely distributed among arthropods, although for most of them it is less important than vision. The auditory sense has clearly developed multiple times in this phylum, for various peripheral structures—tympani, sensilla, Johnston's organs and metatarsal organs—have taken on the role of "ears" in different lineages. As in vertebrates, audition has probably evolved out of the sense of touch.

HEARING IN MOLLUSCS

As late as the 1980s, molluscs were thought to lack an auditory sense, and some scientists speculated as to why cephalopods are "deaf"

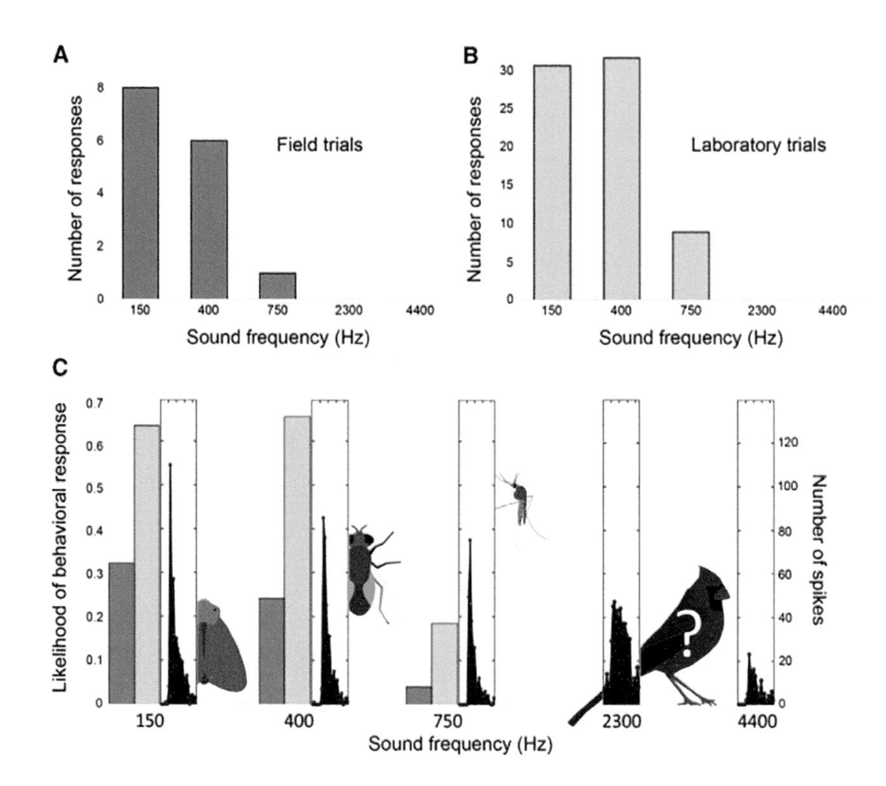

Figure 7.11 Auditory sensitivity of *Deinopsis spinosa*. Both in the field (A) and in the laboratory (B), this net-casting spider responds with a backflip to sound frequencies below 1,000 Hz. These are frequencies characteristic of insects, the spider's prey. (C) Its auditory system is sensitive to higher frequencies as well (purple histograms), but these do not evoke a backflip, perhaps because they are characteristic of avian predators.

Source: From Stafstrom, J. A., Menda, G., Nitzany, E. I., Hebets, E. A., & Hoy, R. R. (2020). Ogre-faced, net-casting spiders use auditory cues to detect airborne prey. *Current Biology, 30*, 5033–5039. Used with permission of Elsevier Science & Technology Journals; permission conveyed through Copyright Clearance Center, Inc.

(Moynihan, 1985). Research by Packard et al. (1990), however, showed that cephalopods—squid, octopus, and cuttlefish—can be classically conditioned to respond to low-frequency vibration transmitted through water. Individual animals (mostly cuttlefish, *Sepia officinalis*) were placed in a water-filled tube with a piston at each end. The vibrating pistons operated in sync, so that when one pushed on the water, the other pulled back. A 10-second period of vibration initially evoked no reaction. Packard then began delivering a brief, mild shock at the very end of the vibration. The shock produced an immediate response: The cuttlefish stopped breathing water for several seconds. But over the course of a series of trials they stopped breathing during the vibration, before the shock was delivered. This conditioned response showed that the animals were detecting the vibration. Sensitivity was greatest at infrasound frequencies, below 20 Hz.

Let's digress for a moment to discuss the physical nature of sound. Sound is a complex phenomenon in which particles of a medium, such as air or water, move back and forth. As they do so, they are at intervals pressed together, raising the local pressure. Scientists can measure either the changes in particle velocity or the changes in pressure, as a sound wave passes by. These are not quite the same thing, in part because velocity is a vector quantity, meaning that particle velocity always refers to speed in a particular direction. Pressure, on the other hand, is directionless.

Our own eardrums respond to pressure—specifically, to the difference in air pressure between the ear canal and the middle ear. The same is true of other terrestrial vertebrates, and insects, like the cricket, that have a tympanum. But how about aquatic invertebrates such as cephalopods, that do not have an eardrum? Does their auditory system also respond to pressure, or to particle velocity?

The experiments of Packard et al. (1990) address this question. In the condition described so far, with both pistons moving in synchrony, the entire column of water in the tube, as well as the cuttlefish itself, moved back and forth. Particle velocity was constantly changing, but water pressure in the tube remained relatively constant. This form of stimulation was detected by the animal, indicating that it could hear fluctuations in particle velocity. But perhaps, it could hear changes in pressure as well? To test this possibility, the researchers changed the phase of one of the pistons, so that they both pushed inward on the tube together, and then both pulled back. This produced substantial increases and decreases in water pressure in the tube, with little overall movement. The cuttlefish gave no response, showing that they are sensitive to particle velocity, rather than to the pressure aspect of sound. (Recent application of this experimental paradigm to fish [Schulz-Mirbach et al., 2020] shows that the complex auditory systems of these aquatic vertebrates respond to both pressure *and* particle motion.)

So how do cephalopods detect sound? Their "ears" are a pair of structures inside the head, near the brain, called *statocysts* (Williamson & Chrachri, 2007). These lymph-filled sense organs bear some similarity to the saccule of fish, described earlier. The inner surface of the cuttlefish statocyst is lined with patches of sensory epithelium, the *maculae*, where receptor cells extend hairlike filaments into the lymph. Hovering near the tips of the hairs are tiny pebbles called *statoconia*. When a sound wave passes through the body of the cuttlefish, the animal's entire body moves back and forth slightly. But because of their inertia, the statoconia lag behind, thus bending the macular hairs and triggering a neural response.

Later work (Mooney et al., 2010) provided electrophysiological confirmation of the role of the statocysts in detecting particle motion. The investigators recorded auditory evoked potentials with electrodes positioned in cartilage near the statocysts of longfin squid (*Loligo pealeii*). The animals were sedated prior to the experimental session. Evoked responses were related to the particle motion aspect of the sound stimuli rather than to pressure, and sensitivity was greatest at low frequencies. Ablation of statocysts eliminated the response.

How low-frequency (including infrasonic) water movement can enable cuttlefish to react to the approach of another animal was shown by Wilson et al. (2018). As a predator such as a fish or dolphin approaches, it pushes water ahead of itself, creating a bow wave that stimulates the cuttlefish's statocysts. At low intensities of stimulation the cuttlefish responds by camouflage, changing its appearance. But at higher intensities it makes an escape response, jetting in the same direction as the particle acceleration and thus away from the predator. However, as some predators draw close, they suck water into their mouths, reversing the direction of particle acceleration in the water. If cuttlefish in the wild responded by jetting in the same direction as this new acceleration, they would be moving toward the predator. It is therefore adaptive that they respond promptly to the initial bow wave.

SUMMARY

The auditory sense plays a significant role in the lives of animals in all three phyla considered in this book. It informs them of changes in their surroundings that represent both dangers and opportunities. It is also important for communication within species, most often to promote mating but also to warn of predators, establish dominance, and regulate other social interactions.

In molluscs, the sense of hearing is relatively crude and is limited to detection of disturbances in the water; it might be considered a form of touch were it not for the fact that the receptors are inside the head rather than on the skin. In arthropods, the auditory sense, while it takes a variety of forms, often involves some frequency discrimination, with high and low frequencies eliciting different behavioral responses. An

example is the ogre-faced spider that detects low- and high-frequency sounds with different receptors: It casts a net in response to the former and freezes in response to the latter.

Frequency discrimination is most elaborate in vertebrates, in whom a complex mechanical and neural system has evolved that carries out a Fourier analysis on sounds. This auditory system, in conjunction with an ability to control the frequency and temporal structure of vocalizations, enables some vertebrates (especially among birds and mammals) to communicate in complex ways.

In some animals, such as bats who use echolocation to hunt and navigate, hearing is arguably the most important sense. However, for many animals, there is another sense that is more vital. For some, such as cephalopods and birds of prey, this other sense is vision. But for others, including many of our fellow mammals, it is smell. To this, and to the closely related modality of taste, we now turn.

8 Taste

The chemical senses, taste and smell, are the oldest methods of acquiring information about the environment. Before there were eyes or ears, primitive organisms were interacting with molecules in their milieu. Classical theories of perception held that we learn about our surroundings because tiny pieces of objects break off and make their way into our bodies. This is not a very good description of how vision, hearing, and touch work, but it is passable as a description of the chemical senses, with molecules being the "tiny pieces."

Based on our experience as humans, the differences between the modalities of taste and smell seem straightforward. First, taste is mediated by receptors in the mouth, whereas smell is mediated by receptors in the nose. Second, something that is tasted, such as a bite of food or sip of liquid, is in direct contact with taste receptor cells. In contrast, something that is smelled, such as a flower, is typically at some distance from the olfactory receptors; the immediate stimulus is a multitude of volatile molecules that drift through the air. A third distinction is a functional one: Taste is concerned mainly with eating, while smell is concerned with eating plus everything else.

But as we will see, none of these differences between taste and smell is clearcut or applies in all cases. The distinctions are blurred to some extent in mammals, more so in other vertebrates, and even more in invertebrates.

Moreover, smell and taste often are activated together, as when we eat pizza; and when this happens, it is difficult to subjectively distinguish their contributions. This combined sensory experience is called *flavor*. It is easy to assume that the enjoyment of, say, chocolate is all a matter of taste, but in fact smell is often the more important contributor, as someone with a head cold quickly realizes.

The overall perception of food being eaten also includes a tactile component, such as the feel of pear on the tongue, the warmth of coffee, or the "burn" of a hot pepper. This is sometimes called the *trigeminal component* of flavor, because it is mediated by the large, three-branched nerve that carries somatosensory messages from the head and face to the brain.

Despite these complexities, recognizing taste and smell as two different senses is warranted. While none of the traditional distinctions between them is absolute, each is valid to some extent. And genetic analysis has shown that receptor molecules for taste and smell generally belong to different molecular families. So despite the caveats, we will take up the two chemical senses one at a time. This chapter is about the sense of taste, and the next one is about smell.

We will start by describing the taste system of humans and other mammals. Later sections of the chapter will describe ways in which the taste systems of fish, amphibians, reptiles, and birds are different from those of mammals. After that we will turn to the quite different taste systems of invertebrates.

DOI: 10.1201/9781003362319-8

TASTE IN HUMANS

THE ANATOMY OF TASTE

TASTE BUDS AND PAPILLAE

The receptor cells for taste are located in *taste buds*, specialized structures found primarily on the tongue, but to some degree in the soft palate and other parts of the mouth as well. Each taste bud contains about 50 taste receptor cells. At the top of each taste bud is an opening called a *taste pore*, where the slender tips of receptor cells come into contact with saliva and *tastants* (substances with a taste) in the mouth. There is a constant turnover of cells within a bud, with new cells being formed around the edges, gradually moving toward the center, and eventually being pushed up out of the bud and sloughed off. This turnover is adaptive, for mechanical and chemical stimulation, and the burn of hot foods and liquids damages the cells over time.

Most taste buds are enclosed within larger structures, the papillae, that are large enough to be seen with the naked eye. These are of four kinds. The most important for taste are the *fungiform papillae*, named for their mushroom shape. They are scattered about on the top of the tongue, especially toward the tip. You can see these if you drink milk and then look in the mirror while sticking out your tongue. Each fungiform papilla has a number of taste buds on its top and sides. Next are the *foliate papillae*, vertical grooves on the sides of the tongue. Largest of all, but fewest in number, are the *circumvallate papillae* (literally, papillae "surrounded by a wall") on the back of the tongue.

A fourth type, the *filiform papillae*, are the most numerous but contain no taste buds. Their name comes from their collective resemblance to a file. They serve a mechanical function, dragging and scraping food, as when we lick an ice cream cone.

FROM THE TONGUE TO THE BRAIN

As tastants dissolve on the tongue, their molecules make contact with the tips of receptor cells in the taste buds. Receptor potentials in these cells are synaptically conveyed to the branches of a number of sensory nerves, especially the *chorda tympani*, which is a branch of the facial nerve (cranial nerve VII) serving the front of the tongue. It takes its name from the fact that it passes through the middle ear on its way to the brain, passing just behind, and thus forming a chord of, the eardrum. The other major taste nerve is the glossopharyngeal (cranial nerve IX), which serves the back of the tongue.

Arriving in the brainstem, these nerves converge on the *nucleus of the tractus solitarius*, and from there taste information is routed through the thalamus to cortical regions that are buried deep within the folds of the cortex, especially the *insula*. While its functions are not fully understood, it is clear that the insula processes information about the internal state of the body—pain, hunger, body temperature, and so on—and associated emotional states. This suggests that the pleasantness or unpleasantness of a taste is an important part of gustatory perception.

BASIC TASTES

It has traditionally been believed that there are four *basic tastes*—sweet, salty, sour, and bitter—of which all other taste experiences are compounded. An indication that these tastes are distinct at the level not just of perception but also of mechanism is that they can sometimes be separately modified. For example, extracts of the vine *Gymnema sylvestre*, which grows from tree to tree in the overstory of forests in India and Southeast Asia, have the ability to selectively impair the sweet taste (Hudson et al., 2018). A tea brewed from its leaves contains an assortment of gymnemic acids, which, if left on the tongue for a few minutes, make sugar taste like sand for about half an hour.

TASTE ADAPTATION

Psychophysical research generally supports the existence of basic tastes. Much of it makes use of the phenomenon of *sensory adaptation*: the gradual decline in subjective intensity of a stimulus that is continuously presented. Consider, for example, what happens when a taste solution such as salt water is flowed across the tongue of a research participant for a minute or two. At first, the solution has a strong salty taste, but as the trial continues, the taste gets weaker and weaker and eventually may disappear altogether.

Donald McBurney and his colleagues used the phenomenon of adaptation to study the nature of the sweet, salty, sour, and bitter tastes, and the relationships among them. They used an experimental paradigm called *cross-adaptation*, in which participants became adapted to one stimulus and then rated their perception of a different stimulus. The adapting solution flowed across the participant's tongue for 1 minute, after which the test solution was quickly substituted. At the end of the trial, the participant gave a numerical rating of the overall taste intensity of the test solution, as well as the individual intensities of its sweetness, saltiness, sourness, and bitterness. Participants were able to make these component ratings reliably, a fact that is consistent with the idea that these four taste qualities are in some sense basic.

In one study (Smith & McBurney, 1969), the adapting stimulus was 0.1 M NaCl, and the test stimuli were 11 other salts, such as KCl and NH_4Br, at concentrations that were judged equal in overall intensity to 0.1 M NaCl prior to adaptation. While NaCl normally evokes just a salty sensation, some of the other salts have a more complex taste, with elements of bitterness, sourness, and/or sweetness. The key result of the study was that adaptation reduced the perceived saltiness of all the other salts. This suggests that a common mechanism is involved in sensing saltiness, even when the stimuli are different. Similarly, adapting to sugar reduced the sweetness of a wide range of sweet substances, including saccharine (McBurney, 1972); and adapting to citric acid reduced the sourness of a range of acids. Bitterness was more complicated: Adapting to quinine reduced the bitterness of some, but not all, bitter substances, suggesting that there are two or more types of bitterness (McBurney et al., 1972).

There was little cross-adaptation between the basic tastes. That is, adapting to something salty did not reduce sweetness, sourness, or bitterness.

However, adaptation to one taste often *increased* the intensity of other tastes. For example, adapting to table salt increased the bitterness or sourness of other salts that had more complex tastes. In fact, after adapting to NaCl, even water has a bitter taste. This is why distilled water tastes bitter to most people: They are adapted to their own salty saliva.

Thus, the mechanisms responsible for sweetness, saltiness, sourness, and bitterness are not entirely independent: They are distinct entities reflecting physiological mechanisms of some kind, but they nevertheless interact (McBurney & Gent, 1979). It has not been possible to fully explain how this happens using psychophysical methods alone.

UMAMI

Meanwhile, the idea that there is a fifth basic taste, as first proposed by Kikunae Ikeda more than a century ago, was gradually gaining acceptance. Working in his lab at Tokyo Imperial University (now the University of Tokyo), Ikeda (1909/2002) found that broth made from seaweed owed much of its hearty taste to the large amount of glutamate that it contained. He named this taste *umami*, meaning savoriness or deliciousness, and noted that it could not be subjectively analyzed into sweet, salty, sour, and bitter components.

The importance of Ikeda's work was not fully appreciated during his lifetime or for decades thereafter (Nakamura, 2011), even though monosodium glutamate (MSG) came to be widely used in cooking. But evidence supporting Ikeda's theory gradually accumulated, causing umami's status as a fifth basic taste to be widely accepted (Beauchamp, 2009).

For example, Yamaguchi (1987) carried out psychophysical research using *multidimensional scaling* to determine what the perceptual space for taste is like. A perceptual space is a theoretical space in which different stimuli are represented by points, and the distances separating those points reflect the perceptual dissimilarities between the stimuli. To determine dissimilarities, Yamaguchi gave each of 150 participants seven taste solutions and asked them which pairs of solutions were very similar, which were slightly similar, and which were very different. Some of the solutions represented basic tastes (sucrose representing sweetness, for example),

while others were mixtures (a mixture of sucrose and table salt, for example). There were 21 stimulus solutions in all, and different participants were presented with different subsets of 7.

A perceptual space based on the overall data set was constructed and featured a tetrahedron with sweetness, saltiness, sourness, and bitterness at its corners. Mixtures of the solutions representing these basic tastes were located inside the tetrahedron or on its edges or sides. For example, a mixture of tartaric acid and quinine was located midway along the edge connecting the sourness and bitterness corners. But importantly, MSG and mixtures containing it were located *outside* the tetrahedron. This indicates that umami cannot just be a mixture of sweetness, saltiness, sourness, and bitterness and therefore by definition is a separate taste.

Ikeda believed that umami's importance lies in its ability to alert us to the presence of protein (of which glutamate is a constituent), much as sweetness alerts us to the presence of carbohydrates. But this aspect of his theory has been controversial, because only "free" glutamate ions can stimulate taste receptors; and while some foods (such as tomatoes) contain free glutamate, the glutamate in protein is in bound form, structurally locked into larger molecules. It can be liberated by cooking (as in Ikeda's seaweed broth), fermentation, or other processes. But when, say, a coyote devours a rabbit, glutamate is not liberated until the protein is digested, a process that begins in the stomach. So how can umami receptors detect the glutamate in protein? As we will see in the next section, researchers have discovered a surprising answer to this question.

THE PHYSIOLOGY OF TASTE

How do tastants activate receptor cells, and how is their identity (as sweet, salty, sour, bitter, or umami) encoded and transmitted to the brain so that perception can occur?

RECEPTOR CELLS AND RECEPTOR MOLECULES

To produce electrical activity in a taste receptor cell, a tastant must somehow cause it to depolarize. This process is mediated by specialized protein molecules, called *receptors*, located in the cell's membrane. It is important to distinguish between taste receptor cells and the taste receptors (molecules) that their membranes contain.

The cells in taste buds are of several kinds, including Type I, Type II, and Type III. Type I are thought to serve only a supporting, glia-like function, rather than playing a sensory role. But Types II and III are both clearly receptor cells. They differ anatomically, in that Type III cells make traditional synapses, complete with synaptic vesicles, onto the nerve endings of chorda tympani and other sensory fibers, while Type II cells make gap-junction connections instead.

Let's consider the receptors and receptor cells for the basic tastes, starting with sour.

SOUR

Angela Huang and her colleagues (2006) used genetic engineering to produce mice who lacked Type III receptor cells and tested the ability of their taste system to respond to various stimuli. The overall response of the chorda tympani nerve to solutions flowed across the tongue was recorded in anesthetized mice. Normal ("wild-type") mice served as the controls. The neural response of the mutant mice to sour tastants (acetic acid, citric acid, etc.) was virtually abolished, but their responses to sweet, salty, bitter, and umami stimuli were equal to those of the wild-type mice. Type III receptor cells are therefore responsible for the sour taste, a conclusion also reached, with different methods, by Yijen Huang et al. (2008).

Chang et al. (2010) showed, by recording from individual Type III cells, that they were vigorously activated when H^+ ions were "uncaged" near the cell. Although there has been controversy on this point, it is now generally agreed that these ions, which are the cations in every acid, are the necessary and sufficient condition for activating the receptor cells for sour. And it has recently been discovered that a transmembrane molecule called OTOP1 is the receptor for sour (see Figure 8.1). It allows the passage of H+ ions into the cell, beginning a cascade of events that results in an electrical response (Teng et al., 2019).

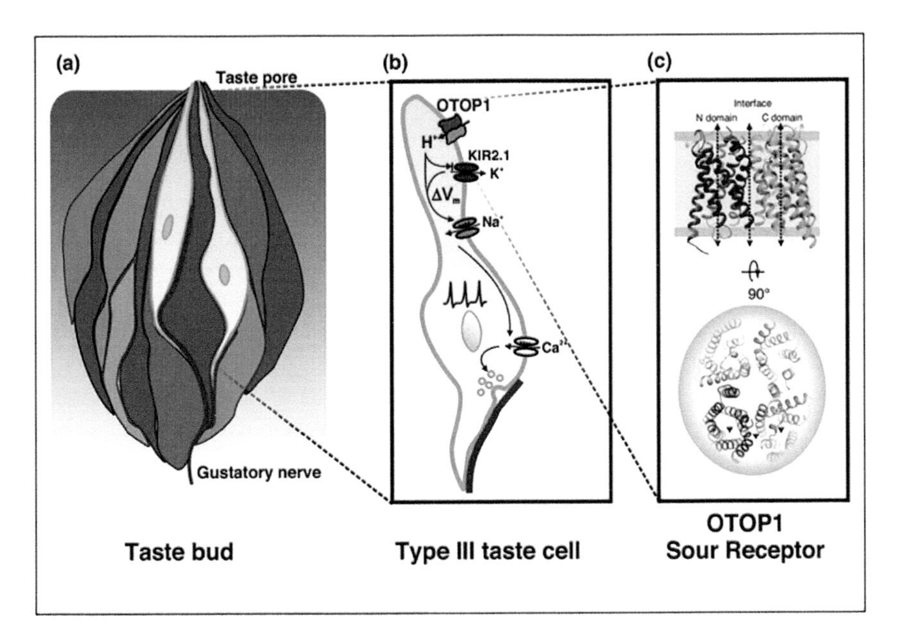

Figure 8.1 Receptoral basis of the sour taste. A few of the cells in the average taste bud detect acids. The middle panel shows an individual cell of this type. Its membrane contains numerous OTOP1 receptors, one of which is shown. This serves as a channel for H^+ ions; their entry is the first step in cell activation. The right panel shows a single OTOP1 receptor, in cross section and in a top view; small arrowheads show possible sites of H^+ entry.

Source: **From Liman, E. R., & Kinnamon, S. C. (2021). Sour taste: Receptors, cells and circuits. *Current Opinion in Physiology, 20*, 8–15. © 2021 Elsevier Ltd. OTOP1 receptor structure based on Saotome, K., Teng, B., Tsui, C. C. (A.), Lee, W.-H., Tu, Y.-H., Kaplan, J. P., Sansom, M. S. P., Liman, E. R., & Ward, A. B. (2019). Structures of the otopetrin proton channels Otop1 and Otop3. *Nature Structural and Molecular Biology, 26*, 518–525; used with permission of Springer Nature BV, conveyed through Copyright Clearance Center, Inc.**

The sour taste is apparently ubiquitous among vertebrates, having been documented in representatives of all classes (Frank et al., 2022). These investigators argue that it was essential to fish as a way of detecting harmful levels of acidity in water, caused for example by the conversion of dissolved CO_2 into carbonic acid. And perhaps the sour taste survived the transition to land because acids can alter the gut microbiome of some terrestrial vertebrates, interfering with digestion. Whatever the reasons, the sour taste appears to be here to stay.

SWEET, UMAMI, AND BITTER

Despite the subjective differences among these tastes, their receptors are chemically related, and very different from OTOP1. Sweet, umami, and bitter receptor cells are all anatomically Type II, and their receptors are not channels for the passage of particular ions, as is the case for sour. Instead, they are G-protein-coupled receptors (GPCRs). A GPCR is a long molecule that weaves, like a shoelace, into and out of the cell through the membrane. When a stimulus molecule (say, a molecule of sugar) interacts with a GPCR outside the cell, the GPCR rearranges itself. This conformational change has effects inside the cell, freeing up G proteins to move about and influence intracellular events, such as the synthesis of second messengers. An electrical response of the cell can result.

Receptors for sweet and umami belong to the T1R family of GPCRs, which includes T1R1, T1R2, and T1R3. Cells responding to sweet substances contain both T1R2 and T1R3, whereas cells coding for umami contain T1R1 and T1R3. The two types of receptors in a given cell typically team up to form a single large protein called a *heterodimer*. The fact that T1R3 is found in both

types of cells suggests that it serves a facilitating role, leaving most of the actual encoding to T1R1 and T1R2 (Xu et al., 2004). The extracellular portion of these receptors has been compared to a Venus fly trap because it has a large cleft that binds readily to tastant molecules.

Cells that respond to bitter substances make use of a different group of GPCRs, the T2R family. This has many members, differing largely in their extracellular portion; a variety of different ones are typically located on an individual cell. This variation is adaptive, for it allows a variety of toxic substances to be perceived as bitter, and thus avoided (Scott, 2004).

SALTY

The receptoral basis of the salty taste remains a puzzle. Channels for sodium, chloride, potassium, and other ions are present in the membranes of most neurons, where they play specific roles in the action potential. Are these same channels reconfigured as receptors in taste receptor cells? Or is there a more broad-spectrum receptor that is activated by a variety of cations and anions? Research on this question is so far inconclusive.

ESTRA-ORAL RECEPTORS

A major finding of research in the last 20 years is that taste receptor molecules are found not just in the mouth but in the stomach and small intestine as well. There are no taste buds there, and the cells that these molecules belong to are not traditional sensory cells. They are, instead, endocrine cells that secrete hormones when stimulated by tastants. For example, when activated by glucose, endocrine cells that contain T1R2/T1R3 sweet receptors secrete glucagon-like peptide-1, a hormone that induces insulin release (Jang et al., 2007). And cells with T1R1/T1R3 umami receptors release cholecystokinin, a hormone that triggers the release of bile and digestive enzymes from the pancreas and gall bladder (Daly et al., 2013). In other words, these receptors help in the process of digestion.

Surprisingly, taste receptors are also found in other organs of the body, such as the lungs and reproductive organs, where their activation may have a variety of health implications, both positive and negative (D'Urso & Drago, 2021). This is a rapidly expanding area of research.

TASTE CODING: LABELED LINE OR PATTERNING?

There has been a long-standing controversy in the field of taste research as to how the identity of a basic taste is encoded. According to the *labeled-line theory*, different receptor cells, and different fibers in the chorda tympani and glossopharyngeal nerves, carry information about the different basic tastes (Chandrashekar et al., 2006). In other words, there are sweet fibers, sour fibers, and so on. According to the *pattern theory*, every tastant activates a wide variety of different fibers, to different degrees, and it is the across-fiber pattern of activity that determines taste (Erickson, 1963).

The truth lies somewhere between these theoretical extremes. Even at the level of taste receptor cells, which show considerable specificity, there is a leakage of information from some cells to others, so that, for example, some Type III cells respond primarily to acids, but weakly—and indirectly—to bitter tastants as well.

At the level of nerve fibers, the range of stimuli to which many neurons respond is even greater. Some respond to only a single basic taste, but some respond to two or three of them. And in some cases, a fiber that responds to only one type of tastant at low stimulus concentrations may respond to more if stronger stimuli are used. If the pattern of responsiveness varied haphazardly from fiber to fiber, the computational challenge facing the brain in determining what is on the tongue would be overwhelming.

In a prescient analysis, Marion Frank (1973) said that our efforts to make sense of patterning must take into account the function of taste, which is to serve a gatekeeping role as animals decide what substances to eat and which ones to spit out. Sweetness usually indicates nutritional value and strongly encourages eating; in addition, animals sometimes ingest salt, which contains ions necessary to body chemistry. On the other hand, sourness, which may indicate spoiled food, is somewhat aversive, and bitterness, the sensory tag of many poisonous substances, is strongly so.

Frank recorded from single fibers in the chorda tympani nerve of anesthetized hamsters, while solutions of sucrose, NaCl, HCl, and quinine were flowed across the animal's tongue. Most fibers responded to more than one of the tastants, and not surprisingly, there was for each fiber one tastant that evoked the strongest response. But the key finding was that most sucrose-best fibers have their second-strongest response to NaCl; NaCl-best fibers gave their second-strongest response to either sucrose or acid; and HCl-best fibers gave their second-strongest response to NaCl (see Figure 8.2). In other words, when response magnitude was plotted as a function of acceptability, the profile of each fiber was a single-peaked function. (There were no quinine-best fibers, and glutamate was not tested because umami was not yet accepted as a basic taste in 1973.)

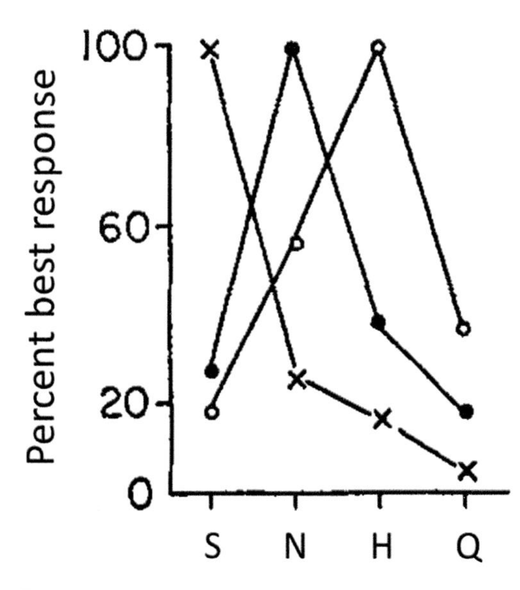

Figure 8.2 Average responses of groups of fibers in the chorda tympani nerve of hamsters. The fibers were divided into three groups based on the stimulus that evoked the strongest response: sucrose (S), NaCl (N), or HCl (H). No fibers responded strongly to quinine (Q). Response magnitudes are expressed as percentages of a fiber's best response.

Source: **From Frank, M. (1973). An analysis of hamster afferent taste nerve response functions. *Journal of General Physiology, 61*, 588–618. Used with permission of Rockefeller University Press; permission conveyed through Copyright Clearance Center, Inc.**

So, in this case at least, the across-fiber pattern is a meaningful one: The relative firing rates of different fiber groups may signal the acceptability of the substance on the tongue. The results of this study paint a simplified picture of taste coding in the hamster, because only four different tastants were used, but they show how an across-fiber pattern can help animals decide whether to eat something or spit it out. With other stimuli or in other animals, across-fiber patterns may carry other types of information.

Another way in which the labeled-line and patterning theories are being reconciled is the fact that an animal's receptor cells, and nerve fibers, may include some "specialist" (e.g., sweet-only) fibers and some "generalist" units that respond to a range of tastants (Frank et al., 2008; Roper & Chaudhari, 2017). The arrangement is similar to that in the visual system of primates, where some retinal ganglion cells and LGN cells encode brightness by responding to all visible wavelengths, while others code for colors in a wavelength-specific way.

Taste coding becomes even more complex when gustatory signals arrive at the cortex. Gustatory cortex occupies a portion of the insula, which is now understood to be concerned with maintaining homeostasis in various systems, partly through hedonic states that prompt certain behaviors. For example, dehydration causes activity in the insula that is experienced as thirst and thereby prompts drinking.

Most experimental work on the insula has been carried out in rats and mice, using single-cell recording and other methods. There is a hedonic gradient within gustatory cortex, with pleasant tastes being represented more anteriorly than unpleasant ones. But neural responses to food in the mouth change over time. Some evidence indicates that an initial wave of cortical activity is the tactile response to an object or liquid contacting the tongue. This is followed by a second wave encoding the tastant's identity and finally by a third wave reflecting its hedonic value (Katz et al., 2001).

Much more research is needed to understand the spatial and temporal patterns of activity in

(say) a mouse's gustatory cortex, how that activity is influenced by the mouse's past experience and current state (Staszko et al., 2020), and how it causes the mouse to eat one caterpillar but spit out another one that contains cyanide or other toxins.

VARIATIONS IN TASTE AMONG MAMMALS

Since the five basic tastes were discovered in a mammalian species (*Homo sapiens*), it is perhaps not surprising that many other mammals also have them. Rats, for example, are rather like us: In a two-bottle test, they show a preference for sucrose and MSG, an aversion to HCl and quinine, and mixed results for NaCl (Inui-Yamamoto et al., 2017).

But the elimination or weakening of a specific taste, due to random mutations in the genes for taste receptor molecules, is an ever-present possibility. If the change does not compromise the animal's adaptation to its environment, it may be passed along to future generations and become a characteristic of the species.

The best studied example of this in mammals is the relative weakness of the sweet taste in felines. As everyone with pets knows, the family cat is not interested in, say, bits of cake, which the dog will gladly accept.

There are many reports, both experimental and anecdotal, that a variety of mammals from rats to horses to baboons are motivated to consume sugar. To determine whether cats really are an exception, Beauchamp et al. (1977) undertook a systematic study in which domestic cats (*Felis catus*) were given a choice between plain water and a sugar solution. Over a 24-hour period, cats were free to drink from bottles containing these stimuli, and the amount drunk from each bottle was measured. A variety of sugars in different concentrations were tested, but the result was always the same: Cats consumed equal amounts of sugar solution and water.

Beauchamp and colleagues next carried out the same experiment with large cats—lions, tigers, leopards, and jaguars—in the Philadelphia Zoo. The solutions were presented in pans, and the test lasted only 5 minutes. Rather than measuring the amount consumed, the experimenters watched the cats and counted the number of times they lapped from each pan. These members of genus *Panthera* showed the same lack of a sugar preference as the domestic cats.

Despite some evidence to the contrary (Bartoshuk et al., 1971), it appears that cats, in contrast to other mammals, are not strongly drawn to sweet tastants. This does not prove that they cannot *detect* sweetness, however. In an electrophysiological study relevant to this possibility, Robinson (1988) recorded the responses of gustatory neurons in the chorda tympani of anesthetized cats while stimulus solutions were flowed across the tongue. Of the 55 units studied, about a third responded best to salt, a third to acid, and a third to both salt and acid. Two units responded to quinine, but none responded appreciably to sucrose. If the cat has sweet fibers, they appear to be few in number.

The lack of sensitivity to sugar actually begins in the receptor cells, where, it will be recalled, the receptor proteins T1R2 and T1R3 form a heterodimer that reacts to sweet tastants. At the Monell Chemical Senses Center in Philadelphia, Peihua Jiang and colleagues (2012) sequenced the gene that instructs the manufacture of T1R2 and found that it is pseudogenized (degraded) in the domestic cat. Any residual ability to detect sugars, as found by some investigators, may be mediated by T1R3 acting on its own.

Jiang et al. (2012) went on to study the gene for T1R2 in a number of other members of order Carnivora. They found that in some, such as the domestic dog, the gene is normal, whereas in others, it is pseudogenized. The researchers suggest that when animal lineages evolve into obligate carnivores, they lose the need for a sweet (carbohydrate-detecting) receptor, and random mutations inactivate the gene for T1R2.

An example studied by Jiang's team is the Asian small-clawed otter (*Aonyx cinereus*): This small, threatened carnivore feeds on crabs, frogs, and other small animals; its T1R2 is nonfunctional, and in two-bowl testing, it showed no preference for sugar-water over water. In contrast, the spectacled bear (*Tremarctos ornatus*), whose diet consists mostly of plants but with an occasional rodent or bird, has an intact T1R2 gene and showed a strong preference for sugar-water in the two-bowl test.

Overall, the results of these and other studies indicate that over the course of evolution, individual tastes may drop out of an animal's gustatory repertoire if they are no longer needed. And sometimes new ones may appear.

TASTE IN OTHER VERTEBRATES

FISH

The sense of taste has been around since the earliest vertebrates. It is present in lampreys, jawless fish that are among the most ancient of vertebrate lineages. Without a jaw, these animals suck in water containing microorganisms, food particles, or, in species that use rasping mouths to attach themselves to other fish, blood. Food is carried into the esophagus on a stream of the lamprey's mucus, while water proceeds to the pharynx, where it is expelled through gills. Taste buds are located primarily on the gill arches. Taste receptor cells in the buds transmit signals to the fibers of cranial nerves, which convey them to a hindbrain structure thought to be homologous to the nucleus of the tractus solitarius (Barreiro-Iglesias et al., 2010).

Recordings from gustatory nerve fibers in larval brook lampreys (*Lampetra planeri*) indicate that their taste receptors respond to salt, quinine, acetic acid, sugar, and especially to animo acids including glutamic acid—in other words, to all of the basic tastes (Baatrup, 1985). They may, of course, also be sensitive to other chemicals that were not tested. But it is impressive that many features of the gustatory system have been conserved over the course of vertebrate evolution.

Genetic analysis of taste receptor molecules in a variety of fish species shows that they have the T1R and T2R families of receptors that are found in mammals. However, there are differences between some receptors of fish and mammals (and other tetrapods). For example, T1R1 and T1R3 in teleosts are quite similar to those in tetrapods, but there are a variety of forms of T1R2 in fish, none of them closely resembling T1R2 in tetrapods. This difference has a functional significance: The T1R1-T1R3 heterodimer is an umami receptor in both fish and tetrapods, but the T1R2-T1R3 heterodimer, which detects sugar in most tetrapods, is another umami receptor in fish (Okada, 2015)! Perhaps this functional overlap allows teleosts to discriminate different amino acids from one another.

Aihara et al. (2008) gave taste tests to medaka fish (*Oryzias latipes*), a small Asian fish that is a popular pet in Japan. Fish in one group were given access to bite-size particles flavored with MSG and other amino acids, while those in another group were offered bitter particles. The fish ravenously consumed the particles with an umami taste, but ate little or none of the bitter particles. The researchers then genetically engineered fish in which T1R and T2R receptors were nonfunctional, and showed that these medakas lacked the taste preferences of wild-type individuals. The results show that the core function of taste—to determine what gets eaten and what doesn't—is the same in fish and mammals, and that even some of the physiological mechanisms are similar.

Catfish are a large order of teleosts in which the sense of taste is highly developed. The most intensively studied species is the channel catfish (*Ictalurus punctatus*), which is native to North America (see Figure 8.3). Like most other catfish, it has long, thin appendages on its face, called *barbels*, that resemble a cat's whiskers. These are not hairs, but sensory organs that contain densely packed taste buds. Catfish have the most taste buds of any animal group, approaching 100,000 in some species. They are found not just

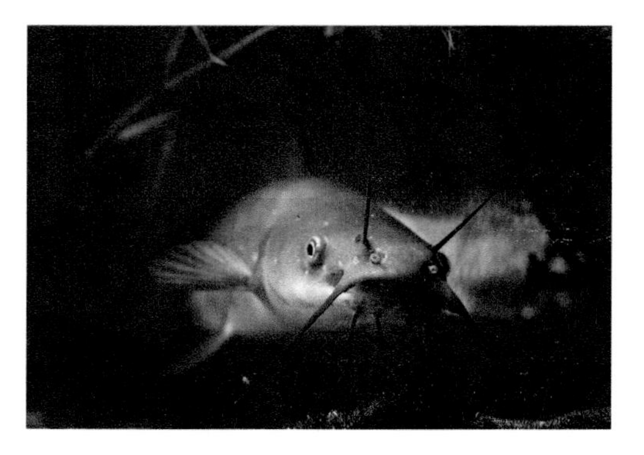

Figure 8.3 A channel catfish, *Ictalurus punctatus*. Note the prominent barbels, sensory structures bearing many taste buds.

***Source:* Used with permission of Shutterstock. Photo Contributor, Aleron Val.**

in the mouth but also on the lips, barbels, and elsewhere on the skin.

The extraoral taste buds are served mainly by a branch of the facial nerve (cranial nerve VII) that is homologous to the mammalian chorda tympani, whereas taste buds deep inside the mouth are innervated by branches of the glossopharyngeal (IX) and vagus nerves (X). The nerves carrying extraoral and intraoral signals go to adjacent regions of the hindbrain, where they appear to mediate different types of feeding behaviors: searching for food, and biting and chewing it, respectively (Caprio et al., 1993).

The channel catfish doesn't fit neatly into our template of five basic tastes. Its gustatory system is heavily weighted toward the detection and discrimination of amino acids. For example, responses to arginine and alanine are conveyed to the brain in largely separate groups of fibers (Caprio et al, 1993). Moreover, it has now been shown that some gustatory fibers respond to bile salts (Rolen & Caprio, 2008). These molecules are produced in the liver of vertebrates to aid in digestion, but small amounts of them are discharged into the water. When detected by the catfish, they indicate that another fish is in the vicinity.

In physiological cross-adaptation experiments, reminiscent of McBurney's psychophysical work, Rolen and Caprio (2008) found that adapting a barbel to amino acids had little effect on the neural response to bile salts, and adapting to bile salts had no effect on the response to amino acids. The implication is that the bile salt receptors are largely separate from the receptors that detect amino acids. But exactly what the bile salt receptors are we do not know. Clearly there is much left to learn about the catfish's sense of taste.

AMPHIBIANS AND REPTILES

Much of the research on the sense of taste in amphibians and reptiles consists of genetic analysis of receptor molecules. Zhong et al. (2021) tallied the genes for T1R and T2R receptors in 14 species of amphibians. They found that all had numerous receptors in the T2R family, presumably mediating a bitter taste in response to a variety of toxic substances. The species with the most types of bitter receptors was the American bullfrog (*Lithobates catesbeianus*) with 178, suggesting a remarkable ability to escape death by poison.

Zhong and colleagues' results for the T1R receptor family were more variable across the 14 species. Only five of the 14 species had receptors of all three types (T1R1, T1R2, and T1R3), suggesting that a majority of these amphibians lack either the sweet taste or the umami taste, or both. Two of the species (both clawed frogs, Genus *Xenopus*) had no T1R receptors at all.

The predominance of bitter receptors in amphibians contrasts markedly with the prevalence of umami receptors in fish. It is not clear what accounts for this transition, coinciding with the emergence of vertebrates onto land.

A similar study, but limited to the T1R receptor family, was carried out in reptiles by Feng and Liang (2018). A clear difference emerged between snakes, on the one hand, and turtles, lizards, and crocodilians, on the other. With one exception, none of the snakes had a single intact gene for a T1R receptor. In contrast, most of the turtles, lizards, and crocodilians had at least two of the three types of T1R receptors. The researchers suggest that since snakes swallow their food whole, they have little opportunity to savor it, and thus the sweet and umami tastes would convey little advantage; there has therefore been no evolutionary pressure preventing random mutations from pseudogenizing the genes for these receptors.

In one of the few behavioral studies of taste in either amphibians or reptiles, Takeuchi et al. (1994) offered flavored food pellets to axolotls (*Ambystoma mexicanum*), critically endangered salamanders that are native to Mexico. The animals consistently snapped at the food pellets, regardless of flavoring, but pellets, once in the mouth, were sometimes swallowed and sometimes rejected. Pellets flavored with sucrose or NaCl were usually swallowed, as were unflavored pellets, but those flavored with quinine or citric acid were often rejected, with a frequency that increased as a function of stimulus concentration. The results indicate that these amphibians possess not only a bitter taste (as expected from genetic studies) but a sour taste as well.

BIRDS

Birds evolved from reptiles (via dinosaurs), and the earliest birds were carnivores. It is therefore doubtful that they had a sweet taste, but very likely that they had an umami taste. And yet many birds living today, such as hummingbirds, are clearly drawn to the sweetness of nectar; they must have evolved sweet receptors. Remarkable research by Yasuka Toda and her colleagues (2021) sheds light on how and when this occurred. Reviewing what was known about the diets of many bird species, they realized that feeding on nectar was a behavior largely confined to songbirds, which constitute about half of all bird species.

Yet not all songbirds drink nectar, or eat fruit, another sugar-containing food. Do those who do not seek out sugary foods still have a sweet taste? To answer this question, the investigators gave a taste test to two species of songbird: the New Holland honeyeater (*Phylidonyris novaehollandiae*: see Figure 8.4), which consumes mainly nectar, and the Atlantic canary (*Serinus canaria*), which eats mainly grains. Given a choice between sugar water and plain water, members of both species showed a clear preference for the sugar.

Now suspecting that all songbirds have a sweet taste, Toda et al. turned to genetic analysis of their taste receptors. Birds don't have T1R2 receptors, which in mammals are necessary for the sweet taste. But birds do have both T1R1 and T1R3, which in other animals combine to form the umami-taste heterodimer. The researchers transfected these receptors, along with a chemical that would luminesce if the receptors were activated, into unrelated cells. This *in vitro* experiment was carried out separately with receptors from a variety of bird species. When the researchers added sugar, cells with receptors from songbirds glowed; cells with receptors from most other birds did not. Both groups of cells also responded to amino acids.

Figure 8.4 A New Holland honeyeater (*Phylidonyris novaehollandiae*), native to Australia. Like other songbirds, it is prompted by its sweet taste to feed on sugary foods, here the nectar of the acorn banksia.

***Source:* Used with permission of Shutterstock. Photo Contributor, Wattlebird.**

The researchers concluded that early in songbird evolution, perhaps 30 million years ago in Australia, mutations in T1R1 and T1R3 caused the heterodimer made of these two proteins to become responsive to sugar, while retaining at least some of its sensitivity to amino acids.

But one puzzle remained. Hummingbirds, whose consumption of nectar shows they have a sweet taste, are not songbirds. How did they get the "sweet mutation"? To answer this question, Toda and her colleagues mixed T1R1 from songbirds with T1R3 from hummingbirds (and vice versa). Neither preparation glowed in response to sugar: The crossed heterodimers were ineffective as sweet receptors. The result means that the sweet taste evolved separately in songbirds and in hummingbirds.

TASTE IN INVERTEBRATES

INSECTS

Invertebrates also have a well-developed sense of taste. Among arthropods, flies—especially the fruit fly, *Drosophila melanogaster*—have been the most studied, so we will take them as our example. Taste receptor cells in flies and other insects are true neurons, with their own axons. They are housed not in taste buds but in hair-like sensilla that are open at the tip, permitting tastants to enter. There are only a few gustatory receptor neurons in a sensillum.

These neurons and the sensilla that contain them are found not only on the fly's proboscis (see Figure 8.5), through which it sucks up food, but elsewhere on its body as well—especially

Figure 8.5 A fruit fly with proboscis extended tastes a potential food item to determine if it should be ingested. Other taste receptors are located in sensilla on the legs and other body parts.

Source: **Used with permission of Shutterstock. Photo Contributor, Jordan Lye.**

on the feet. This enables the fly to determine, by walking around on a potential food item such as a decaying piece of fruit, whether it is worth eating.

Like vertebrates, the fly has an appetitive response to sugars and amino acids but rejects substances that to us are bitter. Acids, too, it mostly avoids. Dilute salt water has a positive hedonic value, but as concentration increases, the solution becomes aversive.

Despite these behavioral similarities, the receptor molecules in the fly are not structurally related to the analogous ones (T1R1, etc.) in vertebrates (Liman et al., 2014). In both phyla, the receptors weave in and out through the receptor cell's membrane, but the sequences of amino acids making up these proteins are totally unrelated. And the receptors for what we may by analogy call the sweet, bitter, and umami tastes are not G-protein-coupled receptors (GPCRs) in the fly, as they are in vertebrates. Gustatory systems that are, at the most basic level, functionally similar in mammal and insect have developed independently. This is a remarkable example of convergent evolution.

MOLLUSCS

We are gradually learning about taste in molluscs, thanks in large part to research on the octopus. While there probably are taste receptor cells in or around the mouth, it is those on the octopus's arms that are being most actively investigated. They are on the rims of the suckers, versatile structures with which the animal can palpate, pick up, and—we now know—chemically examine an object. The rim of each sucker is richly supplied with receptor cells. Some are oval-shaped and are entirely enclosed within the sucker's epithelium. These are mechanoreceptors. Others are more elongated, with a vase-like shape, and have a narrow tip that extends to the surface of the epithelium. Here, fingerlike extensions, cilia and microvilli, are in direct contact with the environment. These are gustatory receptor cells, usually referred to as *chemoreceptor cells,* since their functions extend beyond the evaluation of potential food items (Graziadei, 1962, 1964).

Lena van Giesen et al. (2020) showed, by single-neuron recording in *Octopus bimaculoides,* that the mechanoreceptor and chemoreceptor cells form two distinct groups: The former respond to pressure or movement, but not to chemicals, and their responses are transient; the latter produce more sustained responses and only to chemicals.

Van Giesen and her colleagues found receptor molecules in the octopus's chemoreceptor cells that are unlike those in either vertebrates or arthropods. They are, instead, related to acetylcholine receptors, from which they evolved. But they are relatively insensitive to acetylcholine, responding instead to a large class of organic molecules called *terpenoids* that are produced by both plants and animals.

There are multiple types of these chemoreceptors, and two or more types are typically expressed on the same chemoreceptor cell. These receptors are *ionotropic,* which is to say that they work by opening channels in the cell's membrane through which ions can pass, thus initiating neural activity.

It is possible that additional chemoreceptors, sensitive to chemicals other than terpenoids, are present in octopus suckers but remain to be discovered.

In searching for prey, an octopus may see a shrimp or crab out in the open and move quickly to seize and eat it. But often, prey animals hide in crevices where they cannot be seen. Octopuses spend a lot of time foraging, extending their arms into nooks and crannies in search of these food items (Hanlon & Messenger, 2018). It has long been assumed that they use a combination of touch and taste to locate prey and to distinguish them from non-prey animals and other items.

An elegantly controlled study by Buresch et al. (2022) now shows that this is true. Octopuses (*O. bimaculoides*) were trained to reach through holes in an opaque dome to retrieve a shrimp or crab (see Figure 8.6). Each hole had a curved elbow attached, to prevent the octopus from seeing into the dome. On test trials, the octopus encountered not a live animal but a cylindrical disk, made mostly of agarose. The disk was sometimes unflavored but on other trials was imbued with extract of shrimp or crab, or of sea star (an animal that these octopuses do not eat), or of algae. The octopus touched and sometimes grasped the disk but was not able to pull it out of the dome.

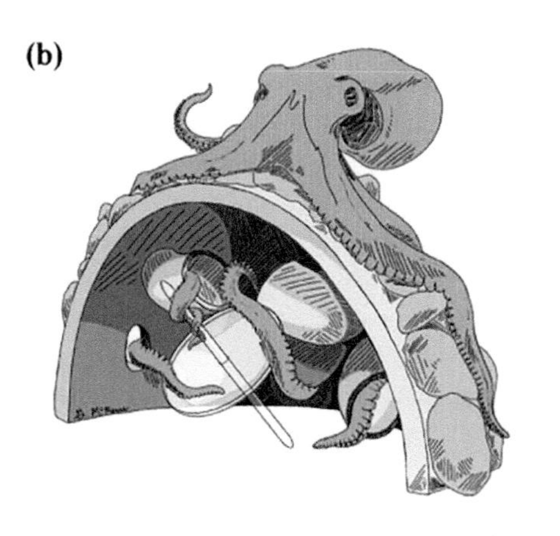

Figure 8.6 Experimental dome used to study contact chemoreception in octopus. The octopus reached into a dome, extending its arms through holes. Its arms searched for prey animals, but on test trials encountered only an agarose disk, held in place by a rod. It touched the disk more if the disk was flavored with shrimp or crab extract.

Source: **From Buresch, K. C., Sklar, K., Chen, J. Y., Madden, S. R., Mongil, A. S., Wise, G. V., Boal, J. G., & Hanlon, R. T. (2022). Contact chemoreception in multimodal sensing of prey by *Octopus. Journal of Comparative Physiology A, 208*, 435–442. Used with permission of Springer Nature BV; permission conveyed through Copyright Clearance Center, Inc.**

The experimenters measured the amount of time an octopus spent touching the disk, and the number of times it curled an arm around it. The results were straightforward: the octopuses spent much more time interacting with the disk when it was flavored with shrimp or crab than when it was flavored with sea star or algae, or not flavored at all.

The octopuses could not have distinguished the disks from one another based on some tactile cue, such as surface texture, because the experimenters filtered the extracts to remove all particles.

But could the octopuses have been using smell rather than taste to distinguish the disks? Did the disks give off odorant molecules that drifted through the water, eventually reaching olfactory receptors, or were the molecules identified while they were still embedded in the surface of the agar? In other words, were the octopuses using *distance chemoreception* or *contact chemoreception*?

To answer this question, Buresch et al. carried out a control test in which dye was mixed with the agarose. When left to soak, none of the dye leached out of the disk, suggesting that the same was true of the animal extracts used in the experiments. This study indicates that octopuses use contact chemoreception, together with touch, to identify hidden prey.

Is this contact sense used in other contexts as well? Indications that it is emerge from an observational study of *Hapalochlaena maculosa,* a venomous blue-ringed octopus native to South Asian waters (Morse & Huffard, 2022). The investigators recorded the behavior of small groups of these cephalopods as they interacted in a laboratory tank. This species is nocturnal, so most of their activity occurred in the dark, recorded under infrared light (see Figure 8.7).

Figure 8.7 A blue-ringed octopus (genus *Hapalochlaena*) in Indonesia's Lembeh Strait. These cephalopods may use contact chemoreception to recognize other individuals.

Source: **Used with permission of Shutterstock. Photo Contributor, Sascha Janson.**

In a typical interaction, one octopus approached and used its suckers to touch another individual and then either quickly moved away or held its ground. Intriguingly, a male was more likely to retreat from a female if during an earlier encounter the two had mated. Morse and Huffard suggest that this reaction may serve to distribute the male's limited number of spermatophores as widely as possible.

The ability to distinguish individuals in this study was not based on vision, since it occurred in the dark, and was not based on olfaction, since it required physical contact between the two animals. It appears to depend, instead, on a combination of contact chemoreception and touch.

Is such chemical recognition of an individual mediated by pheromones? Probably not, because pheromones normally convey generic information such as an animal's sex or hormonal state rather than its identity. But Cummins et al. (2011) have shown that squid (*Loligo pealeii*) can detect and respond to a pheromone when they touch it with their arm suckers. It had previously been observed that male squid become aggressive toward other males immediately after they come in contact with a mass of eggs deposited by a female. The stereotyped behaviors include raising and splaying the arms, chasing and lunging at other males, and grappling with them. Schooling groups of males are calm as they approach and inspect the eggs, but become aggressive as soon as they touch them. This behavioral switch does not occur if the eggs are sealed in a glass flask, where they can be seen but not touched. But Cummins and colleagues found that an extract of egg capsules, painted on the flask, was sufficient to trigger the aggressive behavior.

Next, they used liquid chromatography to fractionate the extract, and bioassayed each of its components separately to identify the active one. They discovered it to be a 10-kilodalton (moderately heavy) molecule, *Loligo* β-MSP, that is secreted by the female as a component of the egg capsules. Females mate frequently while in an egg-depositing phase, and *Loligo* β-MSP prompts males to aggressively compete for these mating opportunities.

SUMMARY

The sense of taste serves as a gatekeeper to the digestive tract. It induces animals to ingest nutritious substances while rejecting potentially toxic ones. Psychophysical evidence supports the existence of five basic tastes in humans and many other vertebrates. Four of these—sweet, salty, sour, and bitter—have long been recognized, and the existence of a fifth, umami (savory, like a protein-rich broth), is now accepted.

The gustatory process begins when substances are taken into the mouth and contact the tips of taste receptor cells, which are usually housed in taste buds. Receptor molecules embedded in the membrane of these cells interact with the tastants, resulting in an electrical response of the cell. Sour receptors work by allowing hydrogen ions into the cell; sweet, bitter, and umami receptors are G-protein-coupled receptors. Researchers are working to determine the nature of salt receptors.

The distribution of these tastes differs among mammals and other vertebrate classes. For example, cats have little or no sweet taste, and among birds, it is present in songbirds but few others. Fish have many types of umami receptors (mostly for different amino acids) whereas amphibians have a large variety of bitter receptors. In fish, taste buds are often found outside as well as inside the mouth; catfish, with more taste buds than any other vertebrate, are the premier example.

In invertebrates as well, taste receptor cells are widely distributed on the body, for example being on the feet as well as the proboscis of many insects. In cephalopods, they have been studied primarily on the arms, where they are found in the rims of suckers. Taste receptor molecules in arthropods and molluscs are structurally different from each other, and from those in vertebrates. But in all three phyla, they mediate appetitive responses to nutritious substances and aversion to poisonous ones, an example of convergent evolution.

At least in cephalopods and perhaps in some other invertebrates, the sense of taste has taken on other roles not related to food evaluation. For example, some octopuses seem to recognize individual conspecifics by touching them with their suckers, an ability that probably depends on a combination of tactile and chemical inspection. For this reason, the sense of taste in invertebrates is now often referred to as *contact chemoreception*.

But diversity of function is much more a characteristic of the other chemical sense, smell, to which we now turn.

9 Smell

In the last chapter, we saw that the sense of taste is used primarily to evaluate potential foods prior to and during ingestion. Encoding tastants as sweet or savory, or as sour or bitter, is the gustatory system's way of prompting acceptance of nutritious foods and rejection of toxic substances. Only secondarily does taste sometimes play other roles in animals, such as social recognition.

The sense of smell is very different. It contributes to the subjective experience of eating by uniting with taste to produce the flavor of foods and, in fact, is often the primary contributor. But smell has a wide variety of other functions as well, from finding a mate to avoiding danger. In many animals, from dogs to bees, it is arguably the most important sense. Humans, too, have a reasonably keen sense of smell; the fact that it plays a much smaller role than vision or hearing in our everyday lives is as much a matter of culture as of biology.

The complex relationships between smells and behavior are made possible by the fact that smell is a *remote*, or distance, sense. A fox who smells a rabbit has many behavioral options, from circling downwind to rushing directly toward a burrow; in contrast, once the rabbit is being tasted, the options are fewer: eating it or (less likely) spitting it out.

In this chapter, we will first consider psychophysical studies with humans, characterizing our olfactory ability to detect and discriminate odors. We experience innumerable odors; classifying them or identifying "basic smells" has proved very difficult. The key breakthrough has, instead, come from neuroscientific research, especially the Nobel Prize-winning work of Linda Buck and Richard Axel, which will be described in this chapter.

In later sections, we will describe some of the many roles that olfaction plays in the behavior of different animals, vertebrate and invertebrate. Some of this discussion will focus on pheromones, secreted molecules that influence the behavior of other members of the same species.

THE HUMAN SENSE OF SMELL

ANATOMY

Air inhaled through the nose enters the nasal cavity, or rather the left and right nasal cavities, since the two are separated by the septum. Together they have a volume of about 34 cm^3. At the top of each chamber is the *olfactory epithelium*, 2.5 cm^2 in size and composed of about 50 million receptor cells interspersed with supporting cells. Turbinate bones break up the flow of air, so gentle eddies are what reach the receptors, not a direct blast from the nostrils.

The olfactory epithelium is coated with mucus, into which cilia extend from the tips of the receptor cells (see Figure 9.1). As odor-laden air sweeps across the mucosa, some of the odorant molecules attach themselves to, and thus activate, receptor molecules (hereafter called simply receptors) on the cilia. This triggers neural activity in the cells.

DOI: 10.1201/9781003362319-9

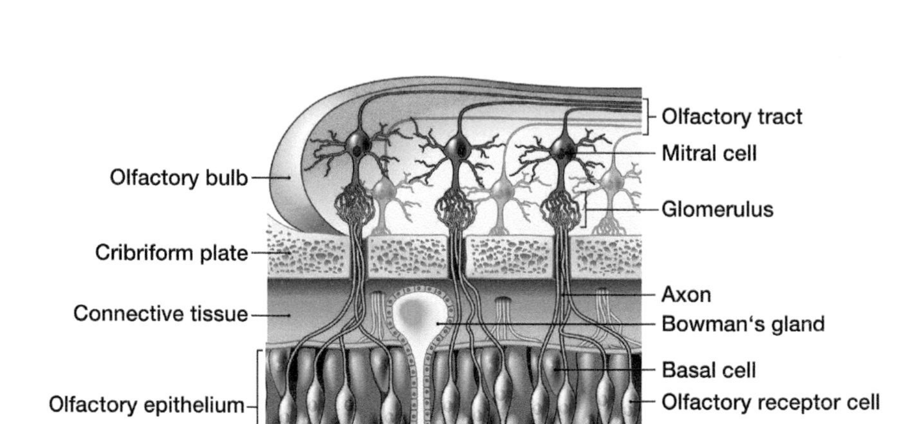

Figure 9.1 First stages of the olfactory pathway. Odorant molecules that reach the top of the nasal cavity flow along the surface of the olfactory epithelium, in mucus supplied by Bowman's glands. Dendrites of olfactory receptor neurons (ORNs) extend cilia into the mucus layer, and receptor molecules on the cilia engage with odorant molecules, thus activating subsets of ORNs. The axons of ORNs carry olfactory signals through a bony sieve, the cribriform plate, into the olfactory bulb, an extension of the brain.

Source: **Used with permission of Shutterstock. Illustration Contributor, Axel Kock.**

Axons of the receptor cells pass through tiny holes in the *cribriform plate,* a sieve-like portion of the skull's ethmoid bone. Once inside the cranium, the axons enter the *olfactory bulb,* a unique brain structure connected by a thick fiber bundle, the olfactory tract, to primary olfactory cortex. Each human olfactory bulb contains thousands of spherical neuropils called *glomeruli* (Maresh et al., 2008). Here, the axons of receptor cells make synaptic contact with second-order neurons, especially *mitral cells,* that transmit the olfactory information to the cortex.

In evolutionary terms, olfactory cortex is one of the oldest cortical regions, as shown by the fact that it has fewer layers than neocortical areas such as visual, auditory, and somatosensory cortex. It is unique in that olfactory information traveling from receptors to cortex does not pass through the thalamus, a complex hub that provides the major input to most cortical areas.

Primary olfactory cortex, in turn, connects with a variety of other brain areas, but especially with the *amygdala,* an almond-shaped subcortical structure that plays a major role in emotional responses. It is perhaps because of the strong links between the olfactory system and the amygdala that our olfactory memories are often tinged with emotion (Herz & Schooler, 2002).

PERCEPTION

DETECTING ODORS

Before discussing how the olfactory system acquires and processes information, we need to consider its capabilities. How good is it at detecting odors and distinguishing among them?

The odorants to which we are most sensitive can be detected when their concentration in air is no more than one part per billion. For other odorants, threshold is thousands of times higher. Differences in sensitivity result from the fact that some odorants dissolve more readily in mucus than others, or bind more readily to receptors.

To analyze what's happening inside the nose when we detect a faint odor, de Vries and Stuiver (1961) measured threshold to a very effective odorant, secondary butyl mercaptan, and interpreted the result in terms of their mathematical model of the nasal cavity. They found that a brief, barely detectable puff of this chemical sends about a billion odorant molecules into the nasal cavity. However, only a fraction of these reach the olfactory mucosa, and even fewer sink into the mucus and come into contact with receptors. The researchers concluded that engagement of a receptor by just one molecule of odorant is sufficient to trigger electrical activity in a cell. This is the theoretical limit of sensitivity, comparable to the fact that one photon is enough to trigger a response in a retinal rod. But, as in the eye, this tiny response of a single cell is not enough to produce a threshold sensation; for this, the combined signals of perhaps 40 olfactory receptor cells are required.

IDENTIFYING ODORS

Beyond the detection of odors is the more complicated issue of identifying them. The number of odors we can distinguish, when they are presented in pairs and asked whether the two are the same or different, is virtually unlimited.

But fitting odors into an organizational framework, analogous to the color circle or the taste tetrahedron, is another matter. Identifying "basic odors" comparable to basic tastes or primary colors has often been attempted, but no scheme proposed so far has been entirely satisfactory. The most famous of these was the smell prism, proposed by psychologist Hans Henning over a century ago. This triangular prism was a theoretical construct representing odor space, with six primaries (basic smells) at the corners: flowery, foul, fruity, spicy, burnt, and resinous. Every other smell was said to be a mixture of primaries, its location on the surface of the prism determined by their proportions.

But Henning's prism didn't pass the smell test. In psychophysical experiments, participants had difficulty analyzing smells into their primary components and sometimes assigned very different smells to the same location on the prism (Geldard, 1972).

More recent, chemically sophisticated classification schemes have proven useful in organizing particular odor domains—perfumes for example—but devising a valid classification system for all odors remains an elusive goal (Gilbert, 2008). This is not because of any lack of effort or ingenuity on the part of olfactory scientists, but because the diversity of smells is too great to be captured by any simple model.

While people have difficulty describing an odor's qualitative properties, they are quite good at determining the location of its source. Georg von Békésy, whose work in audition was described in Chapter 7, found that by comparing the intensity and time of arrival of an odorant in the two nostrils, someone can say whether it is coming from the left or the right. The greater the intensity of the stimulus to one nostril relative to the other, the farther to that side its source is perceived to be. The same is true in the time domain: If odorant-laden air is delivered even a millisecond earlier to one nostril than to the other, it will be localized on the leading side (Békésy, 1967).

BIOLOGICAL BASIS OF VERTEBRATE OLFACTION

THE MAIN OLFACTORY SYSTEM

Our understanding of olfactory stimulation and coding was greatly advanced by the work of Linda Buck and Richard Axel (1991), who shared the 2004 Nobel Prize in Physiology or Medicine. They set out to identify the receptors for smell. Earlier research had suggested that these molecules were G-protein coupled receptors (GPCRs), like those responsible for the sweet, umami, and bitter tastes. To start their search for analogous olfactory receptors, Buck and Axel obtained genetic material from the olfactory mucosa of rats. Next, they used an in vitro technique called the *polymerase chain reaction (PCR)* to produce from this genetic material measurable amounts of a variety of DNA molecules. They found that these PCR products included genetic instructions for a multitude of GPCRs that were unique but had shared features indicating that they were members of a family of receptors. Later research showed that in mice, there are about 1,000 of these distinct olfactory receptor types and in humans about 350.

Further research in Buck's lab (Ressler et al., 1993) examined the relationship between GPCRs and olfactory receptor cells, using a technique called *autoradiography*. Histological sections of mouse olfactory mucosa were treated with radioactively labeled RNA for specific GPCRs, and these RNA molecules were taken up by the receptor cells where they would, in life, be used. The sections were then treated with photographic emulsion and left in the dark for two

weeks, during which radioactive emission from the cells exposed particles of the emulsion, producing dark spots that were later visible under the microscope.

Ressler and her colleagues found that a given RNA label was not uniformly distributed across the mucosa; rather, it was inside specific cells, a small fraction of those in the mucosa. And when adjacent sections were incubated with RNA for different GPCRs, the same cell was never labeled in both sections. In other words, a receptor cell has only one type of receptor molecule.

In summary, the nasal mucosa contains millions of olfactory receptor neurons (ORNs), and each of these has on its membrane a single type of receptor molecule, a GPCR. There are many types of these receptors; they are similar enough in their basic structure to be considered a family, different from the much smaller family of GPCRs that serve as taste receptors. Each of these olfactory GPCR variants is present in thousands of individual receptor cells (Buck, 2005).

The degree of anatomical specificity is even more remarkable in the olfactory bulb (Buck, 2005). Each olfactory receptor neuron in the mucosa sends its axon to only one glomerulus. But this glomerulus receives axons from a large number of ORNs, all having the same type of receptor molecule. In the mouse, for example, there are about 1,000 classes of receptor cells and about 2,000 glomeruli in each bulb, so receptor neurons of a given class send axons to only a handful of glomeruli—perhaps as few as two per bulb.

An individual receptor cell typically responds to several related odorants. And two different receptor cells often respond to overlapping sets of odorants. Consider a hypothetical example in which one class of GPCRs (and therefore one class of receptor cells) responds to odorants A and B, while another responds only to odorants B and C. When one of these three odorants is presented, an animal can in theory identify it on the basis of which class or classes of ORNs respond. In other words, olfactory coding is based on a combination of specificity and patterning.

But what enables an odorant to activate a particular receptor molecule in the first place? No one is quite sure, but the most widely accepted mechanism is that proposed by Amoore (1963) in his *stereochemical theory of odor.* Amoore held that receptor molecules were three-dimensional structures embedded in the membrane of receptor cells, and that a receptor molecule's shape was a geometrical fit to a certain type of odorant molecule. When a molecule of the odorant drifted close to the membrane, it was likely to settle into the niche provided, like a key in a lock. As one line of evidence for his theory, Amoore pointed to examples of molecules that are similar to one another in shape and smell, although they are chemically unrelated.

Receptor cells responsive to a particular odorant are not confined to a single spot in the olfactory mucosa, but neither are they randomly distributed. At least in some animals, including the mouse, the mucosa is divided into zones, with each type of receptor cell occurring (along with other, perhaps similar, types) in only one zone (Ressler et al., 1993). But within a zone, the locations of the individual receptor cells are random. This crude mapping becomes much more precise in the olfactory bulb, where glomeruli processing information from receptor cells with overlapping sensitivity profiles are often close together.

In contrast, sensory coding in the olfactory cortex appears to be less dependent on mapping. A given odorant can cause widespread activity in this cortical area, and different odorants can cause similar, diffuse patterns of activity (Osmanski et al., 2014). How can an animal determine, on the basis of this diffuse activity, what odorant is present?

The results of an experiment by Wilson et al. (2017) suggest a first step toward solving this puzzle. These researchers trained rats to distinguish two odorants. These were special rats, genetically engineered so that their olfactory receptor cells responded not just to odorants but also to light, a technology called *optogenetics.* When their olfactory mucosa was illuminated during behavioral testing, all receptors were temporarily activated, *masking* their responses to odorants.

When the light was simultaneous with the delivery of the odorant, the rats responded randomly: They couldn't smell the odorant. But if the light was delayed just a tenth of a second, blocking all but the earliest component of the neural response to the odorant, the rats' performance was highly accurate. The study shows that the first 100 ms of the response contains all the information needed to identify the odorant.

In fact, odorant identity is *more* precisely represented in the initial phase of the neural response than in later phases (Blazing & Franks, 2020), due to the fact that individual ORNs typically respond to a number of odorants, but more strongly to some than to others. A cell's response to its most effective odorant is generated rapidly, whereas responses to less effective stimuli occur more slowly. The overall neural response therefore grows more muddled and diffuse as time passes.

THE VOMERONASAL SYSTEM

In addition to the main olfactory system just described, many vertebrates have an *accessory olfactory system*, the receptor cells for which are located in a structure called the *vomeronasal organ (VNO)*. The VNO is lower in the nasal cavity than the olfactory mucosa, just above the hard palate that forms the roof of the mouth. The left and right VNOs are close to the vomer, a midline skull bone that is part of the septum dividing the two sides of the nasal cavity.

VNOs are widespread among vertebrates. They are present in many amphibians, reptiles, and mammals, though not in fish or birds; and they are vestigial in apes and people. In most cases, the VNO is open to the nasal cavity, so that molecules must enter the nose to stimulate it; but in snakes, ungulates including cattle and deer, and some other vertebrates, the VNO extends through the hard palate into the mouth, providing another access route for stimulation. Even when it does not do so, a tube called the *nasopalatine duct*, which ascends through the hard palate into the nasal cavity, may allow stimulus molecules entering the mouth to reach the VNO of rodents, bats, and cats (Jacob et al., 2000).

The main olfactory epithelium, at the top of the nasal cavity, can be accessed only by odorant molecules that are *volatile*—light enough to drift through the air. However, many molecules of biological significance are too large and heavy to be volatile. They stimulate the VNO instead, traveling to its receptors in flowing mucus.

Some of these are *pheromones*, chemicals secreted or excreted by an animal that can influence the behavior of a conspecific who detects them. Depending on the species, pheromones can trigger aggression, induce fear, or promote mating. But not every secreted odorant that influences behavior is a pheromone. The term applies only if the affected behavior is unlearned, relatively stereotyped, and occurs throughout a species (e.g., in all adult female mice).

By means of VNO receptors, animals can also respond to pheromones of *another* species, as when a cheetah nuzzling grass can tell the sex of a springbok that passed by recently.

Receptor molecules in the VNO are, like those in the olfactory mucosa, GPCRs. However, they are members of a different family—or actually of two families, VR1 and VR2 (Dulac & Axel, 1995; Herrada & Dulac, 1997). These are found in separate regions of the VNO epithelium, and there are indications that they have different functions, VR1 receptors being primarily involved in reproductive behavior and VR2 receptors in aggression (Silva & Antunes, 2017).

Relying as it does on nonvolatile molecules, it may be wondered why the vomeronasal system is considered part of the olfactory system. The answer is found in the VNO's central connections. Axons of vomeronasal receptor cells terminate in the accessory olfactory bulb, a structure that looks like a smaller version of the (main) olfactory bulb, and contains glomeruli. From here, mitral cell axons travel in the accessory olfactory tract to the brain. But they bypass the olfactory cortex, with many going directly to the amygdala. From here, information from the VNO makes its way to the hypothalamus, a brain region heavily involved in regulating instinctive behavior.

OLFACTION IN MAMMALS

The sense of smell is important to most mammals. Those in whom olfactory structures are most highly developed are called *macrosmatic*, while primates, with a less elaborate olfactory system, are considered *microsmatic*.

This distinction was considered an important one in earlier times, when the chemical senses were seen as less refined than vision and hearing, senses that dominate the everyday lives of humans. But psychophysical documentation of our own considerable olfactory abilities, including detection of low-concentration odorants and localization of their source, has rendered the term microsmatic, at least as applied to primates, obsolete. Nevertheless, there is no doubt that for some mammals smell is the dominant sense.

DOGS

OLFACTORY SYSTEM

Most familiar of these is the dog. Its interest in smells is obvious whenever it is out on a walk or when another animal is encountered. And its olfactory system, beginning with its nose, is optimized for extracting information from the environment. By turning its head or making independent movements of its nostrils, a dog can make selective use of one nostril or the other in attending to an odorant. Which one is used depends on the odorant and its emotional significance for the animal, as Siniscalchi et al. (2011) showed. Dogs in their study were free to approach and sniff an odorant-soaked cotton swab, while a video camera behind the swab recorded their movements during each 3-minute trial. On the first of seven trials with each odorant, the dogs consistently sniffed it with the right nostril. What happened next was different for different odorants. For aversive or threatening smells, such as sweat from their veterinarian, the dogs continued to use their right nostril. But for smells that were presumably pleasant or neutral and required only routine monitoring, such as food or lemon, they gradually switched to their left nostril.

This dichotomy tells us something about olfactory processing in the brain. Unlike most sensory systems, in which information from the left side of the environment goes primarily to the right half of the brain, and vice versa, olfactory pathways in vertebrates remain largely ipsilateral. The left nasal mucosa sends fibers to the left olfactory bulb, which connects with the left olfactory cortex, left amygdala, and so on. So when dogs sniff an unfamiliar object with their right nostril, this suggests that they are formulating an initial reaction to it primarily with the right side of their brain.

Let's return to the path followed by odorant molecules. Once past the dog's nostrils and in the nasal cavity, the airstream divides, with an upper portion channeled upwards toward the olfactory mucosa while the lower portion heads downward toward the lungs. Deflecting and breaking up the upper portion are the *turbinates*, thin sheets of bone that are lined with mucus-covered epithelium, some of which is olfactory epithelium. The dog's main olfactory epithelium has a distinctive yellow-brown color, allowing it to be distinguished from nonolfactory epithelium, which is red-orange (Barrios et al., 2014). Whereas the human nasal cavity has three prominent turbinates, the dog's has ten, curling in different directions like pieces of scrollery. As inspired air encounters the dog's turbinates, it meanders along a circuitous path that gives odorant molecules ample opportunity to settle out and be adsorbed onto the cilia of olfactory receptor cells.

A dog's olfactory epithelium is huge, with a total area up to (depending on the breed of dog) about 150 cm^2, roughly the size of a teacup's saucer; in comparison, a human's olfactory epithelium is only as big as a postage stamp, about 5 cm^2. As a result, dogs have on the order of 30 times as many olfactory receptor neurons as humans—more than a billion ORNs in some cases (Hepper & Wells, 2015).

It is universally recognized that dogs are capable of extraordinary feats of olfactory detection: for example, finding disaster victims, living or dead, who are buried in rubble. This must depend in some way on the size of their olfactory mucosa and their remarkably high number of olfactory receptor cells. But what exactly is the relationship between anatomy and sensitivity?

It will be recalled that under optimal conditions, a single molecule of odorant is enough to trigger neural activity in an individual human ORN (de Vries & Stuiver, 1961). This is the ultimate in cellular sensitivity; no animal can have cells more sensitive than this. There must be other reasons that having many ORNs increases sensitivity. Part of the answer is that perception of an odorant requires that many—perhaps 40—ORNs be activated with near simultaneity; and if an animal has more of them, the chances of this happening are higher. The geometrical complexity of the dog's turbinates, which causes odorants to linger over patches of mucosa, must also increase the number of molecules adsorbed.

However, a comprehensive survey of psychophysical experiments in humans, dogs, and other mammals (Laska, 2017) indicates that dogs are roughly comparable to humans in their detection thresholds for some odorants. Such cross-species comparisons are difficult to make, since different species require different experimental methods. For example, humans can generally attend to a task longer, and with fewer reinforcers, than other animals. Still, it appears that for some plant-based odorants, humans are at least as sensitive as dogs, while for some animal-based odorants, such as constitute the smell of prey, dogs have an edge (Laska, 2017).

However, the main contributor to dogs' olfactory superiority to humans may be not higher sensitivity to odorants that both can detect, but rather, an ability of dogs to detect *more* odorants. This would require not just more ORNs, but more types of them. Do they exist?

An inventory of the boxer dog genome by Quignon et al. (2005) turned up 1,094 genes for olfactory receptor proteins, about 20% of which were nonfunctional pseudogenes. So allowing for differences between breeds, it is reasonable to conclude that dogs have between 800 and 900 different types of olfactory receptors in their nasal mucosa, compared to about 350 in humans. The difference is even greater if we include the dog's VNO receptors, which humans don't have.

A schematic drawing of a dog receptor protein is shown at the top of Figure 9.2. The molecular chain weaves in and out through the membrane: Its extracellular (EC) portions interact with odorant molecules, causing changes in the receptor that affect the disposition of its intracellular (IC) portions, which in turn trigger events leading to neural activation. This GPCR is a long chain of amino acids, represented by small circles in the figure. The molecule shown could be any one of the dog's 800+ olfactory mucosa receptors: Blue circles represent amino acids that are usually the same across receptors; red circles are ones that vary from one receptor type to another. Dog receptors are quite similar to those in the rat (bottom half of figure) and other mammals.

OLFACTORY ABILITIES

This abundance of receptor types is undoubtedly a factor in dogs' ability to sniff out illegal drugs, smell our moods, detect diseases, and follow the track of a missing person. Let's now consider a couple of these abilities in more detail.

Tracking

Hepper and Wells (2005) set out to study the ability of dogs, when encountering a trail of human footprints, to tell by smell alone which way the person was walking. The participants were two-year-old male dogs (retrievers and German shepherds) who had some tracking experience.

To make a track, the experimenters cut out a series of 21 squares of carpet, 49 cm on a side, and laid them end to end on an outdoor concrete surface. A person wearing old hiking boots then walked the track from end to end, taking one step per carpet square. After an hour, a dog was brought at right angles to the midpoint of the trail and told to track, i.e., to follow the trail in the direction taken earlier by the walker. This was done repeatedly, with a fresh trail on fresh carpeting each time. All six dogs did well, turning in the correct direction on at least 90% of trials.

What information were the dogs using in this test? Perhaps they noticed the shape of a footprint on the carpeting, and simply turned in the direction of the footprint's toe. This wouldn't really be tracking at all. To test this possibility, Hepper and Wells carried out a second experiment. Using rubber gloves, they rearranged the order of the carpet squares, making a random sequence but leaving the heal-to-toe orientation of each square unchanged. Retested, the dogs performed at a chance level, showing that they had indeed been following a trail—a sequence of footprints—in the first experiment.

Next, the experimenters made the test harder by removing squares from the beginning and end of the trail. Dogs were quite successful with trails that were 11, 9, 7, and 5 squares in length,

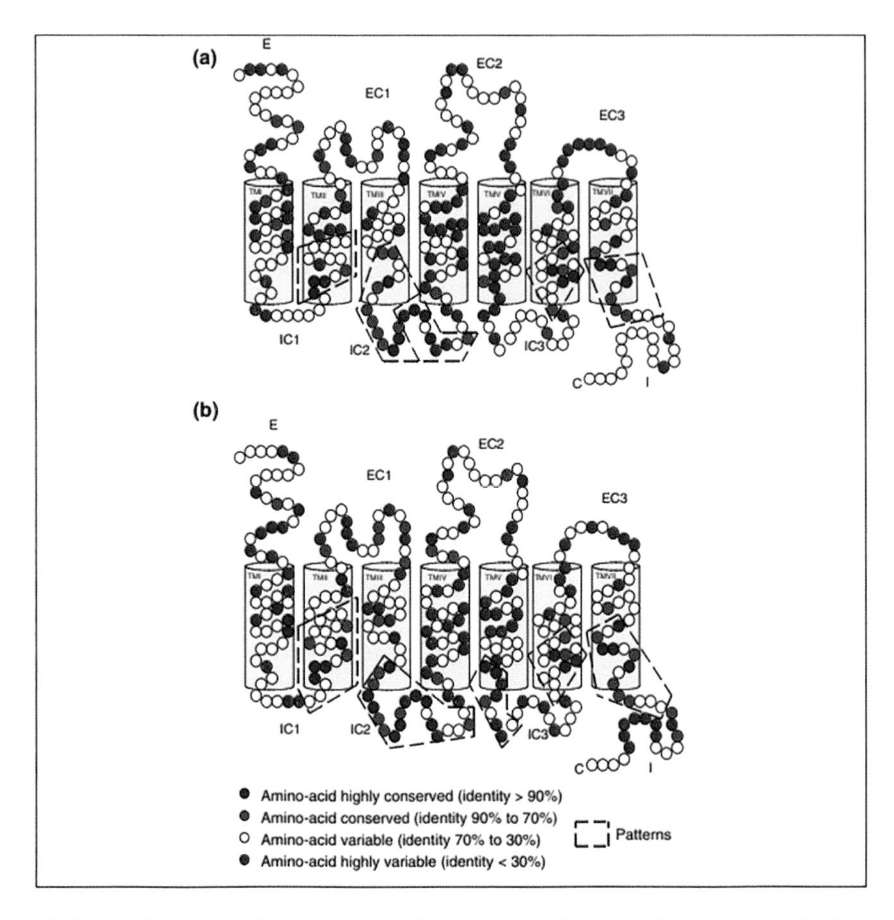

Figure 9.2 Schematic drawing of receptor molecule in the main olfactory epithelium of dogs (top) and rats (bottom). See text for details.

Source: **From Quignon, P., Giraud, M., Rimbault, M., Lavigne, P., Tacher, S., Morin, E., Retout, E., Valin, A.-S., Lindblad-Toh, K., Nicolas, J., & Galibert, F. (2005). The dog and rat olfactory receptor repertoires. *Genome Biology, 6*, R83. © 2005 Quignon et al.; licensee BioMed Central Ltd.**

dropping to a chance level only when the trail was shortened to three squares.

How to explain this ability? The researchers believe that some change in the odor of the footprints over time, perhaps a gradual fading, provides dogs with a clue: They turn in the direction of the stronger-smelling footsteps. But the change in odor intensity must be a small one, for the footprints were made at the rate of about 2 per second. In a series of five footsteps (the shortest series on which the dogs were consistently successful), the time between the first and fifth footsteps was only 2 seconds!

Detecting Disease

It is widely believed that dogs can be trained to detect the presence of health conditions in humans. For example, there are suggestive but relatively uncontrolled reports (reviewed by Dalziel et al., 2003) that some dogs can detect an impending epileptic seizure in their human owner. To evaluate this possibility and determine its possible olfactory basis, Davis (2017) obtained sweat from people with epilepsy who had just had a seizure. By means of gas chromatography, he discovered that this sweat contained volatile organic compounds that were not present in sweat obtained from the same individuals between seizures or from other individuals who did not have epilepsy (Davis, 2017). Chief among these was menthone, a chemical usually derived from peppermint. It is not clear by what chemical pathway menthone is produced in the body.

Following up on Davis's findings, Maa et al. (2021a) combined menthone with two of the other volatile organic compounds to make a "seizure scent" for experimental use. Dogs (mostly golden and goldendoodles) were trained to respond positively to this scent by touching the experimenter's left hand, but otherwise to respond by touching the experimenter's right hand. They were then tested, in a series of trials, with sweat samples from 60 patients with epilepsy. The samples had been collected at various times before, during, and after ictal activity as part of clinical seizure monitoring. The dogs responded positively to samples obtained during the hour preceding seizure onset, or up to 81 minutes after, but they responded negatively to samples obtained from the same patients at other times.

So the dogs were clearly able to detect an impending or recent seizure, but how specific is the odor to this one condition? Might there be other situations in which the "seizure scent" would be present? To explore this possibility, Maa et al. (2021b) carried out an experiment in which five people, none of whom had epilepsy, watched movies designed to stir up particular emotions—a horror movie and a fast-paced comedy—on separate days. Sweat samples were obtained from their palms during each movie, and presented to dogs who had been trained to respond to the "seizure scent." Trainers who interacted with the dogs were given the samples in unlabeled opaque Mylar® bags.

The surprising result was that dogs responded positively, by touching the experimenter's left hand, when sniffing the sweat from people who were watching the horror movie, but not from those who were watching the comedy. Apparently, the odorants in the sweat of someone who is experiencing fear are similar or identical to those in the sweat of a person who is having an epileptic seizure. Maa et al. (2021b) hypothesize that menthone is an alarm pheromone given off by someone in fear and is secreted at the time of an epileptic seizure as well. Even if a person with epilepsy is not aware of an impending seizure, subtle physiological changes that are not consciously perceived may be sufficient to trigger release of this alarm pheromone.

Further research is needed to fully evaluate Maa's hypothesis. But the larger point of this research is that we must be cautious in characterizing a dog's olfactory skills. Dogs are remarkably capable of perceiving compounds that are, to us, faint or odorless, and can be trained to report their presence. But exactly what they are detecting and how specific the odor is to a particular medical condition, are questions that can only be answered by carefully controlled experiments.

The same cautionary note is sounded by Elliker et al. (2014), in a test of whether dogs can learn to detect prostate cancer from the smell of a patient's urine. Aware that dogs can pick up subtle, unintended cues from humans, these researchers used rigorous double-blind procedures in their experiments: Neither the dogs, nor the trainers who tested them, nor the investigators who handled the samples, knew which were from patients and which were from control individuals.

On each trial, the dog was brought into a room containing a plastic array with four holes, spaced 0.75 m apart. A flask containing a sample was positioned beneath each hole. A sample from a cancer patient was positioned beneath one of the holes, and samples from three controls were beneath the other three holes. The location of the cancer sample was randomly varied from trial to trial. The dog was rewarded for sitting or lying with its nose over the hole above the cancer sample.

During training, dogs gradually learned to distinguish the smell of specific prostate-cancer patients' urine from that of controls. Samples from more than a dozen patients were repeatedly presented during this training period. Thinking that the dogs had identified a distinctive odor diagnostic of prostate cancer, the experimenters proceeded to a testing phase in which the urine of unfamiliar patients was presented, along with samples from new controls. Dogs were now presented with each sample only once. Surprisingly, the animals now performed at a chance level. Nothing they had learned previously had generalized to the new samples.

Elliker et al. concluded that their canine subjects had succeeded during training by memorizing the smell of each patient's urine on an individual basis, rather than by discerning a common odorant diagnostic of cancer. This was

a remarkable feat of olfactory memory, but not the one that would be of the greatest benefit in medical diagnosis.

It is, of course, possible that with different training methods, or different types of samples, or different medical conditions, a different result would have been obtained. But this study shows how important it is for claims that dogs can detect various medical conditions by smell to be critically evaluated.

MOUSE PHEROMONES

Earlier in the chapter, we saw that the main olfactory epithelium (MOE) of mice is highly developed, with three times as many classes of olfactory receptor molecules as the human olfactory epithelium, randomly distributed within separate zones the functional significance of which is unknown.

THE DUAL OLFACTORY HYPOTHESIS

In this section, we turn to the other half of the mouse's olfactory system, the vomeronasal system. The vomeronasal organ is located lower and farther forward in the nasal cavity than the MOE, making it more likely to come into contact with large and heavy molecules. Since many mammalian pheromones are of this nonvolatile variety, the VNO plays a major role in pheromone detection. For example, *darcin*, a pheromone with effects described below, has an atomic mass of 18,893 Daltons (Roberts et al., 2010).

The role of the VNO in detecting nonvolatile pheromones has been examined in experiments by Fernando Martínez-García and his colleagues (2009) at the University of Valencia. They measured the olfactory preferences of female mice who had never been exposed to male mice. They were allowed to explore trays containing bedding that had been soiled by other mice. A female test mouse was put in a cage with two trays, and was free to go back and forth between them, rummaging in the urine-soaked bedding. The time she spent at each tray was recorded over a 5-minute period.

When the bedding in one tray had been soiled by female mice and the other by male mice, the test mouse spent most of her time at the tray with male-soiled bedding. But when the trays were covered with perforated lids that allowed only volatile odorants to reach her, she showed no preference. Control experiments showed that female mice could detect volatile male odorants, but were not attracted to them.

To determine the role of the vomeronasal system in these results, Martínez-García et al. lesioned the accessory olfactory bulb in some test animals. The lesioned females were not attracted to male-soiled bedding. These experimental results imply that the vomeronasal system, of which the accessory olfactory bulb is a part, mediates the attraction of intact female mice to male pheromones. More broadly, the findings are consistent with the *dual olfactory hypothesis* (Martínez-García et al., 2009) that the vomeronasal system is largely responsible for detecting and responding to nonvolatile pheromones, while the main olfactory system detects volatile odorants, most of which (in mammals) are not pheromones.

But these functional distinctions between the VNO and the MOE are neither absolute nor universal across species. For example, in rodents, some VNO receptors of the VR1 type are activated by volatile pheromones (Silva & Antunes, 2017). And in humans, who lack a functioning vomeronasal system, volatile compounds in sweat have been reported to act as pheromones (Stern & McClintock, 1998), although this finding remains controversial (Wyatt, 2020). More research is needed to determine how completely, and how widely among vertebrates, the idea of a separate system for the detection of pheromones applies.

DARCIN, A MOUSE PHEROMONE

Pheromones are common among mammals but are especially numerous in mice. This is fortunate for scientists, since the tools and methods of genetic analysis are well established in these animals. For example, there are inbred strains of mice, such as the C57 strain, whose members have virtually identical genomes. This makes it possible to test the effect of a single genetic manipulation in a large number of mice whose other genes are the same across individuals.

Standardized experimental paradigms for studying behavior have also been developed for mice, such as the *resident-intruder test* of aggression. In this test, a male mouse is put in a cage along with a female, and the two reside there for several days. Then, the female is removed and another male (the "intruder") is put into the cage. Some resident males tolerate the intruder, but others chase and attack him, biting and grappling with him. The intruder may fight back but usually gets the worst of it. If a mouse has been castrated, he is fairly docile and does not score high on the resident-intruder test.

The urine of male mice contains 21 different sex-specific proteins. Castrated mice do not secrete these male urinary proteins (MUPs). However, if urine from an intact male mouse is swabbed onto the back of a castrated mouse just before he takes part in the aggression test, he demonstrates the full panoply of stereotyped aggressive behaviors (Kaur et al., 2014). The transformative, abrupt effect of the applied urine suggests that it contains one or more aggression-inducing pheromones.

To identify the specific molecules involved, Kaur et al. fractionated the urine, purified the individual MUPs, and used them in the resident-intruder test with castrated mice. Only two of these ligands, MUP3 and MUP20, elicited aggressive behavior. Subsequent research has focused primarily on MUP20, also called darcin (Roberts et al., 2010).

Before definitively classifying MUP20 as a pheromone, it must be established that the aggression that it triggers is an unlearned response. Since laboratory mice are typically raised within a colony to ensure normal emotional development, even castrated mice will have been exposed to MUP20 previously. This means that they may have learned to associate its presence with the sights, sounds, or direct experience of, aggression. This protein only activates receptor cells in the vomeronasal organ (Stowers & Kuo, 2015).

Mice from BALB/c, an inbred laboratory strain widely used in research, lack MUP20. In an isolated colony of these mice, no one has ever been exposed to this ligand. And when presented with it for the first time in their lives, males promptly become aggressive. This experiment proves that darcin is a pheromone (Stowers & Kuo, 2015).

And its effects are not limited to provoking aggression. Darcin is also attractive to female mice, as Roberts et al. (2010) showed. The experimenters painted darcin in one spot on the ceiling of a testing arena, and female urine in another spot. Females were brought into the arena and their behavior observed. By standing on their hind legs and stretching upwards, they were able to touch either patch with their noses, causing robust vomeronasal stimulation. They showed an interest in both patches but spent twice as much time contacting darcin as they spent contacting female urine.

In a further experiment, Roberts and colleagues showed that volatile odorants from a male mouse are not, on initial presentation, particularly attractive to a female. However, if she is exposed to these same volatiles in combination with darcin, she will subsequently be attracted to the volatiles alone. But she will not be attracted to volatiles from a different male mouse. In other words, she learns, under the influence of darcin, to prefer the smell of one male (in the wild, her mate) to that of his potential rivals.

Darcin is just one of many mouse pheromones, but this account of some of its actions hints at the overall complexity of pheromonal signaling in this small mammal.

Finally, it should be noted that chemicals released by mammals of one species may have strong, unlearned effects on mammals of other species. For example, 2-phenylethylamine, a substance present in high concentration in the urine of carnivores, causes avoidance in mice, and activates a specific population of mouse olfactory receptors (Ferrero et al., 2011). Such a chemical is called a *kairomone*, an interspecies signal that benefits the receiver—in this case, the mouse, who scurries to safety when warned by smell that a raccoon, fox, cat, or other mammalian predator is nearby.

OLFACTION IN BIRDS

There was a traditional belief among ornithologists (including Audubon) that birds, given their generally sharp vision and acute hearing, have little need for a sense of smell. This idea never made much sense, since it is inconsistent with bird anatomy: Birds have olfactory bulbs that receive fibers from nasal mucosa and send axons to a region of the forebrain that is homologous to the olfactory cortex of mammals. But a groundswell of behavioral research, beginning in the 1970s, was needed to sweep the notion away. It turns out that many birds have a good sense of smell, which they use in a variety of ways, including for navigation.

PIGEONS

A pioneer in this field was Floriano Papi of the University of Pisa, who proposed that pigeons can find their way home by attending to odors carried on the wind. For example, if a breeze from the south frequently carries the smell of lilacs to the aviary, and a bird is then transported from the aviary to a remote site where there is a heavy lilac smell, it would conclude that it was south of home and would fly north when released. To test this idea, Papi et al. (1973) raised homing pigeons (*Columba livia*) in two aviaries: one open to the wind, the other protected from wind by plastic sheeting. Each pigeon was shuttled between the two aviaries every three days. When in one of the aviaries, the pigeons wore a plastic mask that prevented them from breathing through their nostrils and therefore from smelling anything (Figure 9.3).

Figure 9.3 Pigeon wearing a mask that interfered with its sense of smell by preventing it from breathing through its nostrils. The masks had a soft inner lining that allowed them to be slipped on and off easily by the experimenters.

Source: From Papi, F., Fiore, L., Fiaschi, V., & Benvenuti, S. (1973). An experiment for testing the hypothesis of olfactory navigation of homing pigeons. *Journal of Comparative Physiology, 83*, 93–102. Used with permission of Springer Nature BV; permission conveyed through Copyright Clearance Center, Inc.

The birds were randomly assigned either to an experimental group that wore masks when in the open aviary or to a control group that wore them in the closed aviary. The point of this design is to equalize the amount of mask wearing in the two groups, while allowing only the control group to smell the wind. Later, the birds were taken to sites at least 10 km from the aviary and released.

Of the experimental birds, 44% were delayed in returning to the aviary or became lost, whereas this happened to only 23% of the control birds, a significant difference. The results are consistent with Papi's hypothesis, since birds were disadvantaged if they had never been exposed to odor-bearing wind. The study shows, however, that smell is just one of the sensory channels that contribute to navigation, since a majority of pigeons in both groups found their way home.

In a further test of the olfactory navigation hypothesis, Papi et al. (1974) exposed pigeons for more than a month to a current of air that intermittently flowed through the aviary. Sometimes, this artificial "wind" blew from north to south and sometimes the reverse. When blowing from the north, it carried the smell of turpentine, and when from the south, it carried the smell of olive oil. Later, the birds were transported eastward and exposed to one odor or the other before being released.

According to Papi's hypothesis, pigeons who smell olive oil at the release site will think they have been transported to the olive-rich south and will fly north when released, in an attempt to return to the aviary, whereas those who smell turpentine will head south. Exactly these results were obtained.

As a result of these and later studies, it is now well established that pigeons have a robust sense of smell and can use it when navigating. However, many questions about olfactory navigation remain, such as when it is used, whether it contributes to maps, compasses, or both, and how olfactory information is integrated with navigational cues provided by other senses (Gagliardo, 2013). The importance of olfaction's role in navigation relative to other sensory modalities no doubt depends on a number of factors, such as the weather and the pigeon's distance from home. The birds can rely on different modalities interchangeably, depending on the amount of information they provide (Gagliardo et al., 2001).

SEABIRDS

Olfaction has also been extensively studied in the procellariiforms, an order of birds that includes petrels and albatrosses (see Figure 9.4). These seabirds use odors arising from the ocean to guide their foraging, for example the odor of dimethyl sulfide given off by algae. High concentrations of this compound can indicate the presence of a food chain that includes fish and squid, favored prey of the birds (Nevitt, 2008).

Foraging by day, these tube-nosed birds return at night to their nests in burrows or rocky crevices on islands or the seacoast. In finding their own nest in the dark, an ability to recognize its olfactory signature would certainly be adaptive. Bonadonna et al. (2004) have demonstrated that blue petrels (*Halobaena caerulea*) have this ability, which is likely present in other procellariiforms as well.

Procellariiforms are monogamous and mate for life. The identifiable smell of the burrow presumably derives from each bird's own scent and that of its partner, for their oily plumage has a musky smell that no doubt rubs off on the nest itself. An individual bird's odor signature is distinctive, enabling the two members of a mating pair to recognize one another, as Bonadonna and Nevitt (2004) discovered in Antarctic prions (*Pachiptila desolata*).

Procellariiforms lay only one egg during the breeding season, and the parents take turns incubating it; one forages at sea while the other sits on the nest. Leclaire et al. (2017) addressed the question of whether they can recognize the smell of their own egg. Working with a colony of blue petrels on one of the Desolation Islands in the southern Indian Ocean, the investigators briefly removed incubating females from their nests, along with their eggs, and brought them to the laboratory. Each adult was placed in the stem of a Y-maze, from which it was free to move forward into one of the maze's two arms. In one arm, randomly determined, a gentle flow of air brought to the parent the scent of her own egg; a similar flow in the other branch carried the scent of a conspecific egg at the same developmental stage.

Figure 9.4 Procellariiforms like these Laysan albatrosses (*Diomedea immutabilis*) have prominent tubular nostrils, testament to their olfactory abilities. Birds of this order use olfaction both to forage on the ocean and to recognize individuals where they nest on land.

Source: **Used with permission of Shutterstock. Photo Contributor, Wayne Lynch/All Canada.**

Some of the birds remained in the stem of the maze, but of those who made a choice, 82% moved forward into the arm from which came the smell of their own egg, showing that they were able to identify it—perhaps because, like the nest itself, an egg picks up the odor of its parents. After being tested, parents with their eggs were promptly returned to their respective burrows, where they resumed incubation.

To see whether a procellariiform's ability to recognize the smell of its own burrow is present early in life, O'Dwyer et al. (2008) tested Leach's storm-petrel chicks (*Oceanodroma leucorhoa*) on an island in Nova Scotia. Within a colony, each chick has its own burrow. The researchers removed individual chicks from their burrows and carried them in cloth bags to their nearby laboratory. Samples of nesting material from the chick's burrow and from a neighboring burrow were also taken to the lab and placed in adjacent plastic containers that had a chick-sized opening in the side. The chick was placed facing the entrances of these two artificial burrows and given the opportunity to choose between them. In the darkened lab, the chick spent several minutes examining the two compartments by extending its head into each of them before settling down in one. Each chick was tested only once and then returned to its own natural burrow.

Two-thirds of the chicks tested chose the compartment with their own nesting material, indicating that they were able to identify their own smell and perhaps the odors of their parents. O'Dwyer et al. note that birds of this species remain in their burrow until they fledge, at which point they fly away, never to return; so their ability to recognize the smell of individuals will not be used for homing until later in life. In the meantime, it may be used for other social purposes, such as kin recognition or mate choice.

OLFACTION IN REPTILES

TURTLES

Among reptiles, turtles have been used the most in behavioral research because of their trainability. For example, Manton et al. (1972) trained green sea turtles (*Chelonia mydas*) to press an underwater lever when a pulse of odorant solution was released into their tank and to refrain from pressing the lever when the pulse contained no odorant. The researchers found that the turtles could detect a variety of chemicals, such as isopentyl acetate and cinnamaldehyde. To determine whether these detections depended on smell rather than taste, Manton and colleagues made the turtles temporarily anosmic by treating their nasal cavities with zinc sulfate, the same procedure used in a study of navigation in homing pigeons mentioned in Chapter 5. This eliminated the ability of the turtles to detect any of the tested odorants, confirming that they had been relying on olfaction.

Recent research by Chiyo Kitayama and her colleagues (2020) at the Ogasawara Marine Center in Tokyo sheds additional light on the olfactory abilities of green sea turtles, by measuring their spontaneous reactions to odorants rather than experimenter-defined operant responses. The investigators were particularly interested in the behavioral relevance of a malodorous substance secreted by *Rathke's glands*, organs that have been characteristic of turtles since Jurassic times (Weldon & Gaffney, 1998).

Rathke's glands are exocrine glands, which is to say that their secretion is released into the environment. It emerges through ducts that pass through the shell where the carapace and plastron meet. One pair of ducts is located slightly posterior to the turtle's front legs, and another pair is just anterior to the back legs (Ehrenfeld & Ehrenfeld, 1973). Following the death of one turtle, an adult male, Kitayama's lab was able to obtain its Rathke's glands and prepare an extract for use as an olfactory stimulus.

In their behavioral experiments, Kitayama et al. (2020) placed juvenile green sea turtles, one at a time, in a water-filled circular arena where they could swim around, periodically lifting their head out of the water. A fan blew odorants downward onto the arena, and the researchers measured the activity level of the turtle: how often it moved from one quadrant of the arena to another during a five-minute trial.

In a control condition, the fan was on but no odorant was presented. But when the odor of shrimp was presented, the young turtles increased their activity, presumably searching for this favored food. In a third condition, the odor of Rathke's gland extract was presented to the young turtles. Their activity level dropped substantially, a behavior that reminded the experimenters of how mice react in the presence of a predator.

This study shows that green sea turtles can smell airborne odorants and can make distinctions among them. Moreover, the secretion of Rathke's glands from a mature male evoked a stereotyped, unlearned reaction in the young turtles, suggesting that it may qualify as a pheromone.

CHEMICAL CRYPSIS IN A SNAKE

An animal's survival depends not just on how well it can detect the smells of prey and predators but also on how well it is smelled by them. *Crypsis*—concealment from others—is a biological strategy that is most familiar in the visual realm but sometimes extends to olfaction as well. Chemical crypsis is illustrated by a reptile very different from the turtle: the puff adder, *Bitis arietans* (see Figure 9.5).

This viper is native to Africa, where it kills more people than any other snake. It is named for its ability, when threatened, to puff itself up and then exhale slowly, with an extended hiss that discourages would-be attackers. But it prefers to wait in hiding, ambushing prey that do not know they have come within striking distance. This strategy succeeds because the adder's coloration blends in with its surroundings (see Figure 9.5) and also because, as Ashadee Kay Miller and her colleagues (2015) at the University of Witwatersrand have shown, other animals cannot smell it.

Figure 9.5 A puff adder (*Bitis arietans*) in a defensive position, ready to lunge. This African viper is named for its ability to puff itself up and hiss when threatened by predators. But often it goes undetected thanks to visual and chemical crypsis.

***Source:* Used with permission of Shutterstock. Photo Contributor, EcoPrint.**

In their experiment, dogs were presented with a cloth that had been left in contact with a puff adder or other snake for 40 minutes to acquire its odor. Once familiar with any "target odor" on the cloth, individual dogs were then taken to a circular array of six jars, one of which contained a cloth with the target odor, and were instructed to select it from among the others by sitting in front of it. The other jars contained odorless cloths or cloths that had acquired environmental odors such as the smell of vegetation. All jars were covered with screening to prevent the dog from touching the cloth.

For five of the six snake species tested, the dogs easily picked out the target odor, but for cloths that had been in contact with a puff adder, the dogs performed at a chance level (see Figure 9.6). Double-blind procedures were used.

An animal's body odor usually depends, at least partly, on byproducts of essential metabolic processes. How are puff adders able to reduce their odor? Miller et al. found a clue when they discovered that the shed skin of an adder had a smell that was very salient to dogs (see Figure 9.6). Perhaps a puff adder's skin is less permeable than that of other snakes, trapping odorants inside until shedding releases them.

Whatever the explanation, this study shows that puff adders have evolved chemical crypsis, a way of avoiding olfactory detection by predators and presumably by prey as well.

OLFACTION IN AMPHIBIANS

Class Amphibia is critical in understanding the evolution of olfaction in vertebrates, because it represents a bridge between aquatic animals (fish) and tetrapods (most reptiles, birds, and mammals). When young, amphibians are aquatic, but after undergoing metamorphosis, most are terrestrial. This transformation requires changes in many organ systems, including the

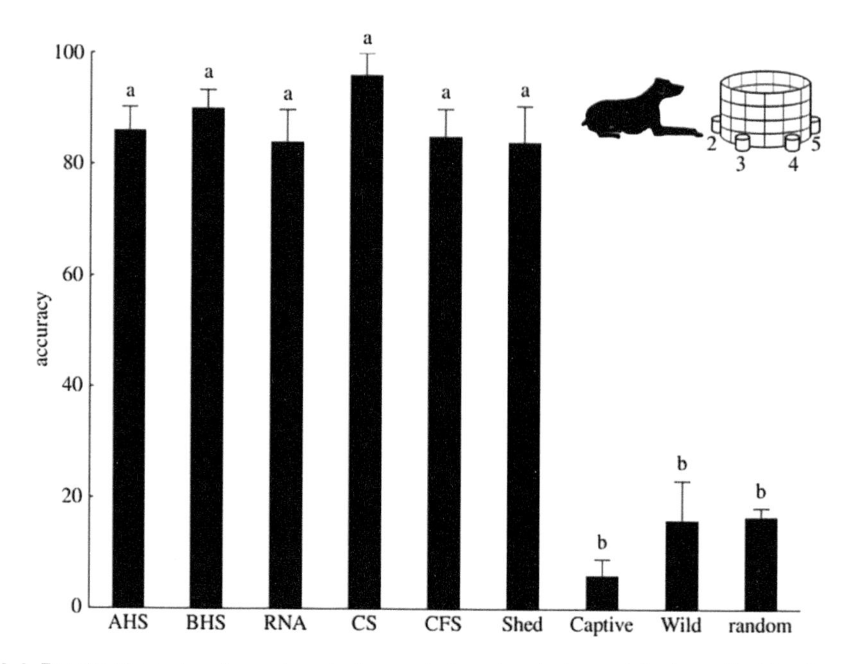

Figure 9.6 Results of a series of experiments demonstrating chemical crypsis in the puff adder. Dogs were trained to select from an array of jars the one containing a cloth that had been in contact with a snake. The dogs responded accurately when detecting the odors of five active hunting snakes (aurora house snake, brown house snake, rhombic night adder, corn snake, and common file snake), indicated by initials beneath the bars. But they guessed randomly when attempting to detect the odor of puff adders, whether these were wild or captive. However, a shed skin from a puff adder was readily detected. Bars marked with an *a* are statistically equivalent, as are those marked with a *b*.

Source: **From Miller, A. K., Maritz, B., McKay, S., Glaudas, X., & Alexander, G. J. (2015). An ambusher's arsenal: Chemical crypsis in the puff adder (*Bitis arietans*). *Proceedings of the Royal Society B, 282*, 20152182. Used with permission of The Royal Society (U.K.); permission conveyed through Copyright Clearance Center, Inc.**

olfactory system. In tadpoles, odorant molecules must be soluble in water if they are to reach the mucosa; in air, however, volatility is the critical parameter (Weiss et al., 2021). This means that different odorants are conveyed to the receptors in, say, tadpoles and adult frogs. In both stages of life, molecules reaching the receptor cells will only be smelled if they activate receptor molecules, so different receptors are required before and after metamorphosis.

As a result, in amphibians evolution has produced two fields of olfactory receptor cells: the vomeronasal mucosa that is dominant in larval forms, and the main olfactory epithelium, which increases in importance during metamorphosis. The vomeronasal system persists in fully terrestrial vertebrates, where (as we have seen) its primary role is the detection of heavy molecules of limited volatility. While the vomeronasal system in terrestrial vertebrates is often associated with pheromones, its main distinction is as a detector of nonvolatile molecules. Many pheromones just happen to be heavy proteins.

During metamorphosis, receptor molecules emerge in the main olfactory mucosa, while some vomeronasal receptors are lost (Syed et al., 2017). Coordination of the different aspects of metamorphosis is critical, for if a tadpole grows legs and steps out on land before or after the olfactory changes occur, it will not be appropriately sensitive to the odorants in its current environment.

Unfortunately for amphibians in our current world, wastewater and groundwater are often contaminated with chemicals that can disrupt the endocrine systems controlling metamorphosis, including its programmed changes in the olfactory system. For example, Heerema et al.

(2018) found that exposing American bullfrog (*Rana catesbeiana*) tadpoles for 48 hours to water containing a small amount of thyroid hormone made them insensitive to the smell of aquatic predators, probably by prematurely reducing their vomeronasal receptors; water containing a cocktail of other endocrine disruptors had the same effect.

Avoiding predators is only one of the uses of the sense of smell by amphibians. Chief among its other functions are finding food, finding a mate, and navigating. An example of the last of these is the journey the brilliant-thighed poison frog, *Allobates femoralis,* makes in adulthood while gently cradling its tadpoles in its mouth. Eggs are deposited and hatch in leaf litter, but to survive, the tadpoles must be taken to water, sometimes hundreds of meters from the oviposition site, where they will develop for two to three months prior to metamorphosis. Pools are common in the tropical rainforest, but many are ephemeral. If the parent (usually the male) deposits the tadpoles in a pond that is destined to evaporate quickly, they will die.

One way olfaction contributes to this frog's ability to select an appropriate pond was discovered by Serrano-Rojas and Pašukonis (2021), in an experiment conducted at the Nouragues Ecological Research Station in French Guiana. Two rival theories were tested: frogs might select pools that already contain tadpoles and thus give off their scent; or frogs might choose pools that smell of decaying leaf litter, since a stagnant pool has clearly been around for a while. Plastic pools of three types, all containing well water, were set up in the forest in areas of high frog density. Some pools were surrounded with wet, decaying leaf litter to give them a stagnant odor; the other two were surrounded with dry leaf litter, and to one of them, the experimenters added tadpoles at a variety of developmental stages.

The result was clear: 84% of the tadpoles deposited by a parent over the course of a day were in the stagnant pools; the numbers deposited in the tadpole-occupied and unoccupied pools were small and about equal. In other words, frogs were strongly drawn to the stagnant pools, perhaps because their smell indicated that they would continue to exist while the tadpoles matured.

This finding shows the subtle yet crucial role that smell plays in the everyday life of amphibians, making them vulnerable—often in unsuspected ways—to small changes in their environment.

OLFACTION IN FISH

THE OLFACTORY MUCOSA

Fish are too diverse for any single description of their olfactory system to apply to all. But certain basic features of the system are common. We will take as an example the zebrafish (*Danio rerio*), a small and unassuming fish that is widely used in research, in part because its transparency in the larval stage makes it possible for scientists to observe its internal organs.

Instead of having the separate main olfactory and vomeronasal epithelia characteristic of tetrapods, the zebrafish has just one olfactory epithelium on each side of the midline. It is a complex, multilobed structure called the olfactory rosette, inside the nasal cavity. Although the rosette is compact, its many folds give it a large surface area.

In fish, the nasal cavity is isolated from the mouth, through which water flows on its way to the gills, and is therefore something of a cul-de-sac. How can water flow efficiently within this cavity to bring odorants to the olfactory mucosa?

Light has been shed on this puzzle by research on *larval* zebrafish, whose olfactory mucosae are pits on the snout, and thus visible to researchers (Reiten et al., 2017). The scientists found that nonsensory epithelial cells surrounding the olfactory mucosa have on their surface numerous cilia capable of active movement.

Reiten and colleagues found that these motile cilia beat in unison, sweeping back and forth at about 24 Hz. The beating is asymmetrical, like that of oars, and causes water in front of the animal to flow toward the snout and into the olfactory pits. Odorants in the water are swept in a medial-to-lateral direction across the olfactory mucosa, activating ORNs. The importance for olfaction of this cilia-produced flow was shown by comparing the robust neural

response to an odorant in normal zebrafish larvae with the negligible response obtained from mutants whose nasal cilia did not move (Reiten et al., 2017).

The researchers also observed this cilia-produced flow of odorants across the olfactory mucosa in larval salmon, suggesting that the mechanism is widespread; it may occur in adult fish as well, though inside the nasal cavity and thus out of view.

Let's turn now to the structure of the olfactory epithelium itself, studied in adult zebrafish. It houses two anatomically distinct types of olfactory receptor neurons (Yoshihara, 2009). Some have their cell body deep within the epithelium and extend a long dendrite toward the surface, where it gives rise to several cilia that are probably not capable of active movement. Cells of the other type have their somata in the epithelium's superficial layers, so that the dendrite extending to the surface is short, and topped with numerous *microvilli* instead of cilia. Microvilli are shorter than cilia and have less of an internal structure. There are receptor cells of a third type, but these are not very numerous.

The ciliated receptor neurons and those with microvilli are apparently forerunners of the receptor neurons in tetrapods' main olfactory epithelium (MOE) and vomeronasal organ (VNO), respectively; for those in the MOE are, by and large, ciliated, while those in the VNO have microvilli. Consistent with this view, the receptor molecules in the zebrafish's ciliated cells are mostly of the OR type, while those in the cells with microvilli are primarily of the VR2 variety (Yoshihara, 2009).

There are other parallels between the olfactory systems of zebrafish and, say, mice (Yoshihara, 2009). In both cases, an individual ORN generally houses only a single type of receptor, although there are exceptions. And zebrafish ORNs with a given receptor all send their axons to a small number of glomeruli within the olfactory bulb, just as in tetrapods. There is a rough—or perhaps just not well understood—odorant map on the zebrafish's bulb; and from its subregions, signals are sent to different parts of the forebrain.

CENTRAL OLFACTORY PATHWAYS

LATERAL OLFACTORY TRACT

Two distinct pathways, the lateral and medial olfactory tracts, carry these signals from a fish's olfactory bulb to its forebrain. Their functional roles were clarified in a series of experimental studies by Kjell Døving and his colleagues at the University of Oslo.

Electrical stimulation of the lateral olfactory tract (LOT) in cod (*Gadus morhua*) triggers what appears to be feeding behavior (Døving & Selset, 1980). The fish swim to the bottom of the tank and assume a head-down position, then move backward over the gravel as if searching for food. This stereotyped reaction resembles normal foraging behavior in this species.

In contrast, *lesioning* the LOT in crucian carp (*Carassius carassius*) interferes with these behaviors (Hamdani et al., 2001). In this study, carp were assigned to one of three groups. In one, the LOT was cut while leaving the medial olfactory tract (MOT) intact; in a second group, the MOT was cut but the LOT left intact; and in a control group, neither tract was lesioned. To test the animals, Hamdani and colleagues surreptitiously added a small amount of salmon extract to the tank through a tube. The control animals and those with an MOT lesion promptly began opening and closing their mouths, roaming the tank, and biting at its floor and walls. The fish with the cut LOT showed none of these feeding behaviors.

Taken together, these experiments indicate a close association, at least in some fish, between feeding behavior and the signals carried by the LOT.

MEDIAL OLFACTORY TRACT

The MOT is subdivided into medial and lateral portions that appear to have largely different functions. Weltzien et al. (2003) showed that lesioning the lateral portion of the MOT in male crucian carp eliminated a key sexual behavior in which, prior to spawning, the male follows the female with his head close to her anal papilla.

Lesioning other olfactory tracts did not produce this reduction in courtship behavior.

As for the medial portion of the MOT, it appears to be responsible for olfactory triggering of an alarm reaction. Fish become alarmed when they smell a predator. In many fish, an additional alarm stimulus is a chemical in the skin cells of the prey species, which is released into the water when conspecifics are savaged by a predator.

Alarm behaviors vary with species. Zebrafish (Speedie & Gerlai, 2008) swim rapidly and erratically on the bottom of the streambed. This zigzagging may serve to stir up debris in which the fish can hide from predators. If multiple zebrafish are present, they move closer together.

To study alarm reactions in the laboratory, Hamdani et al. (2000) extracted alarm substance from the skin of crucian carp and recorded the responses of other carp when it was introduced into their tank. In some animals, the researchers cut the medial portion of the medial olfactory tract (mMOT), leaving all other pathways intact, and in other animals, they did the reverse. Carp with an intact mMOT had a normal reaction when they smelled the alarm substance, but the frequency of alarm behaviors dropped by 85% in individuals whose mMOT had been lesioned. This study shows that the mMOT plays a key role in olfactory triggering of alarm.

Integrating these findings with results from anatomical fiber tracing, Hamdani and Doving (2007) proposed that in many fish, three neural channels extending from the olfactory mucosa to the brain contribute to feeding, sexual behavior, and alarm reactions, respectively. This simple but elegant theory advances our understanding. However, some types of olfactory behavior in fish remain to be integrated into this framework, such as the role of smell in natal river imprinting and homing (Dittman & Quinn, 1996).

FOREBRAIN

The partial separation of neurons responsive to different odorant categories, seen in the fish's olfactory bulb and olfactory tracts, continues in the forebrain. Single-neuron recordings from the channel catfish (*Ictalurus punctatus*) show that cells in the lateral forebrain respond best to amino acids and nucleotides, while cells in the medial forebrain respond strongly to bile salts (Nikonov et al., 2005).

Amino acids and nucleotides indicate the presence of food. Nikonov et al. suggest that the portion of the fish's lateral forebrain that is responsive to them may roughly correspond with the olfactory cortex of mammals; here may begin the processing that leads to hunting for prey.

In contrast, bile salts, which are produced in the liver and excreted in feces, can have hormonal effects on fish that detect them. Some may serve as pheromones, delivering social information (such as reproductive status) to other individuals; and some may act as kairomones, alerting catfish to the presence of a predator. The ability of these emotionally significant odorants to drive neurons in the medial forebrain suggests that this brain region may be homologous to portions of the mammalian amygdala, a structure involved in emotional processing (Nikonov et al., 2005).

PLASTICITY

In some fish, changes in olfaction may occur as a result of changes in behavior or metabolic state. Consider *Astatotilapia burtoni*, an African cichlid fish native to Lake Tanganyika. In this species, olfactory changes occur in both males and females, and in both cases, their immediate cause appears to be increased nutritional need resulting from a reduction in eating. But the reason for the reduction is different in males and females.

The female mouthbroods her fertilized eggs, which is to say that she carries them in her mouth until they hatch. During this time, she does not eat. Once the fry swim away, she is able to eat normally until she spawns again.

In males, the reduction in eating occurs as part of a change in status within the male dominance hierarchy. A few of the males in a community are dominant, but most are in a subordinate mode. Subordinate males spend their time fleeing from dominant males, swimming in the company of females, and eating what food is available. Dominant males, in contrast, are highly territorial, chasing subordinate males

Figure 9.7 A male *Astatotilapia burtoni* in dominant mode. Changes in status brought about by social factors affect the coloration, behavior, and olfactory sensitivity of this African cichlid fish.

***Source:* Used with permission of Shutterstock. Photo Contributor, Anney Lier.**

away. And they mate frequently, capitalizing on the reproductive opportunities afforded by their high status, which they advertise with bright colors (see Figure 9.7). Their busy lifestyle doesn't allow much time for eating.

Maruska and Fernald (2010) found that males can transition from subordinate to dominant mode in a short time. They used males that had once been dominant but were then reduced to subordinate status by the presence of a large male. When the experimenters removed this dominant male from the tank, the subordinate quickly took advantage of his changed situation. He became territorial, his coloration changed, his testes enlarged, and he began mating in a half terra cotta pot provided by the researchers as a spawning area. And he spent less time eating.

Nikonov et al. (2017) measured the olfactory sensitivity of these fish by recording their electroolfactogram (EOG). This electrical wave, generated by the olfactory mucosa, is the summated response of ORNs. The researchers obtained the EOG with a small electrode on the surface of the mucosa of the anesthetized animal, while flowing solutions of alanine and arginine, two amino acids associated with food, across it.

The mucosa responded to the amino acids, but the response was much larger in some individuals than others. A large response was evoked in dominant males and in mouthbrooding females—the two groups who had been eating little (the males) or nothing (the females) for days or weeks. Responses were smaller in subordinate males and in females between periods of mouthbrooding. The researchers hypothesize that the olfactory changes may serve to prepare females to eat voraciously once they are free to do so, and to alert males to small amounts of food they can eat quickly without having to forage widely.

The overall point is that in this species of fish and no doubt in others, the olfactory system is dynamic, its sensitivity changing as a result of metabolic and perhaps other factors (Maruska & Butler, 2021).

EFFECTS OF OCEAN ACIDIFICATION

Among those other factors are anthropogenic ones, including climate change. The effects of climate change are so numerous and complex that entire journals are now devoted to the subject. A major driver of climate change is an increase in atmospheric CO_2. Much of this gas dissolves in oceans and other bodies of water, where it is converted into carbonic acid. The result is a drop in pH, reflecting increased numbers of hydrogen ions (H+). The nasal mucosa is composed of sensory neurons in direct contact with the water, so it would not be surprising if it showed effects of this acidification.

To study effects of acidification on the sense of smell, Munday et al. (2009) measured the olfactory preferences of larval clownfish (*Amphiprion percula*) 11 days after hatching. They had been raised either in ordinary seawater with a CO_2 concentration of about 390 ppm or in seawater to which additional CO_2 had been added, increasing its concentration to 1,000 ppm. This is a level of CO_2 that will be common in the oceans by the year 2100 if present trends continue. Both groups of fish participated in tests in which they were placed individually at one end of a flume, facing a partition that divided it into left and right halves. Water flowed toward them on each side of the partition: plain seawater on one side, and seawater with odorant added on the other side. By choosing to swim to the left or right, the fish demonstrated a preference for, or avoidance of, the odorant.

In the wild, these fish live in a symbiotic relationship with sea anemones, on reefs near islands with vegetation. In the larval stage, they seek out islands with certain types of trees, preferring the odor of tropical rainforest trees such as *Xanthostemon* to that of oily *Melaleuca* trees that grow in swampland. And when tested in the flume, larvae in the control group (raised in ordinary seawater) made the expected choices, swimming upstream in *Xanthostemon*-scented water and avoiding *Melaleuca*-scented water.

However, fish raised in CO_2-enriched water behaved differently. They found the odor of *Xanthostemon* mildly appealing, but surprisingly, they were more strongly attracted to the smell of *Melaleuca*. The experiment suggests that ocean acidification doesn't just decrease a fish's overall olfactory sensitivity but can also change qualitative aspects of olfactory perception—how something smells.

Munday et al. (2010) found that such changes can be dangerous. Fish raised in water with a CO_2 level of 850 ppm were attracted by the odor of a predator, the grouper *Cephalopholis cyanostigma*. This startling attraction to predator odor was confirmed in larval fish from another species, the damselfish *Pomacentrus wardi*.

To assess the ecological importance of the abnormalities caused by larval exposure to elevated CO_2, Munday and colleagues released damselfish larvae on small patches of reef in the Great Barrier Reef. The larvae were at the *settlement stage*, when they choose and settle into an adult habitat. Half had been raised in plain seawater, and the other half in water to which additional CO_2 had been added. Each fish was tagged so that it could be identified and observed by the experimenters.

Fish in the two treatment groups differed considerably in behavior. Those raised in untreated seawater behaved cautiously, cowering in holes in the reef most of the time. In contrast, those raised in lower-pH water ventured boldly out of their holes and swam about in the open. When the experimenters looked for them the next afternoon, only 7% of the larvae from this group could be found, the rest having presumably been eaten by predators whose smell they were drawn to. But 90% of larvae in the control group were still on the reef, living cautiously.

Ocean acidification threatens fish survival in many ways. This important study shows that disruption of the olfactory system is one of them.

Porteus et al. (2018) used physiological methods to analyze the effects of elevated CO_2 in European sea bass (*Dicentrarchus labrax*). The fish were anesthetized, and an electrode was inserted into their olfactory nerve. It recorded the summed activity of ORNs while odorants, dissolved in seawater, were flowed across the rosette. On some trials, the odorants were dissolved in plain seawater and, on other trials, in seawater with an elevated level of CO_2.

The researchers found that increasing the level of CO_2 reduced the amplitude of the nerve's response to some odorants, such as glutamate (an amino acid) and scymnol sulfate (a shark bile acid), but not to others, such as serine and cysteine. By affecting the response to different chemicals differently, elevated CO_2 could distort the fish's perception of an olfactory "cocktail" such as the smell of a predator.

Porteus and colleagues (2021) also offered a hypothesis as to how elevated CO_2 might affect responsiveness to odorants. They note that, for a receptor to be activated, an odorant molecule must bind with it, if only briefly. Part of what draws them together is electrical attraction between positive parts of the odorant molecule and negative regions of the receptor molecule, or vice versa.

When excessive CO_2 is dissolved in the water, a large number of hydrogen ions are released, free to move about and cling to larger molecules, including odorants and olfactory receptors. By so doing, they will interfere with the binding of odorants to receptors, to a degree that depends on their specific shapes and charges. The investigators hasten to point out that this *protonation* is probably just one of the mechanisms involved.

Ocean acidification has many effects on sea life. The studies discussed here show that some of these involve the olfactory system. This is not surprising, since olfaction is the sense through which animals are most directly affected by the chemistry of their environment. To us, these changes may appear subtle and complex, but for fish, they are a threat to survival—of the individual and in some cases of the species.

OLFACTION IN ARTHROPODS

Olfaction is one of the most important senses of arthropods. Their vision and hearing are remarkable in some ways, but overall not as keen as those of vertebrates. Their sense of smell, in contrast, gives very precise information, for it often involves recognition of specific molecules and triggers specific responses to them.

THE FRUIT FLY

We will start our discussion with the fruit fly, *Drosophila melanogaster.* It has been the object of intense research, especially in the field of genetics, for more than a century (Benton, 2022), and its entire genome is now known.

OLFACTORY SYSTEM

Drosophila smells with two pairs of organs on its face: the antennae (near the eyes) and the maxillary palps (near the mouth). All four are bushy structures, crowded with hairlike processes, the *sensilla,* which contain the dendrites of the olfactory receptor neurons (see Figure 9.8).

These dendrites have on their surfaces the olfactory receptor molecules that interact with odorants. This may sound similar to the situation in vertebrates, but in fact, the receptors themselves are quite different. It will be recalled that in the *metabotropic* receptors of vertebrates, an odorant interacts with the receptor's extracellular portion, causing changes in its intracellular portion. These changes in turn trigger chemical processes within the ORN that lead to excitation. In contrast, insects have *ionotropic* receptors, activation of which directly opens ion channels in the ORN's membrane.

A curious feature of insect olfactory receptors is that each is a complex of several molecules: the highly specific receptor itself, and one or more flanking *co-receptors,* which are more broadly tuned and appear to play a variety of supporting roles (Benton, 2022).

These differences are fundamental enough to show that the olfactory systems of vertebrates and arthropods are distantly, if at all, related. This means that the similarities between them—described below—are the result of convergent evolution rather than of shared inheritance.

The axons of ORNs project to the antennal lobe of the brain, roughly analogous to the vertebrate olfactory bulb. The antennal lobe contains glomeruli, dense areas of neuropil containing both cell bodies and a multitude of fibers, including local connectors, axon terminals of

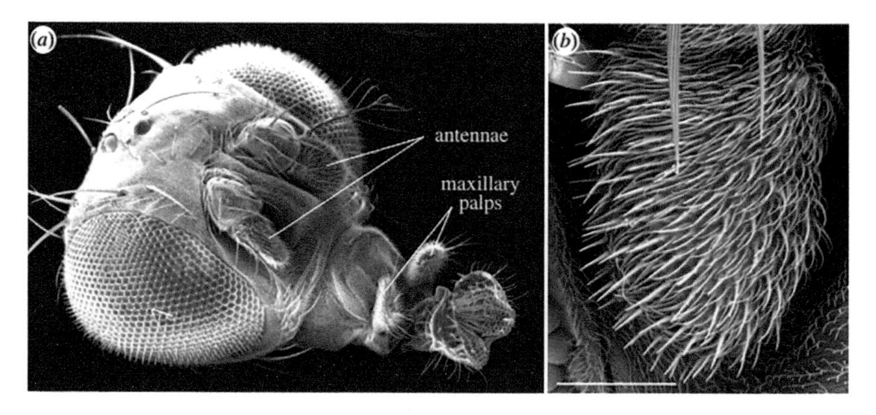

Figure 9.8 The olfactory organs of *Drosophila melanogaster*, seen in scanning electron micrographs. On the left is the animal's head, showing its antennae and maxillary palps. On the right is a close-up of an antenna, showing the sensilla that house the dendrites of olfactory receptor neurons.

Source: **From Benton, R. (2022).** ***Drosophila*** **olfaction: Past, present and future.** ***Proceedings of the Royal Society B*****,** ***289*****, 20222054. Used with permission of The Royal Society (U.K.); permission conveyed through Copyright Clearance Center, Inc.**

ORN axons, and axons of glomerular cells projecting to more central brain areas. Importantly, a glomerulus receives inputs from a single class of ORNs, or perhaps in some cases a handful of similar classes.

Glomerular axons travel to the insect forebrain, where they terminate in either the lateral horns or the mushroom bodies. The lateral horns are regions of the protocerebrum where olfactory inputs connect with downstream circuits that initiate different stereotyped behaviors; the mushroom bodies have a more integrative, plastic circuitry that enables them to play an important role in olfactory learning and memory.

DROSOPHILA PHEROMONES

A Pheromone Secreted by Males

Drosophila use their sense of smell to detect both environmental odorants, such as those that emanate from rotting fruit, and volatile pheromones, which serve a variety of social functions. The best studied of its pheromones is 11-cis vaccenyl acetate (cVA), which is produced only by males. Secreted from the tip of the male's abdomen, cVA drops to the ground where it forms olfactory landmarks that attract other Drosophila to the scene (Mercier et al., 2018).

When many males are crowded together, all releasing cVA, the amount of this volatile odorant in the air rises. And at high concentrations, cVA promotes aggression among males (Wang & Anderson, 2010). But at lower concentrations, it is cVA's effects on sexual behavior that are more notable.

Experimental analysis of these sexual behaviors has been greatly facilitated by the identification of the olfactory receptor for cVA, *Or67d*, which is found only in a single class of ORNs and is the only receptor they contain. By creating a mutant fly that lacks this receptor, Amina Kurtovic and her colleagues (2007) at the Research Institute of Molecular Pathology in Vienna have been able to see how flies behave when they cannot smell cVA.

The investigators measured courtship, a complex set of behaviors in which a male follows a female, touches her body, flaps his wings in song, and contacts her genitalia with his mouthparts. When a wild-type (i.e., genetically normal) male and female are put together for a 10-minute test, the male courts the female about 80% of the time. If instead the male is put in the testing arena with another wild-type male, he will court him, but only infrequently (10%).

Kurtovic et al. discovered that the behavior of a mutant male, unable to smell cVA, is different. When placed with a virgin female, his courtship index is unchanged by the mutation; when placed with a wild-type male, however, his courtship index rises to almost 30% as a result of the mutation. The implication is that the smell of cVA suppresses male–male courtship, perhaps a mild expression of its aggression-promoting role.

The smell of cVA also has an effect on the sexual behavior of females. When placed together with a wild-type male, a wild-type virgin female was more than twice as likely to copulate as a mutant female. Overall, the results indicate that cVA is a dual-purpose pheromone that promotes mating in females while inhibiting it in males (Kurtovic et al., 2007). An added complexity is that during mating, a male transfers some cVA to the female, thus perhaps reducing the motivation of other suitors to mate with her (Ejima et al., 2007).

The different effects of cVA on males and females are puzzling, since flies of both sexes detect it with the same receptors and process these ORN signals with comparable, if not quite identical, glomeruli in the antennal lobe. Datta et al. (2008) discovered, however, that when the axons of these glomerular cells reach the lateral horn, they ramify, ending in a number of branches; and in males, they give off an extra branch that extends ventrally. This genetically programmed ventral branch may contribute, through circuits it selectively activates, to sex differences in the fly's behavioral response to cVA.

A Pheromone Secreted by Females

Another Drosophila pheromone, Z4–11Al, is secreted only by females but attracts flies of both sexes. It activates a specific class of ORN, the ab9A neuron, that has a surprising property (Lebreton et al., 2017). Each of these neurons has two receptors, Or69aA and Or69aB, which are closely related. These two isoforms are virtual twins, but with slight differences that give them different sensitivities. Whereas Or69aB is

primarily sensitive to Z4–11Al, Or69aA is mainly responsive to other chemicals found in fermenting fruit and yeast, Drosophila's preferred foods. The ORNs therefore respond either to the pheromone or to food odorants, or especially to both. They provide a mechanism by which flies of both sexes are drawn to a location offering both mating opportunities and the nutritional resources needed for reproduction.

Wasp Pheromone, Fly Kairomone

Olfaction is also used after mating, as female flies seek out places to deposit their eggs. These will hatch into larvae before growing into adults. But the larvae face danger from parasitoid wasps, especially those of the genus *Leptopilina.* The wasp inserts an egg into the body of a Drosophila larva using her sharp ovipositor. As this egg hatches and passes through developmental stages, it eats the fly larva from the inside, finally emerging in adult form from the husk of the dead fly larva.

One of the female fly's few resources in protecting her eggs from this fate is that she can smell wasps, mainly because of the wasp sex pheromone *iridomyrmecin,* although other odorants play a supporting role. *Drosophila* detects iridomyrmecin with the ORN called ab10B, by means of a specific receptor. But as in the case of the neurons studied by Lebreton's team, ab10B neurons are home to two receptors: One detects iridomyrmecin, while the other detects additional chemicals produced by the wasp (Ebrahim et al., 2015). Taken together, this cocktail of odorants strongly activates ab10B neurons, signaling unambiguously that a wasp is nearby. Female *Drosophila* prefer to deposit their eggs in a location that does not have this wasp smell.

Iridomyrmecin is a pheromone for the wasp, but not for the fly, because it is of a different species. In relation to the fly, it is a *kairomone,* a chemical released by an organism that provides useful information to individuals of a different species.

Drosophila melanogaster has some 60 odorant receptors, each with its own story, known or unknown. But the few examples given here illustrate the sorts of ways—sometimes singly but often in combination—in which they contribute to the Umwelt of these small but complex insects.

THE HONEYBEE

Fruit flies can perhaps be described as sociable, occasionally congregating on, say, a grape leaf to mate and interact in other ways, while distancing themselves from flies of other species (Soto-Yéber et al., 2018), but they are not considered *social insects* because they do not have a well-defined community structure, like many ants, bees, and termites. One of the most highly organized of these *eusocial* insects is the honeybee, *Apis mellifera,* whose dance language was described in Chapter 5. Bees use all their senses to communicate with one another about the location of food, but in maintaining the social structure of the hive, olfaction is paramount.

Their many pheromones carry numerous signals to one another, in a remarkably complex network of chemical communication which is by no means fully understood. But some of its behavioral features have been known for a century, and the last few decades have seen rapid advances in understanding the chemistry of bee pheromones.

The most important feature of honeybee social structure is the division of the hive into three castes of bees, which are morphologically distinct and have different social functions. These three castes are the queen, the workers, and the drones. The drones are the only males, and their role is finished once they have mated with a new queen; they die soon after mating, or if not, are eventually driven from the hive.

So we will focus on the behavior and pheromones of queens and workers. Only the queen mates, at the beginning of her reign; she mates with dozens of drones and stores their sperm for use throughout her life. Her fertilized eggs can develop into workers or future queens. The workers spend their lives attending to the queen's personal needs, nurturing the young, cleaning and defending the hive, and foraging. All of these activities are controlled by pheromones.

The most important pheromones are those secreted by the queen herself. One of these is the *queen mandibular pheromone (QMP)* secreted by glands near her mouthparts. Like many pheromones, it is a cocktail of several chemicals (Bortolotti & Costa, 2014). One of these chemicals, 9-ODA, was first reported in 1960, but later research turned up four other components (Slessor et al., 1988). All are considered parts of QMP, rather than separate pheromones, because their combined effect is much greater than the sum of their individual effects.

QMP orchestrates many of the activities of the hive. The best studied of these is retinue formation, the *retinue* being the queen's personal entourage of workers who feed, clean, and stroke her. This function of QMP is enhanced by four other substances, secreted in various locations on the queen's body, and the combination of all nine semiochemicals is called *queen retinue pheromone* (Keeling et al., 2003).

To measure the contribution of these four additional substances, Keeling and colleagues carried out a behavioral test, or *bioassay*, in which a plastic object roughly the size and shape of a queen was placed in a Petri dish containing a group of workers. A dimple in the top of the dummy queen accommodated a small amount of test substance. The number of times bees contacted the dummy over a 5 minute period, touching it with their mouthparts or antennae, was tallied. The nine-component blend elicited about three times as many contacts as QMP alone, although the four additional components were ineffective when presented without QMP. The investigators conclude that queen retinue pheromone is a true multiglandular pheromone of at least nine components, the most complex pheromone known to be involved in producing a specific behavior.

Localized measurements of 9-ODA indicate that it and presumably other constituents of QMP are constantly being produced by the queen. Workers in her retinue pick these substances up on their antennae and mouthparts. When they then come into contact with other workers, small amounts of QMP are passed along. In this way, the pheromone becomes widely distributed within the hive, influencing the behavior of workers in a variety of ways (Naumann et al., 1991).

For example, QMP can influence how workers raise the "brood" of immature bees. Eggs laid by the queen have, for a time, the potential to develop into either queens or workers. Which path they follow depends on what the current workers do. A future queen is fed a richer diet (including royal jelly) than a future worker. When the present queen is in the prime of life, no new queens are born. But if she is old or sick, the amount of QMP she produces decreases. This causes workers to start feeding royal jelly to some of the larvae, and do other things to prepare for the birth of new queens. These changes can be induced experimentally by removing the queen from a hive. And when Butler and Gibbons (1958) administered QMP to such a queenless colony, the new behaviors were halted. The study shows that QMP suppresses the rearing of queens.

Worker bees have pheromones of their own, the best known of which is the alarm pheromone (Pernal, 2021). There are actually several alarm pheromones, but the most powerful is one secreted by glands adjacent to the stinger. At the approach of an intruder, workers secrete this pheromone and create air currents with their wings to disperse it. And if they sting the intruder, the glands are torn loose with the stinger, releasing additional pheromone that induces other workers to join in the attack. Africanized bees ("killer bees") are especially aggressive in defending their hive primarily because of differences in the amount and effectiveness of their alarm pheromones.

Pheromones are generally considered to fall into two groups. *Releaser pheromones* quickly trigger behavioral responses. An example is the alarm pheromone just described. It summons other workers to immediately join in an attack. *Primer pheromones*, in contrast, work more gradually and indirectly, up- or down-regulating genes and affecting endocrine and metabolic systems. These changes make the animals receiving them better suited to playing their behavioral roles. An example is the queen mandibular pheromone: Transmitted throughout the hive, it conditions different groups of workers for their respective tasks. However, the distinction between releaser and primer pheromones is not a sharp one. Some pheromones have both types of actions. The QMP, though mainly a primer pheromone, has some releaser properties, as shown by its ability to trigger retinue behavior within a few minutes.

This brief survey mentions only a few of the many pheromones used by honeybees. A fuller discussion can be found in reviews by Bortolotti and Costa (2014) and Pernal (2021).

It should be noted that olfaction in honeybees is not only a matter of pheromones. They smell floral and other odors and use them in navigating to a food source. Bees even learn to associate different odors with different locations: Catching a whiff of a remembered smell back at the hive days later, they will immediately return to the now odorless place where that smell was earlier paired with food (Reinhard et al., 2004).

BOLAS SPIDERS

In previous sections, we have considered both pheromones and kairomones. These are two categories of *semiochemicals*: substances secreted or excreted by an organism that influence other organisms (Wyatt, 2014). To qualify as a pheromone, a substance must influence others of the same species. A kairomone, on the other hand, influences members of one or more other species, providing them with some sort of advantage, such as information about a nearby predator. A third type of semiochemical is one that influences members of other species, but in a harmful way. That is, it conveys disinformation, involving deception of some kind. Such a substance is called an *allomone*.

An example of an allomone is the cocktail of organic molecules secreted by bolas spiders such as the widely distributed *Mastophora hutchinsoni*. Females capture prey using a sticky blob at the end of a silk thread, called a bolas because of its resemblance to the traditional South American hunting weapon of the same name.

Haynes et al. (2001) found that the bolas plays a key role in a complex behavioral sequence. When on the hunt, the spider releases a cocktail of odorants that mimics the female sex pheromone of a prey species, such as the bristly cutworm moth *Lacinipolia renigera*. Attracted by the smell, male moths flutter to the area. Their wingbeats are perceived by the spider, who proceeds to the next stage in the sequence: making the bolas. When a moth approaches, she pulls back the bolas and flicks it toward him, so that it swings against and immobilizes him. She paralyzes him with a bite and wraps him in silk for later consumption.

In summary, most insects and spiders live in an Umwelt filled with odors, with semiochemicals playing a prominent role. Even plant odors can influence an insect's behavior—for example, the smell of a flower that attracts bees and butterflies—and can therefore be considered semiochemicals. And sometimes, these signals can be deceptive, as in the case of fragrant orchids that provide no food to insect pollinators (Vereecken & McNeil, 2010). The olfactory landscape is complex, the richest and most challenging component of most arthropods' sensory worlds.

HERMIT CRABS

BEHAVIOR

We will conclude this discussion of olfaction in arthropods with a look at hermit crabs. This group of crustaceans includes both aquatic and terrestrial species. A defining property is that, having a soft and vulnerable abdomen, they appropriate and live in empty mollusc (usually snail) shells. In Chapter 6, we described the care with which they inspect a new shell before "trading up" from an old one that they have outgrown. This behavior becomes more complicated in a social context, where a series of crabs, differing incrementally in size, may wait by an empty shell that is too big for any of them. When a crab that is large enough to use the empty shell shows up and abandons its old one, all the crabs in the vacancy chain move up in rapid succession (Rotjan et al., 2010).

This behavior is largely a visual one, but vision is not their only well-developed sense, for hermit crabs also have an elaborate olfactory system, beginning with olfactory receptors located on the more anterior of two pairs of antennae.

Like other decapods, hermit crabs use sex (and perhaps other) pheromones to send olfactory signals to one another (Okamura et al., 2017), but an equally important function of their sense of smell is to locate food. Rittschof and Sutherland (1986) tested terrestrial hermit crabs (*Coenobita rugosis*) with solutions of various food odorants. These were poured into plastic cups that were embedded up to the rim in a sandy Costa Rican beach where crabs walked about. The crabs were drawn to the odors of mango, coconut, and other fruits, to the odor of rotting gastropod flesh, and even to the presumably unfamiliar odor of horse feces.

In a separate test, paper towels were saturated with stimulus solutions and placed on the beach. Rittschof and Sutherland noted whether the crabs were attracted to them, and whether, when they came in contact with the paper towels, they made eating movements, bringing their claws to their mouth. They were not attracted to sucrose, or to the juice of a freshly killed limpet, but if they stumbled upon paper towels soaked in these solutions, they made vigorous eating motions. The nature of this response suggests that it is induced by the sense of taste, also called *contact chemoreception*, mediated by receptors on the legs and other body parts. Solutions of melon juice and honey had a double action, attracting the crabs *and* inducing eating behavior, implying that they were both smelled and perceived through contact chemoreception.

OLFACTORY SYSTEM

The anterior antennae (or *antennules*) of crabs have on their terminal segment a dense cluster of sensory filaments, the *aesthetascs*, each of which contains the dendrites of olfactory receptor neurons. The crab flicks the array of aesthetascs back and forth to "sniff" its environment. At the microscopic level, the details of this process differ depending on whether the crab is aquatic or terrestrial, but in both cases, the function is to bring fresh samples of odorant into contact with the aesthetascs (Waldrop & Koehl, 2016).

Research on other crustaceans has shown that olfactory receptor molecules on the ORN dendrites have a curious blend of properties. In their immediate response to odorants, they are ionotropic, opening pores in the dendritic membrane, and yet the result of this action appears to be metabotropic, involving G-protein-mediated signaling (Corey et al., 2013).

The brains of crustaceans, like those of many other arthropods, have three main subdivisions: the protocerebrum, the deuterocerebrum, and the tritocerebrum. In a terrestrial hermit crab, the tritocerebrum is small, and the olfactory system is extensively represented in both the protocerebrum and the deuterocerebrum (Harzsch & Hansson, 2008). The axons of ORNs in the antennules go to the large *olfactory lobes* in the deuterocerebrum, where they end in glomeruli. There are about 800 glomeruli in each olfactory lobe, suggesting that there are about this many classes of olfactory receptors.

Axons of glomerular cells leave the olfactory lobes and travel anteriorly into the protocerebrum, where they end in the layered neuropil of the left and right *hemiellipsoid bodies*. These are reminiscent of the mushroom bodies of insects, but whether the similarity reflects a degree of homology, or only convergent evolution, remains to be determined (Harzsch & Hansson, 2008).

Neurogenesis, the birth of new neurons, occurs in specific brain areas of many animals. In decapod crustaceans, the group that includes lobsters and crabs, it is characteristic of the central olfactory pathway, where it occurs in parts of the olfactory lobes and the hemiellipsoid bodies. It has been suggested (Schmidt, 2007) that the additional neurons enable sensory plasticity, allowing these arthropods to learn about novel odors as their olfactory environment changes.

OLFACTION IN MOLLUSCS

BEHAVIOR

In this section, we will consider the sense of smell in molluscs, focusing on cephalopods, the group in which most of the behavioral work has been done. The advanced cephalopods—squid, octopus, and cuttlefish—have well-defined olfactory organs that are on the face near the eyes. The available evidence indicates that their olfactory systems are similar to one another in other ways as well.

Cephalopods use vision, touch, and the chemical senses to identify food. In clear water and bright light, vision plays a major role; and when prey are hiding in a crevice, touch and contact chemoreception work together. But when the prey is out of reach and out of sight, smell is essential.

The value of smell to a hungry octopus was demonstrated by Maselli et al. (2020). They first presented individual octopuses (*O. vulgaris*) with a buffet of food items: anchovies, clams,

and mussels. The octopuses consistently ate the anchovies first, showing that this was their favorite among the choices offered. Next, the researchers presented each octopus with an array of three jars, each containing one of the three foods. The jars were opaque, but perforations in their lids allowed the smell of the prey to reach the octopus. The cephalopod participants seized the jar containing the anchovies, unscrewed the lid, and ate the contents, before turning less enthusiastically to the other jars. Perhaps not surprisingly, the octopuses were not fooled when the experimenters attached photographs of less desirable foods to the jar of fragrant anchovies.

To get an idea of what odors a cephalopod may detect in everyday life, Boal and Golden (1999) allowed individual cuttlefish to settle onto the glass floor of a tank. Looking up from below the tank, the experimenter could observe the rate at which the animal ventilated. Water flowed into the tank from another tank, to which odorants were added. When dead shrimp were placed in the upstream tank, the cuttlefish began ventilating faster within a few seconds. The same reaction occurred when another cuttlefish was placed in the upstream tank, or water that had previously held loggerhead turtles, potential predators of cuttlefish. Several other odorants were also effective. The experiment shows that cuttlefish have an olfactory system capable of detecting a variety of ecologically significant odors that occur in its natural environment.

OLFACTORY SYSTEM

Anatomical study of the olfactory organs of the squid *Lolliguncula brevis* reveals an epithelium populated with olfactory receptor neurons (ORNs) having a variety of morphologies (Emery, 1975b). These are true neurons, with one or more dendrites extending to the sensory surface and an axon that travels to the olfactory lobe of the brain. The ORNs are classified into five types, based on whether there is one dendrite or many, whether the cell has cilia-lined invaginations, and other properties.

Mobley et al. (2008) set out to compare the odorant sensitivities of the different types of squid ORNs by applying a chemical, agmatine, that enters neurons through ion channels whenever the cells are activated by olfactory stimuli. The accumulation of agmatine in some cells but not others reveals which ones responded to the odorant. The experimenters used only a few odorants, but these were sufficient to show that the different classes of ORN differ in their sensitivity profiles.

Gilly and Lucero (1992) measured the occurrence of a particular behavior, jetting, in restrained but awake squid (*Loligo opalescens*). Among the substances that triggered the jetting response when applied to the olfactory organ was squid ink. This response is adaptive, for squid release ink when they are threatened, and an individual whose neighbors are threatened will likely benefit from jetting out of the area. Melanin is the pigment that makes squid ink black, and L-dopa, which is a precursor of melanin as well as of dopamine, also evoked a strong jetting response.

Surprisingly, single-cell recordings from a sample of ORNs showed that they were hyperpolarized by these chemicals, which is to say that neural firing was reduced below its spontaneous level (Lucero et al., 1992). This means either that a reduction in neural activity can trigger the jetting response or that other ORNs, not captured in the experimenters' sample, give a positive response, perhaps mediated by dopaminergic receptors.

Some other chemicals, such as propylparaben, also evoked jetting, whereas others did not. In a control experiment, jetting responses to propylparaben did not occur when the olfactory organ was temporarily anesthetized, proving that they are triggered by smell rather than by vision or touch (Gilly & Lucero, 1992).

In summary, the squid's olfactory system detects a variety of chemicals and is capable of initiating behavioral responses to them. But our understanding of cephalopod olfactory sensitivity and coding is still at an early stage. Andouche et al. (2021) have taken an important step by using genetic analysis to document the diversity of receptor molecule classes in cephalopods and other molluscs. Many of these receptors are ionotropic, like those in arthropods.

A particular co-receptor, IR25, has a long evolutionary history, variants of the gene for it being present in both molluscs and arthropods, and even in nematodes. Andouche and colleagues turned up evidence of this receptor not just in the cuttlefish's olfactory organs, but in its suckers and its fins. These last results show how blurred are the distinctions in molluscs among olfaction, taste, and a more generalized chemical sense.

SUMMARY

The sense of smell is for many animals the most important modality. In vertebrates, it is mediated by olfactory receptor neurons (ORNs) in the olfactory mucosa, inside the nasal cavity. Receptors in the membrane of ORN cilia interact with odorant molecules, causing electrical responses that are conveyed by ORN axons to the olfactory bulb and from there to the olfactory cortex, amygdala, and/or other brain structures.

The receptors themselves are metabotropic and constitute a large family of molecules distinct from the much smaller group that mediate taste. Most ORNs contain receptors of just a single type. The olfactory bulb contains distinctive neuronal clusters called glomeruli, each of which receives fibers from ORNs of a particular type. However, most receptors are responsive to multiple odorants, so identifying an odorant must involve a comparison of signals from different classes of ORNs, and thus from different glomeruli.

In addition to the main olfactory epithelium, many vertebrates, including amphibians, reptiles, and some mammals, have an accessory olfactory epithelium that is lower in the nasal cavity and therefore more accessible to heavier, nonvolatile odorant molecules, including some pheromones. Receptor cells here have microvilli rather than cilia. The two types of cells are interspersed in the olfactory mucosa of fish, but become separate in amphibians.

Vertebrates differ widely in their olfactory abilities and the roles this sensory modality plays in their everyday lives. This chapter explored olfaction in dogs, mice, turtles, frogs, pigeons, zebrafish, and other vertebrates.

Arthropods also have a well-developed sense of smell, with receptors that are ionotropic rather than metabotropic. Pheromones and other odorants play important roles in, for example, the ability of crustaceans to find food, flies to regulate courtship, bees to keep their hive running smoothly, and spiders to lure moths to their death.

Olfaction has been studied in cephalopods, where it is used to locate prey and to avoid predators, for example by jetting in squid. Olfactory organs near the eyes contain ORNs with ionotropic receptors. However, much remains to be learned about olfactory sensitivity and processing in Phylum Mollusca.

With this description of the olfactory sense, we come to the end of the traditional "five senses" attributed by Aristotle to humans. Even for our species, the list is incomplete, since it does not include a variety of sensory experiences such as thirst, balance, and muscular effort. And a list composed of vision, hearing, touch, taste, and smell is even more inadequate to capture the range of senses possessed by animals. In the next chapter, we will describe three of these sensory modalities that we do not have.

10 Senses We Don't Have

In this chapter, we will consider three senses. Two of them involve forms of energy—electricity and infrared radiation—to which we humans are for all practical purposes insensible, because of our high thresholds. We can feel the warmth of sunlight, and the stabbing pain of electric shock, but at the very low intensities of IR radiation perceived by copperheads, and of electricity perceived by sharks, we experience nothing. We don't feel that we are at a serious disadvantage here, since these sensory abilities have evolved in specific animals to address needs that we don't face.

MAGNETIC SENSE

However, there is a sense that we lack entirely, although it plays an important role in the lives of a wide range of animals, including many fish, birds, and insects. And if we had this sense, we would make good use of it. It is *magnetoreception*, the ability to detect magnetic fields. We will begin our discussion with this sensory modality.

The most important magnetic field for animals is that of the Earth itself, the planet being a giant magnet. Lines of magnetic force stream outward from the South Pole, travel northward, and dip into the Earth again near the North Pole. At any location, the local magnetic lines have both an azimuth (roughly northward, although there are local variations) and an inclination (tilting upward or downward).

MAGNETORECEPTION IN VERTEBRATES

Magnetoreception is widespread among vertebrates, being present in all five classes: fish, amphibians, reptiles, birds, and a few mammals. We will focus on two animals in which it has been heavily researched: turtles and migratory songbirds, especially the European robin. Work with turtles has shown how innate magnetic maps enable hatchlings to set out on long, complicated journeys, and how adults are able to find their way to a specific location based on its magnetic signature. Work with birds has addressed similar questions, and also probed deeply into the physiological mechanisms underlying magnetoreception. Describing research on both turtles and birds will give us an understanding of the main issues in the field of vertebrate magnetoreception.

Toward the end of our discussion, we will take a quick look at magnetoreception in the other vertebrate classes: fish, amphibians, and mammals.

TURTLES

One of the most thorough behavioral analyses of magnetoreception in any animal is the body of work on the loggerhead sea turtle (*Caretta caretta*) by Ken and Catherine Lohmann and their colleagues at the University of North Carolina.

DOI: 10.1201/9781003362319-10

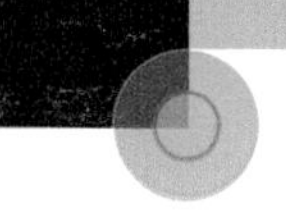

Figure 10.1 A hatchling loggerhead sea turtle, having just emerged from its nest in the sand, moves quickly toward the relative safety of the ocean.

Source: **Used with permission of Shutterstock. Photo Contributor, Salty View.**

Infant loggerheads hatch from eggs buried in sand close to the ocean; they claw their way to the surface, scramble down the beach to the water, and begin a long migratory journey (see Figure 10.1). In a series of studies, the Lohmanns collected hatchlings on a Florida beach as they were starting to emerge from their underground nests, transported them to a nearby laboratory, and tested them in darkness. All turtles were released at the end of each study.

For testing, individual hatchlings were placed in a seawater pool, wearing a harness that was tethered to a movable overhead boom. The direction in which the turtle swam was electronically recorded every 10 seconds. Turtles in a baseline condition swam eastward, the direction from their birthplace to the ocean.

Then, the researchers activated magnetic coils outside the pool, changing the magnetic field impinging on the animal. In one study (Lohmann & Lohmann, 1994), they manipulated a single parameter of the magnetic field: its *inclination*. Inclination is the degree to which the magnetic lines of force deviate from horizontal. These lines jut abruptly out of the ground at the south pole and tilt outward from the Earth as they travel northward across the southern hemisphere; they are roughly horizontal as they cross the equator; and then in the northern hemisphere, they tilt downward again. The turtles in this study were at all times in the northern hemisphere, where there is a correlation between the tilt of the magnetic field at a given location, and the location's *latitude*.

Before looking at the results of this study, let's look at the migratory path followed by loggerhead turtles from the southeast coast of the United States once they are in the Atlantic Ocean. They travel clockwise in a vast, roughly circular path called a *gyre*, created by the Gulf Stream and other currents (see Figure 10.2). In some parts of the gyre, they can rest and drift on the current, but at other points they must swim vigorously to stay on course. For example, as they approach Europe from the west, there is a "fork in the road," with some currents branching off and heading northward. If turtles are caught in this northerly drift, they will be carried into very cold water and die; they must paddle southward to stay in the gyre. At this danger point, the inclination of the magnetic field is about 60°. A similar, but less deadly, branching of currents occurs in the southwest corner of the gyre, as the turtles move toward the region of their birth: here, where magnetic field inclination is about 30°, a drift could carry them into the Gulf of Mexico rather than to the east coast of Florida. To prevent this, they must swim northward.

To see whether magnetic inclination influences turtles' swimming (in the laboratory pool) in a way that would promote survival in the ocean, the investigators adjusted inclination and measured the animals' responses. When inclination was 60°, turtles swam south-southwest, a direction that would insure their remaining in the gyre; and when inclination was 30°, they swam northeast, avoiding a not altogether unpleasant, but ultimately futile, sojourn in the Gulf of Mexico. Inclinations above 60° or below 30° evoked no consistent response from the turtles, perhaps because animals so far to the north or south of the gyre are unlikely to survive in any case; any resources used to maintain sensitivity to these extreme values would be wasted.

In summary, this study shows that hatchling loggerhead turtles can perceive and discriminate different magnetic field inclinations. The fact that they are able swim in an appropriate direction means that they have some sort of *magnetic compass*. Perhaps turtles take as "north" the direction in which the magnetic field is tilted downward.

Using similar methods, Lohmann and Lohmann (1996) found that young loggerheads can also perceive the overall *intensity* of the magnetic field. Intensity, like inclination, tends to increase with distance from the equator, but the

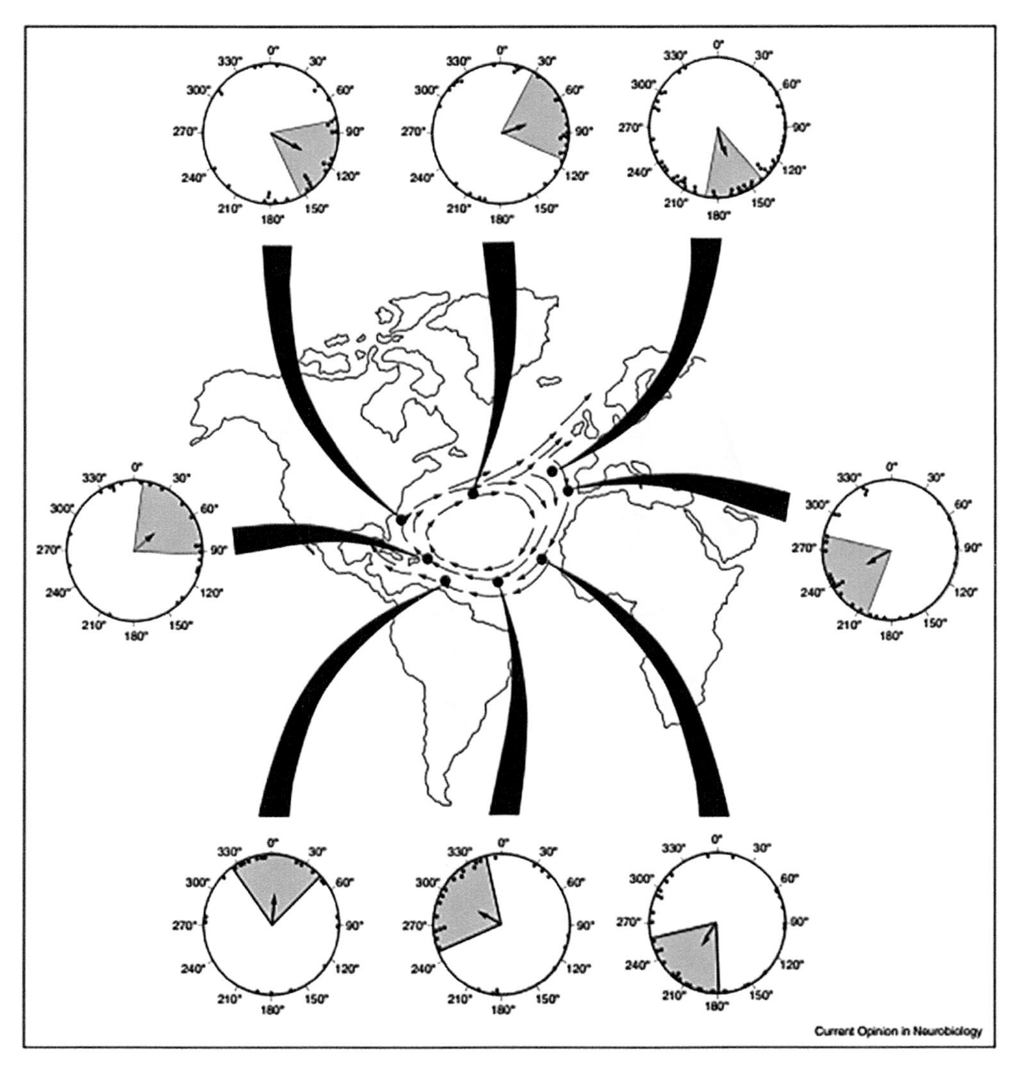

Figure 10.2 When exposed in the lab to magnetic fields replicating those occurring at specific locations in the North Atlantic subtropical gyre (black dots on map), hatchling loggerhead sea turtles swam in directions that would keep them in the gyre if they were actually in the ocean. In the orientation diagrams, small dots indicate the directional response of individual turtles, and the arrow shows the mean response.

Source: **Reprinted with permission from Lohmann, K. J., Putman, N. F., & Lohmann, C. M. F. (2012). The magnetic map of hatchling loggerhead sea turtles. *Current Opinion in Neurobiology*, *22*, 336–342. Used with permission of Elsevier Science & Technology Journals; permission conveyed through Copyright Clearance Center, Inc.**

two parameters are only moderately correlated, so being able to perceive both gives a turtle additional information about its location. In fact, each point along the turtle's route is characterized by a unique combination of inclination and intensity—its *magnetic signature*. It is perhaps by taking both parameters into account that the animal is able to discriminate locations that differ in longitude, even if they are at the same latitude (Putman et al., 2011).

In subsequent studies, hatchling turtles in a laboratory tank were presented with magnetic fields exactly duplicating those they would encounter at locations in their oceanic voyage. In each case, the animals swam in the direction that, interacting with the pull of the current, would keep them in the gyre (see Figure 10.2). The fact that the promptings of their magnetic sense are useful throughout their far-flung

journey suggest that the links between magnetic fields and directional responses are part of an overall spatial framework, an inherited *magnetic map* (Lohmann et al., 2012).

Loggerheads are long-lived turtles, often living beyond the age of 50. Hatchlings are only about 5 cm in length, but as they mature, they grow to a length approaching a meter. As they develop, they travel widely and return regularly to preferred feeding areas. At about age 30, they reach sexual maturity and mate in coastal waters. After mating, females seek out the location where they were born to lay their eggs, a behavior called *natal homing*.

While multiple senses may contribute to the success of natal homing, magnetoreception plays a major role, as shown in a population study by Brothers and Lohmann (2015). The study makes use of changes over time in Earth's magnetic field. It is well known that the field reverses every so often, with the magnetic north pole becoming the south pole and vice versa; the last of these reversals occurred about three-quarters of a million years ago. Less well known is the fact that smaller changes in the field, called *secular variation*, occur very frequently. For example, the roughly east-west line running through locations where inclination has a particular value, say 57°, may move slightly northward or southward from year to year; other isolines, such as that for 58°, may move in the same direction but by a different amount. The result is that two isolines may be more widely separated in some years than in other years.

Brothers and Lohmann compared a record of these changes with the nesting patterns of female loggerheads along the east coast of Florida, which has many beaches. They found that between isolines that were closer together than in earlier years, the density of nests was higher than average, whereas it was lower than average between isolines that had become more widely separated.

The authors interpreted their findings to mean that female loggerheads are imprinted at birth on the magnetic signature of the beach where they were born. Thirty years later, this lifelong memory draws them back to a beach with this same signature, whether it is actually the same beach or, because of secular variation, a beach north or south of their birthplace. Females imprinted with magnetic signatures between, say, 57° and 58° inclination will nest on a beach where the inclination is between these values. This will result in a high density of nests if the 57° and 58° isolines have moved closer together over time, but a low density if the isolines have moved farther apart.

The importance of this study is the support it gives to the idea that loggerheads are imprinted with a magnetic signature of their birthplace.

But learning of magnetic signatures is not confined to hatchlings, as shown by the fact that partially grown turtles choose foraging areas and return to them at intervals, guided at least in part by magnetic signatures. For this study, Lohmann et al. (2004) captured green turtles (*Chelonia mydas*) at their feeding grounds on the Florida coast, transported them to a nearby laboratory, and tested them in a larger version of the circular tank surrounded by magnetic coils used in their research with hatchlings. When presented with the magnetic signature of a site several hundred kilometers north of the testing site, the tethered turtles swam southward, and swam northward in response to the signature of an equally remote site south of their current location. The animals, in other words, appear to be trying to return to their foraging site. The results imply that sea turtles have available for use, if not for conscious perception, a cognitive map that contains information about north-south changes in magnetic signatures.

The extensive work with turtles helps provide us with an understanding of how magnetoreception can be used in navigation, giving animals directional information (a magnetic compass) and information about their spatial location within a larger framework (a magnetic map). But how are magnetic fields detected in the first place? This is a challenging question that has not yet been fully answered. The answer is undoubtedly different in different animals, and some animals appear to use more than one method.

BIRDS

The question has been most thoroughly researched in birds, and so it is to them that we now turn. We will use as our primary example the European robin (*Erithacus rubecula*) because of the long tradition and large body of research associated with this widely loved songbird. Many robins migrate, spending the warmer months in northern Europe and the winters in southern portions of the continent. Their migration is usually nocturnal, guided in part by a stellar compass of the kind described in Chapter 5. But an additional method of guidance—magnetoreception—is also at work.

This can be shown by observing the behavior of robins confined in a windowless room during their migratory season. Especially at night, they face the direction in which they would fly off, if at liberty to do so. In a classic study using this approach, Wolfgang and Roswitha Wiltschko (1972) placed individual robins in an octagonal cage with eight radially arranged perches. The room's artificial lighting maintained a normal light/dark cycle, and contact of a robin with each of the eight perches was electronically recorded throughout the night. Large coils outside the cage enabled the Wiltschkos to control the magnetic field to which the birds were exposed.

The experiments were conducted in spring, at a time when the robins would normally have been migrating toward the northeast. And with the magnetic field adjusted to normal values for the laboratory's location, the birds, on average, faced in that direction.

This result suggests that the robins were responding on the basis of the magnetic field, but doesn't prove it: They might have been relying on subtle cues in some other sensory modality—faint sounds or smells, for example. To test the role of the magnetic field directly, the experimenters reversed its direction. In this condition, birds spent the night facing southwest, proving that they were relying on their magnetic sense.

Later work confirms the results of this early study. In place of the eight perches used by the Wiltschkos to measure a bird's preferred direction, more recent research has used a simpler device called an *Emlen funnel* (see Figure 10.3). The bird is placed on a small platform at the bottom of the funnel, which is lined on the inside with coated paper. Birds with migratory restlessness hop and scramble up the funnel in a particular direction, making scratch marks on the paper that are later measured by the experimenters.

In one recent study, Packmor et al. (2021) used a different migratory songbird, the Eurasian reed

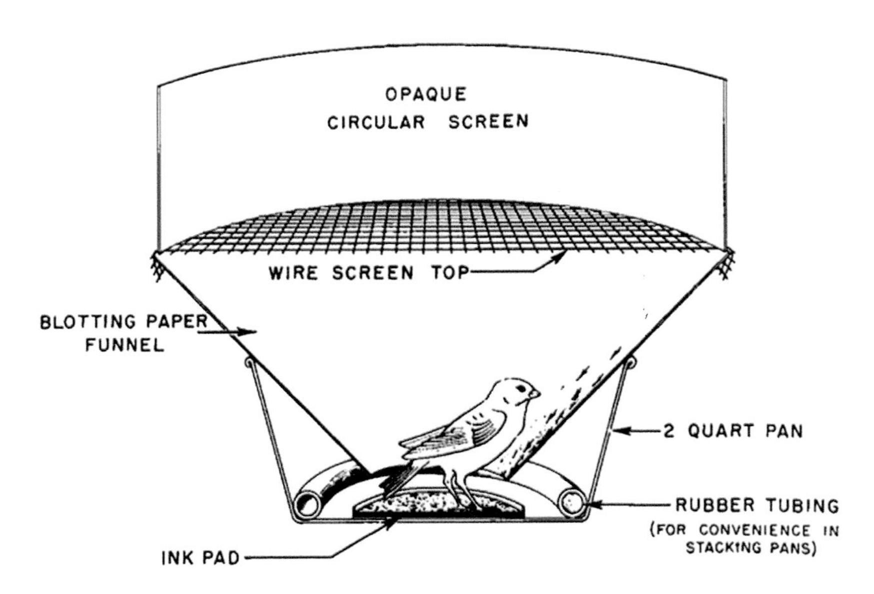

Figure 10.3 An Emlen funnel. During migration season, a restless bird hops repeatedly onto the sloping sides of the funnel, leaving marks that indicate the direction in which it would fly off if allowed.

Source: **Reprinted with permission from Emlen, S. T., & Emlen, J. T. (1966). A technique for recording migratory orientation of captive birds. *The Auk*, *83*, #3, 361–367. Reprinted by permission of Oxford University Press on behalf of the American Ornithological Society; permission conveyed through Copyright Clearance Center, Inc.**

warbler (*Acrocephalus scirpaceus*), to study the effect of a tiny but powerful magnet glued to the bird's forehead. The rationale for the experiment was that the magnetic field given off by the magnet would swamp any magnetoreceptors located in the bird's face, preventing it from responding to the weaker geomagnetic field. Each bird participated in three conditions: one with the magnet, a second with a sham magnet of the same size and weight, but no magnetic properties; and a control condition with nothing attached to the face. In all three conditions, a translucent cover was placed over the funnel so that the bird received only diffuse light. All testing was done during the autumn migration season.

In the control condition, the birds oriented normally, facing southeast. In the sham condition, they did the same, showing that any annoyance at having something glued to their forehead did not interfere with their ability to orient. With a magnet attached, however, the birds were disoriented, clambering up the funnel in random directions. The experiment indicates that birds have a magnetic sense and that it helps them orient for their migratory journey. At the end of this study, all attachments were removed and the birds were released near their capture site, in time to continue their migration.

The Role of Magnetite

But how are birds actually detecting the magnetic field? One contributing factor is thought to be the presence of magnetite (Fe_3O_4) and other iron oxides in sensory neurons inside the beaks of robins and other birds. In the dendrites of these nerve cells, particles of the compounds are arrayed in rows that may serve as miniature magnets, depolarizing the dendrite when torqued by a magnetic field (Falkenberg et al., 2010). The neurons belong to the ophthalmic branch of the trigeminal nerve, which primarily mediates sensations of touch and pain in the upper face. (The ophthalmic branch is named for the fact that it travels near, and serves tissues surrounding, the eye; it does not carry visual information.) Electrophysiological recordings in another migratory songbird, the bobolink, show that these nerve cells primarily encode the intensity of the local magnetic field (Semm & Beason, 1990).

Many birds, including robins, gradually acquire a magnetic map of the area where they live or through which they migrate (Lohmann et al., 2022). Since Earth's magnetic field varies from place to place, birds may use its intensity or inclination at their current location—perceived with magnetite-containing neurons—to help determine where they are on the map (Semm & Beason, 1990; Wiltschko & Wiltschko, 2013).

If there are large magnetic field variations within a given area, caused for example by iron deposits in the soil, signals from beak magnetoreceptors may actually be disorienting: Wiltschko et al. (2010) found that pigeons whose upper beak had been locally anesthetized, eliminating these signals, flew home from a magnetic anomaly more quickly and directly than control birds.

Further evidence that magnetite-containing neurons play a role in navigation comes from research in which robins received a brief, strong magnetic pulse that recalibrated their magnetite (Holland, 2010). The pulse was oriented from west to east; that is, it was perpendicular to Earth's magnetic field. Birds were caught, pulsed, and then released. The direction in which they flew away from the release point was monitored from an observation tower.

Robins in a control group, which did not receive a pulse, flew away from the release point in a northwesterly direction, consistent with their normal migration route through the area in which the research was done. Pulsed robins, in contrast, flew off toward the northeast, indicating that magnetite, changed by the pulse, was sending altered signals to the brain (see Figure 10.4). Studies with other birds show that this aftereffect dissipates after a few days.

In summary, robins navigate partly on the basis of signals from magnetite-containing neurons. But this system responds primarily to the strength and inclination, rather than the azimuth, of magnetic fields, and is therefore not a very precise compass; most researchers think that its more important contribution is to the robin's cognitive map. Moreover, it is not essential to the robin's magnetic sense, since birds in a dimly lit room without a view of the sky can usually orient normally, even if sensory signals from the beak are blocked (Zapka et al., 2009).

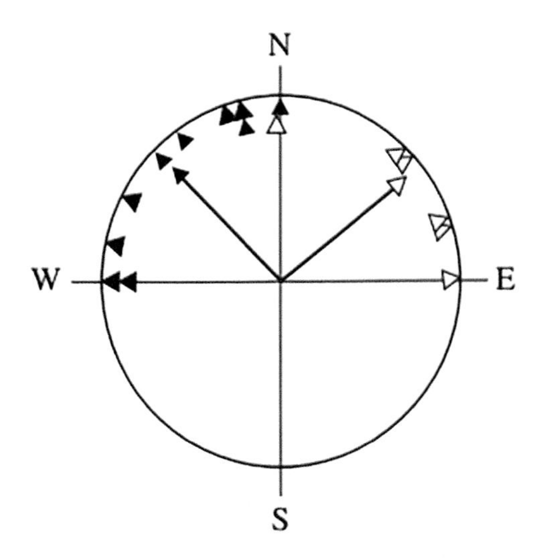

Figure 10.4 The effect of a magnetic pulse on the migratory orientation of robins. The directional response of each bird is represented by a triangle. Control robins who received no pulse oriented in their normal migratory direction (filled triangles). In contrast, the orientation of birds who received a magnetic pulse perpendicular to the geomagnetic field was shifted by about 90° (unfilled triangles). Arrows show the mean bearing of each group.

***Source:* From Holland, R. A. (2010). Differential effects of magnetic pulses on the orientation of naturally migrating birds. *Journal of the Royal Society Interface*, *7*, 1617–1625. Used with permission of The Royal Society (U.K.); permission conveyed through Copyright Clearance Center, Inc.**

However, if the beak is anesthetized and they are tested in total darkness, the birds are completely disoriented (Stapput et al., 2008). These results imply that there is a second component of the robin's magnetic sense, and that it requires light.

The Role of Cryptochrome

There is now evidence that this second process involves a chemical called *cryptochrome* that is located in the bird retina (Hore & Mouritsen, 2022). A cryptochrome molecule, like a molecule of visual pigment, has a portion that absorbs light and is changed by it. This portion of the cryptochrome molecule is called its *chromophore*. When the chromophore absorbs a photon, it reacts by seizing an electron from a neighboring part of the molecule. This unpaired electron makes the chromophore what chemists call a *radical*. Other electrons shift within the cryptochrome molecule so that a portion distinct from the chromophore is deprived of one of its electrons; the unpaired electron left behind makes this portion of the molecule a second radical.

This *radical pair* gives the activated cryptochrome molecule special properties. All electrons spin. If the two electrons making up the radical pair spin in opposite directions, this arrangement is called a *singlet*; if they spin in the same direction, they constitute a *triplet*. The radical pair spontaneously changes from the singlet form to the triplet form, and back again, millions of times per second. Either form of chryptochrome can initiate a cascade of chemical steps that leads to neural activity. But cryptochrome in its singlet form is also able to settle back into the nonradical ground state.

What does all this have to do with magnetoreception? The answer is that a magnetic field can change the proportions of the singlet and triplet forms of cryptochrome by promoting the conversion of the singlet to the triplet form. This reduces reversion to the resting state, thus permitting more neural activation to occur.

Evidence that cryptochrome has the magnetic sensitivity needed for it to play a role in magnetoreception comes from a study by Xu et al. (2021). Moreover, cryptochrome is present in the retina when and where it needs to be to aid in bird migration, the activity that relies most heavily on magnetoreception. By measuring mRNA used in the production of cryptochrome, Günther et al. (2018) showed that in European robins, more than twice as much cryptochrome is being made during migration seasons (fall and spring) as during summer and winter; in contrast, chickens, which do not migrate, produce cryptochrome at a steady rate year round.

Remarkably, the cryptochrome is located in the outer segments of specific receptors: long-wave-sensitive single cones and double cones (which also contain long-wave-sensitive pigment). This co-localization of blue-sensitive cryptochrome with red-sensitive visual pigment is consistent with the idea that these cones have two separate functions—magnetoreception and color vision—that are used at different times.

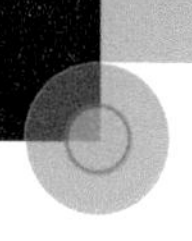

When flying at night, robins are using magnetoreception to navigate; during the day, when they are resting between flights, they use color vision to look for food and perhaps for other robins. It is not clear how the robin's visual system recognizes which type of message these cones are sending at a given time, but the very complex wiring of the avian retina (Günther et al., 2021) and the fact that cryptochrome signals are routed to a special region of the robin's brain (Zapka et al., 2009) suggest that this determination is made at an early stage of processing.

It will be recalled from Chapter 5 that night-migrating birds such as the robin have a star compass, which they must learn to use by observing the rotation of the starry sky and (if they are in the northern hemisphere) designating the center of its rotation as North. Does the magnetic compass also depend on learning, and how does it interface with the star compass? To answer these questions, Alert et al. (2015) hand-raised newly hatched robins indoors, without a view of the sky but with exposure to Earth's magnetic field.

As migration season approached the following spring, the young birds were divided into two groups: an experimental group that was moved to an outdoor aviary with a wide view of the sky, and a control group that continued to live indoors in a windowless room. After two weeks—enough time for the experimental group to develop their star compass—all the birds were again tested using Emlen funnels, without a view of the sky. The experimental birds now had a functioning magnetic compass, but the control birds did not. These results show that the magnetic compass does not develop on its own; it needs to be calibrated against the star compass.

In summary, the European robin and other birds appear to have at least two sensory mechanisms for detecting magnetic fields: a cryptochrome-based process that contributes mainly to a magnetic compass, and a magnetite-based process that contributes mainly to a magnetic map (Wiltschko & Wiltschko, 2019). We have considerable evidence of their existence, but the details of how they operate are not well understood. We will now consider the pathways that these two types of magnetic information follow in the brain.

CENTRAL MAGNETOSENSORY PATHWAYS IN BIRDS

The Magnetite Pathway

Magnetite receptors in the beak are neurons whose axons contribute to the ophthalmic branch of the trigeminal nerve. This branch travels to the brainstem, where its fibers terminate on cells in the trigeminal nuclei. Some of these second-order cells respond to magnetic stimulation, as shown by Heyers et al. (2010). These investigators divided European robins into two groups, and placed each bird in a cage surrounded by magnetic coils. Those in the experimental group were exposed for 3 hours to a constantly changing magnetic field, while those in the control group had no magnetic stimulation at all, the Earth's magnetic field being canceled out by the coils. At the end of the exposure, the birds were sacrificed, and their brains were prepared for study under the microscope.

The experiment depends on a particular gene that is upregulated in active neurons, creating more of a protein called ZENK. By looking for differences in the amounts of ZENK in the two groups of robins, Heyers and colleagues could identify groups of cells that had been activated by the magnetic stimulation. They found that such cells were located in specific portions of the trigeminal nuclei.

What happens next to the magnetic information conveyed by these cells has been discovered by Kobylkov et al. (2020). Working with another night-migrating songbird, the Eurasian blackcap (*Sylvia atricapilla*), the researchers used a combination of procedures to identify, as Heyers and colleagues had done, cells in the trigeminal nuclei activated by magnetic stimulation, and learn to what other brain regions these neurons send their axons.

This experiment revealed that a particular part of the forebrain, the *telencephalic frontal nidopallium*, receives a projection from the cells in the trigeminal nuclei that are activated by magnetosensory inputs. The nidopallium is a region of the bird's brain that is widely thought to be involved in both cognition and motivation, so the identified subregion is well positioned to play a role in enabling birds to develop magnetic maps that will help them reach their navigational goals. Other trigeminal pathways, concerned with somesthetic rather than magnetic information, project to other brain regions.

The Cryptochrome Pathway

Magnetic information that depends on cryptochrome begins in the retina and travels in a subset of optic nerve fibers to the dorsal geniculate complex in the thalamus, a structure roughly equivalent to the mammalian lateral geniculate nucleus. Magnetosensory signals are relayed from here to a specialized forebrain area called *cluster N* (Heyers et al., 2007). Cluster N is part of the visual Wulst, a large region of the bird's forebrain devoted to visual processing, as described in Chapter 2.

The importance of cluster N for the magnetic compass was shown by Zapka et al. (2009). The authors lesioned this structure by injecting ibotenic acid, a neurotoxin, into it on both sides of the brain while the birds were anesthetized. Subsequent behavioral tests, inside dimly lit wooden huts with no view of the sky, showed that the birds were no longer capable of orienting properly on the basis of the magnetic field. However, they were still able to use their star compass when given a view of the sky, showing that the cluster N lesions had not caused a general visual disturbance or a cognitive impairment regarding compasses. Interestingly, other animals in whom the ophthalmic branch of the trigeminal nerve had been cut (but cluster N left intact) showed no impairment of their magnetic compass. This last result supports Zapka and colleagues' view that the magnetite pathway contributes mainly to the mapping component of bird navigation, rather than the compass component.

The Vestibular Pathway

Recording from individual neurons in the brainstem of pigeons, Wu and Dickman (2012) made the surprising discovery that there are magnetoresponsive cells in the *vestibular nuclei*, brainstem structures distinct from the trigeminal nuclei and the lateral geniculate nucleus. Neural activity was influenced by the azimuth, inclination, and intensity of the magnetic field, and each cell responded most vigorously to a specific direction, which varied from cell to cell.

The receptors that provide the input to these cells remain a mystery. They appear to be located in the inner ear, where there is no light, so the receptors cannot be of the cryptochrome variety. Additional research is needed to determine whether they contain magnetite or use some other mechanism of activation.

Still another unresolved question is whether these different streams of magnetoreceptive information converge somewhere in the brain, and how their signals are functionally integrated as birds navigate by means of magnetic compasses and maps (Malkemper et al., 2020).

ANTHROPOGENIC DISRUPTION OF BIRD MAGNETORECEPTION

It might seem that the ability of robins to perceive the Earth's magnetic field would be immune to anthropogenic factors, but such is not the case, as Engels et al. (2014) discovered when they attempted to measure the preferred direction of birds on a university campus during spring migration season. The birds were enclosed in wooden huts with no view of the sky, and were therefore expected to use their magnetic compasses. But surprisingly, their directional preferences, measured with Emlen funnels, were haphazard.

Given the campus's urban setting, the investigators suspected that electromagnetic noise, for example from electronic equipment and AM radio signals, might be interfering with the birds' perception of Earth's magnetic field. To test this possibility, they electrically shielded the huts with grounded aluminum plates. This maneuver greatly reduced electromagnetic noise within the huts without affecting the steady magnetic field. And remarkably, it enabled the robins to use their magnetic compass. It is not yet known whether the noise interferes with the magnetite component or the cryptochrome component of the robin's magnetic sense. But the results of this study raise the possibility that electromagnetic noise of human origin poses a risk to birds that are using magnetic compasses as they attempt to navigate past heavily populated areas.

Now that we have taken a detailed look at magnetoreception in a reptile and a bird, let's briefly consider what is known about this sensory modality in representatives of the other vertebrate classes.

FISH

The ability of fish to detect the geomagnetic field, and use it in navigation, has been studied for more than half a century (Naisbett-Jones & Lohmann, 2022). Salmon have been the focus of much of this work, because of widespread interest in their challenging final migration, as they swim and leap upriver to reach the stream where they were hatched, and spawn there. But salmon actually navigate at all stages of their lives. As *fry* just beyond the hatchling stage, they emerge from the gravel nests in which they spent their first days, and navigate up- or down-river to a lake where they can slowly mature; next they migrate downstream to the ocean, in which they must find their way for several years; and then they begin their arduous final journey up the same river.

Collectively these journeys rely on a combination of sensory mechanisms, including a celestial compass and olfactory memory. But magnetoreception also plays a role, as discovered by Quinn (1980). This investigator caught sockeye salmon fry (*Oncorhynchus nerka*) and placed them in the center of a laboratory tank. The tank had four arms that extended in different directions; once a fish swam into an arm, baffles prevented its return to the central chamber. After 45 minutes, Quinn tallied the number of fish in each arm to determine their preferred directions; he found that, on average, they swam in the direction in which they would normally traverse the lake. And when the magnetic field in the tank was turned 90°, the preferred orientation of the fry changed in the same direction and by the same amount. This result proves that salmon have a magnetic compass.

Putman et al. (2014) showed that magnetic information is also used by young salmon when they enter the ocean for the first time. Chinook salmon (*Oncorhynchus tshawytscha*), less than a year old, were reared in a hatchery at the mouth of a river. At the age when they would normally begin their oceanic journey, individual fish were placed in large buckets in which their orientation could be observed. Magnetic coils exposed the fish to magnetic fields corresponding to either the northern or the southern edge of their population's normal feeding grounds in the Pacific Ocean. Each of these magnetic signatures had appropriate values of both inclination and intensity.

Individuals presented with the magnetic field appropriate to the northern edge of their foraging range oriented southward, while those exposed to the southern magnetic field faced northward. In other words, the fish responded in a way which would keep them, if they were actually in the ocean, within their normal foraging area. Control experiments using unrealistic combinations of field inclination and intensity produced haphazard responses, suggesting that it is to particular combinations of the two parameters that the fish respond. The results show that these animals have not only a compass but also a magnetic map of sorts; and since they have never undertaken a journey of any kind, let alone an oceanic one, the map must be innate.

But what physiological mechanisms mediate the magnetic sense of fish? In rainbow trout (*Oncorhynchus mykiss*), there is evidence for the existence of magnetite magnetoreceptors. Walker et al. (1997) demonstrated that these fish have a magnetic sense, by training them to detect a magnetic anomaly—a small area of increased magnetic field strength—in the center of a tank. Fish were rewarded with a squirt of liquid fish food for hitting a response bar, but only when an anomaly was present at the same location. Over the course of several sessions, the trout gradually learned to distinguish between the presence and absence of the anomaly.

The authors subsequently made electrophysiological recordings from various sensory nerves in the head, and found some nerve fibers that responded to changes in magnetic field intensity. These fibers belonged to the ophthalmic branch of the trigeminal nerve, and anatomical tracing revealed that they originated in the fish's snout, close to but not in the olfactory mucosa. Later work by the same research group (Diebel et al., 2000) found chains of magnetite crystals associated with these nerve endings, and measurements with a microscopic probe indicated that the chains had magnetic properties.

These findings in trout, taken together with other evidence, indicates that the magnetic sense of a variety of fish depends, at least in part, on a magnetite system. And the anatomical similarity of the trout's putative magnetoreceptors, with their chains of magnetite particles, to similar receptors in the beaks of birds discovered by Falkenberg et al. (2010), suggests that a magnetite-based system is both very old and very widespread among vertebrates.

AMPHIBIANS

The amphibian whose magnetic sense has been most extensively studied is the Alpine newt (*Ichthyosaura alpestris*). This lively, orange-bellied animal is widespread in Europe; it spends much of the year on land but returns to ponds and streams during breeding season. It appears to have both a magnetite-based map capability and a magnetic compass that is light-dependent and therefore suspected to involve cryptochrome. While this newt does not swim oceans or fly between continents, it does wander for kilometers before returning to its home pond, and in these short journeys it has good use for magnetic navigation.

The fact that the newt's compass is light-dependent means that an experimenter can turn the compass on and off by placing the animal in light or in total darkness. In a series of ingenious experiments, Diego-Rasilla and Phillips (2021) showed that Alpine newts periodically use their magnetic compass, presumably in conjunction with a map, to take stock of their location. The investigators collected newts from ponds in Spain's Cieza Mountains, and transported them in light-tight containers to a testing site several kilometers away. Once at the testing site, they were held under diffuse natural light until testing. Tests were carried out in a circular arena, under dense foliage with no view of the sky. A newt was released in the center of the arena, and was free to move about; in fact, newts walked directly from the center to the wall of the arena. The direction of this initial walk was taken to indicate the direction in which the newt would have journeyed if not confined in the arena.

Some animals were tested on the day they were collected, while others were tested on the following day. Those held overnight walked consistently in the direction of the pond from which they had been taken, while those tested on the same day they were taken walked in random directions. This result has several possible explanations. For example, the newts in the latter group might have still been stressed by their capture at the time of testing. Or perhaps the newts kept overnight at the testing site had greater opportunity to update their position than the latecomers.

To sort out the possibilities, Diego-Rasilla and Phillips carried out a second experiment, with two new groups of newts. Two testing sites were used. Half the newts were transported from the pond to one testing site, and the others were taken to the other site. There they remained overnight, exposed to diffuse natural light. In the morning, each group was transported in a light-tight container to the other site, and tested after a two-hour period of acclimatization. The results were startling. Each newt acted as if it were still at the place where it had spent the previous evening: That is, it oriented in the direction of its home pond from that false test site, rather than from the actual test site.

Apparently the newts had taken stock of their location while at the false test site, and retained this memory despite a change in location before testing. To determine exactly when they encoded this information, the investigators carried out a third experiment in which newts were kept in total darkness—which is to say, with their magnetic compass deactivated—at the false test site, being briefly exposed to natural light only during short time windows. The results showed that the crucial period is a half-hour or so of evening twilight, when the geomagnetic field is strongest and least variable. Diego-Rasilla and Phillips propose that at that time the newts use their compass and mapping abilities in combination, so that their location can be determined and remembered for 24 hours. Twilight may be an optimal time for navigational updating in other animals as well (Willis et al., 2009), given that it allows compass readings to be compared with a very salient visual cue to direction, the location on the horizon of the setting sun. In the following section, we examine the possibility of such twilight calibration in a mammal.

MAMMALS

There are occasional published reports (e.g., Baker, 1980; Chae et al., 2019) of magnetic sensitivity in humans, but so far, the evidence for this is not compelling. One of the few mammals for which there *is* strong evidence of magnetoreception is the bat. To determine whether bats, like newts, can compare visual and magnetic compass information at twilight, Holland et al. (2010) carried out an elaborate experiment.

They caught greater mouse-eared bats (*Myotis myotis*) at their home cave in northern Bulgaria and transported them to the laboratory in cloth bags that prevented vision of their surroundings. In the late afternoon of a subsequent day, the bats were placed in transparent containers allowing them a full view of the sky and horizon, where they remained for an hour and a half that included sunset. During this time, half the bats (the experimental group) were exposed to a magnetic field turned 90° clockwise, so that a compass placed in the field pointed east; the other half (control group) experienced an unaltered magnetic field.

Following these experimental manipulations, the bats were driven (again, in cloth bags) to a release site some 25 km north of their home cave, fitted with tiny radio transmitters, and released. The experimenters were able to track each bat's nocturnal flight as it set out to reach home. The bats were fed mealworms before their departure, so as to reduce their interest in foraging and make homing their primary motivation.

Animals in the control group headed south, directly toward home. But those in the experimental group headed *east*. To explain their results, Holland et al. (2010) hypothesized that every evening at sunset, the bats calibrate their magnetic compass with respect to the direction of the setting sun, their "gold standard" for west. For the control group, this may have required some minor tweaking, but nothing more. But bats in the experimental group had to turn their internal compass counterclockwise, compensating for the clockwise turn of the artificial magnetic field. Later that night, all bats were tested in the natural magnetic field, but with their latest sunset calibration still in effect. So when those in the experimental group attempted to fly south, they actually flew east.

But do the results really have anything to do with sunset? Perhaps the distorted flight path of the experimental bats is simply a behavioral rebound from the abrupt changes in their magnetic field. To answer this question, the researchers repeated the experiment using new groups of bats, but carrying it out later at night once all traces of sunset were gone. Now the experimentally altered magnetic field had no effect: Both experimental and control animals flew accurately toward home. The data support the theory of Holland and colleagues that bats who leave their caves at sunset, or shortly thereafter when the horizon is still glowing in the west, calibrate their magnetic compass by comparing it with solar cues.

But research on bat magnetoreception is still in its early stages, and the underlying sensory mechanisms are not well understood. Magnetic-pulse experiments indicate, in the case of the big brown bat, that magnetite is involved (Holland et al., 2008), but where it is located, whether a cryptochrome channel is also involved, and where in the bat's brain these signals are processed, are questions that remain unanswered.

MAGNETORECEPTION IN ARTHROPODS

CRUSTACEANS

As denizens of the ocean floor who venture out to forage at night and then return to their den, lobsters would seem likely to benefit from a magnetic sense. In fact, these crustaceans do have one, as shown by Boles and Lohmann (2003). Caribbean spiny lobsters (*Panulirus argus*) were captured in the coastal waters of the Florida keys, and transported by boat to testing sites. During this boat ride the lobsters were in covered, opaque containers that prevented them from seeing their surroundings, and the boats followed circuitous routes to foil any attempt by the lobsters to keep track of their journey. The next day the animals, temporarily blindfolded by rubber caps on their eyestalks, were tested in a circular arena where the direction in which they walked was measured. Remarkably, the lobsters consistently walked in the direction of the place where they had been captured. The results imply that the lobsters were using a magnetic sense, combining information about both orientation (a magnetic compass) and location (a magnetic map) to achieve what the investigators call *true navigation*.

To test this interpretation, researchers carried out a second experiment in which two groups of lobsters were tested at a single testing site near the capture site, but were subjected to different magnetic fields. One artificial field duplicated that naturally occurring at a site 400 km north of the testing site; the other mimicked the signature of a location 400 km to the south. The two groups of lobsters walked in opposite directions: the first group southward, the second group northward. Since the magnetic field was the only difference between the two groups, it is clear that their navigational behavior was being determined by their magnetic sense.

Ernst and Lohmann (2016) later showed that the preferred walking direction of lobsters can be modified by a brief, strong magnetic pulse, a result indicating that their magnetic sense depends, at least in part, on magnetite-based receptors.

INSECTS

Although a magnetic sense is thought to be present in a variety of insects, compelling evidence for it remains elusive.

Bees

Karl von Frisch (1967) labored for decades to demonstrate magnetic sensitivity in honeybees using a series of clever experimental designs, but consistently obtained negative results. However, some later researchers reported intriguing but indirect evidence of bee magnetoreception, such as the observation that bees' timekeeping appears to be influenced by daily fluctuations in the geomagnetic field (Gould, 1980). But a critical analysis by Baltzley and Nabity (2018) of a psychophysical attempt to measure bees' magnetic thresholds (Kirschvinck et al., 1997) shows that the issue is still unresolved.

Ants

In several species of ants, there are indications of magnetic sensitivity, but it is not clear how they might use this information. Let us consider as an example the fire ant, *Solenopsis invicta*, known for its painful sting. It is native to South America, but humans have spread it to other continents.

Anderson and Vander Meer (1993) studied the effect of a magnetic field on the behavior of fire ant foragers as they leave their nest, come upon a food source, and bring news of it back home. As this two-way foot traffic increases, a pheromone trail is laid down.

Anderson and Vander Meer's experiment was conducted in darkness, except for dim red light that is invisible to ants. The testing arena was a tray on which a petri dish containing an ant colony was placed. Once the ants were acclimated to these surroundings, a dead cockroach was placed on the tray 22 cm from the nest. At the same time the magnetic field was reversed for some colonies, while staying the same for colonies in a control group.

The researchers recorded the time from the introduction of the roach to the formation of a clearly discernible trail. When the orientation of the magnetic field remained unchanged throughout the trial, the ants took about 15 minutes to lay down a trail. However, when the field changed just as the foragers set out in search of the roach, it took half an hour for a trail to be laid down. Altering of the field somehow disrupted their performance, suggesting that magnetoreception may play a role in the ants' short-range navigation by helping them maintain a constant bearing, i.e., walk in a straight line.

The bodies of bees, ants, and other insects have been found to contain small, widely distributed particles of magnetite and other iron compounds, which may contribute to magnetoreception in some cases. But cryptochrome is also commonly present in insects, and there is stronger evidence for its involvement.

Butterflies

An example is provided by research on the monarch butterfly (*Danaus plexippus*), which migrates annually between the northern United States and Canada, and its wintering grounds in Mexico and California. Clouds of these bright orange insects could once be seen moving southward during their fall migration, but habitat destruction (especially affecting milkweed, their only food during the caterpillar stage) and pollution have taken a heavy toll, causing the International Union for the Conservation of Nature (IUCN) to declare the species endangered in 2022.

Monarchs use several types of information in navigating, including a sun compass. But their ability to stay on course during overcast conditions led Guerra et al. (2014) to suspect that they have a magnetic compass as well. To test this possibility, the researchers observed individual monarchs in a miniature flight simulator surrounded by magnetic coils. In a normal magnetic field, the tethered butterflies (tested during fall migration season) attempted to fly southward, but when the field's vertical component was reversed so that field lines tilted down as they extended southward, the butterflies oriented northward. And when the field's vertical component was set to zero, the butterflies showed no consistent orientation. In other words, they were using an inclination compass.

During these experiments the flight simulator was lit by diffuse white light including all wavelengths. When the researchers tested control conditions that filtered out violet light (wavelengths around 400 nm), the butterflies became disoriented, as if their magnetic compass had stopped working. This led Guerra et al. to suspect that the compass depended on a radical-pair mechanism mediated by cryptochrome.

To test this possibility further, Wan et al. (2021) used CRISPR, a genetic engineering tool, to create mutant monarchs lacking cryptochrome. These individuals could not detect an inclination reversal of the magnetic field; thus, the hypothesis that cryptochrome mediates the monarch's magnetic compass was supported.

MAGNETORECEPTION IN MOLLUSCS

Experimental study of the marine gastropod *Tritonia diomedea* shows that it is sensitive to Earth's magnetic field and responds to it in a surprising way. This slug-like nudibranch lives off the Pacific coast of North America, where it moves slowly about, seeking prey and mates while staying out of reach of its nemesis, the ever-hungry sunflower sea star *Pycnopodia helianthoides* (Wyeth & Willows, 2006).

Lohmann and Willows (1997) found, in an experiment conducted over the course of a few days, that *Tritonia* placed in a circular arena consistently oriented eastward. Reversing the horizontal component of the magnetic field—resulting in a combination of axial and inclination components that was unfamiliar to the slugs—disrupted this behavior, causing them to orient randomly. These results showed that the animals have a magnetic sense, but left unanswered the question of exactly why they faced east in the natural magnetic field. In attempting to replicate their initial finding, the researchers discovered that on succeeding days slugs always showed a clear directional preference, but it was not always eastward. In fact, the direction in which they faced changed gradually over the course of a lunar month. Apparently these animals use the geomagnetic field as a compass to help them move in a large counterclockwise circle over the course of a lunar cycle, perhaps as an efficient foraging strategy.

For researchers, an exciting feature of *Tritonia* is its relatively simple nervous system. Its brain consists of three pairs of interconnected ganglia (pedal, cerebral, and pleural), and within these can be found some neurons that are large enough (up to 1 mm), and consistent enough in location from animal to animal, for them to be individually identified and studied. The activity of some of these neurons occurs in synchrony with the swimming movements by which *Tritonia* flees from a predator (Willows, 1971).

Lohmann et al. (1991) found a pair of giant neurons, one in the left pedal ganglion and one in the right, that respond to changes in the magnetic field. Their responses are sluggish, occurring about 10 sec after the magnetic field change. This neural activity does not occur if the connections between the nudibranch's brain and body surface are cut, suggesting that the brain cells receive an input from magnetoreceptors in the periphery.

Later work (Wang et al., 2003, 2004) uncovered two additional pairs of pedal ganglion neurons that respond to magnetic field changes and have axons that carry efferent messages to peripheral structures. They appear to be part of a pathway by which the geomagnetic field can influence *Tritonia*'s locomotion.

Taken as a whole, these studies indicate that researchers are closing in on an understanding of how the magnetic sense of a mollusc contributes to its ability to navigate.

MAGNETORECEPTION: A SUMMARY

So far in this chapter, we have reviewed compelling evidence for the existence of magnetoreception in members of each of the three phyla—vertebrates, arthropods, and molluscs—covered in this book. Behavioral experiments, particularly in vertebrates, have dissected the contributions that the magnetic sense makes to animals' navigational abilities, whether they are on long-range migrations or shorter journeys.

Progress in understanding the physiological basis (or bases) of the magnetic sense is also being made, but more slowly. There is a good body of evidence for two subsystems. One of these involves magnetite, and is anatomically associated (in vertebrates) with the trigeminal component of the somatosensory system. It is widely believed, but remains unproved, that magnetite particles misaligned with the ambient magnetic field exert torque on nearby neural structures, activating them.

The other subsystem, studied especially in birds, involves cryptochrome in visual receptors. When molecules of this chemical are excited by light, they become susceptible to the influence of a magnetic field that can change the proportions of different states of the molecule, indirectly resulting in neural activity.

ELECTRICAL SENSE

PASSIVE ELECTROSENSE

All animals produce small amounts of electricity. Some animals detect the electricity generated by other animals, but do not use their own electricity to probe or affect their environment. These animals have what is called *passive electrosense*.

SHARKS

Following up on earlier demonstrations that some fish react to electrical currents, Adrianus Kalmijn (1971) undertook an elegant series of experiments to determine whether an electrical sense plays an important role in the hunting behavior of *Scyliorhinus canicula*, the small-spotted catshark. Individuals about half a meter in length were collected in the English Channel and the North Sea, transported to Kalmijn's lab at the University of Utrecht, and tested in a seawater-filled wading pool with a sandy bottom.

The European plaice, a small flatfish, was used as prey. When introduced into the pool, it hurried to the bottom and hid under a layer of sand, unobserved by the lethargic sharks dozing nearby. Once aroused, however, the sharks began swimming just above the sand, searching for something to eat. When a shark got within 15 cm of the plaice, it turned downward, sucked in sand and expelled it through its gills, thus uncovering the fish. Then it seized the plaice its jaws and proceeded to eat it.

How did the shark know where the plaice was hiding? Perhaps it saw the sand move or smelled the small fish. To test these possibilities, Kalmijn enclosed the plaice in a compartment with walls and roof made of stiff, opaque agar, and covered the whole thing with sand. This eliminated visual and olfactory cues, but the shark was still able to locate (although not to seize) its prey. By default, sensing of the very weak electrical current emanating from the plaice itself became the

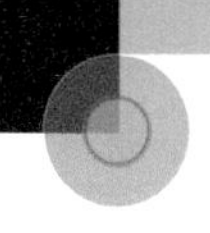

most likely explanation for the shark's hunting skill. To put this possibility to a crucial test, the experimenter covered the agar chamber with plastic wrap, an electrical insulator. Now the shark swam by, oblivious to the small fish hiding below. Finally, Kalmijn showed that the shark would attack electrodes hidden in the sand, even preferring them to a chunk of whiting clearly visible nearby.

There are several types of electroreceptors, but the most common in sharks and their relatives are the *ampullae of Lorenzini,* named for seventeenth-century Italian ichthyologist Stefano Lorenzini, whose careful descriptions of these mysterious structures piqued the curiosity of later researchers as to their function. Each ampulla is a gently rounded sac connected by a tube to the body surface. The tube ends in a pore, which is often visible to the naked eye. A shark may have hundreds of ampullae, located in rows around the mouth and on other parts of the head. The sac and the duct are filled with an electrically conductive jelly. Embedded in a connective tissue matrix at the end of the sac are receptor cells that make synaptic contact with sensory neurons (Andres & von Düring, 1988).

But what do these receptors sense? This question was the subject of much speculation until they were definitively shown to be electroreceptors by Murray (1962). Working with small-spotted catsharks as well as several species of rays, this investigator recorded from nerve fibers receiving signals from the receptors in ampullae. Small amounts of voltage from a battery were delivered to the surface opening of an ampulla's tube by means of a cotton-wick electrode. When the stimulating voltage was negative, nerve fibers increased their firing rate; when it was positive, the frequency of nerve impulses dropped below the axon's spontaneous level.

Later research has provided information as to how electricity excites the receptors. Like neurons, they have a resting potential, the inside of the cell being negative relative to the external milieu. When negative voltage, transported along the highly conductive jelly filling the tube, reaches a receptor cell's cilium, it reduces the voltage level outside the receptor, thus producing a slight depolarization. The cell membrane of the receptor is *excitable,* thanks to voltage-sensitive calcium channels. These now open, allowing positive calcium ions to enter the cell and produce a much larger depolarization—one sufficient to trigger action potentials. These nerve impulses cause the release of transmitter substances at synapses of receptor cells on sensory neurons, affecting their firing rate. Other types of ions further tune the response (Bellono et al., 2018).

PLATYPUS

An electric sense in vertebrates is found primarily in fish and amphibians, but surprisingly, also in the semi-aquatic platypus (*Ornithorhynchus anatinus*) and its fellow monotreme, the echidna, both native to Australia. Like other mammals, these hairy, warm-blooded animals provide milk for their young, but they also have several characteristics more typical of reptiles. One of these is that they lay eggs. Another is that males have a spur on each ankle through which the platypus (though not the echidna) can inject venom into rivals as it fights for access to females.

The electroreceptors of the platypus have been studied in some detail (Andres & von Düring, 1988). They somewhat resemble the ampullae of Lorenzini but have evolved from sweat glands and their ducts, rather than from lateral line receptors. These glands secrete mucus which fills their ducts and probably plays a conductive role similar to that of the gelatinous substance in the ducts of ampullae.

DOLPHIN

One other mammal has been shown to be electroreceptive: the Guiana dolphin (*Sotalia guianensis*), which swims in the coastal waters of South and Central America. Its electroreceptors are of still another variety, having evolved from vibrissae (whiskers) (see Figure 10.5). Like other electroreceptors, these consist of sensitive cells to which external voltage can be efficiently conveyed (Czech-Damal et al., 2012).

The Guiana dolphin and the platypus, like the catfish shark, presumably use their passive electrosense to detect a nearby source of voltage—usually another animal—and obtain some information about its size, location, and

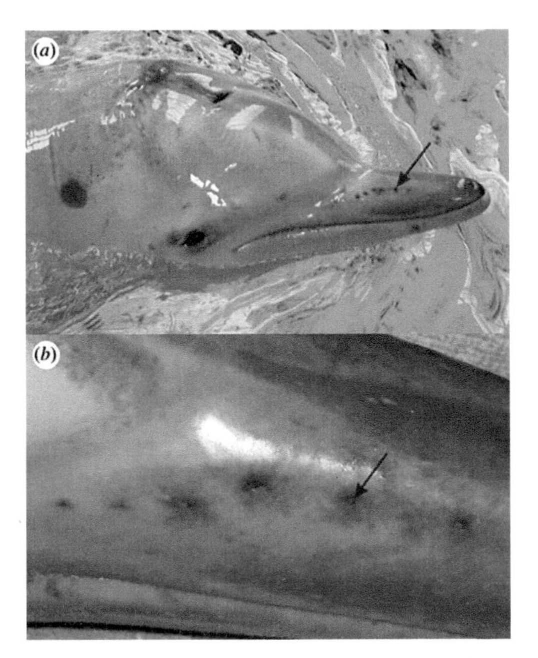

Figure 10.5 The Guiana dolphin, *Sotalia guianensis*. Arrows point to its vibrissal crypts, which house electroreceptors.

Source: **Reprinted with permission from Czech-Damal, N. U., Liebschner, A., Miersch, L., Klauer, G., Hanke, F. D., Marshall, C., Dehnhardt, G., & Hanke, W. (2012). Electroreception in the Guiana dolphin (*Sotalia guianensis*). *Proceedings of the Royal Society B*, *279*, 663–668. Used with permission of The Royal Society (U.K.); permission conveyed through Copyright Clearance Center, Inc.**

movement, based on the number of electroreceptors stimulated and the spatiotemporal pattern of their activation.

ELECTRIC FISH

In some fish, however, a more sophisticated form of electrosense has evolved. Rather than simply receiving exogenous electrical information, these animals produce specialized electrical discharges that serve to probe or influence the environment. They are called *electric fish*, and their enhanced sensory ability is called *active electrosense*.

There are two types of electric fish: those who are *weakly electric* use their low-voltage signals to locate objects and to communicate with conspecifics, whereas those who are *strongly electric* produce more intense discharges that stun or otherwise harm other animals. Weakly electric fish are more numerous and we will discuss them first.

WEAKLY ELECTRIC FISH

We will focus on one well-studied species of weakly electric fish, the elephantnose fish *Gnathonemus petersii*, that inhabits the Niger and other West African rivers (see Figure 10.6). It is named for a trunk-like extension of its mouth called a *Schnauzenorgan*, which plays an important role in its electrosense. This fish lurks unseen in murky water, using its electrosense to perceive its environment. Between its body and its tail is its electric organ, which produces the *electric organ discharges (EODs)* that are key to its electrosense.

The EODs are brief pulses of electricity that fill the water surrounding the fish with an electric field, and the fish can sense this field with electroreceptors that are widely distributed on the body surface, being highest in density on the head. If the fish is in midstream, far from other animals and objects, the rather featureless field will be refreshed, unchanged, with every EOD pulse. But an object entering the field will distort it, casting a sort of electrical "shadow" onto the fish's body. This is an area where the local amplitude of EODs is altered. The shadow can be measured by researchers using a recording electrode that is moved from place to place on the body surface.

Figure 10.6 An elephantnose fish (*Gnathonemus petersii*).

Source: **Used with permission of Shutterstock. Photo Contributor, boban_nz.**

By comparing the properties of these shadows with the ability of fish to perceive the objects that cast them, Gerhardt von der Emde (2004) has advanced our understanding of how active electrosense works. He first set out to determine whether elephantnose fish can perceive the distance to an object. In his laboratory at the University of Bonn, he presented two objects at different distances from an elephantnose fish, and rewarded it with a small worm for going to the object that was farther away. The experiment was conducted in darkness, so in making its choice the fish had to rely on its electrosense. Fish quickly mastered this task, even when the two objects were of different sizes.

How were they able to perceive distance? The task is a challenging one, for the electrical "shadow" (image) of an object depends not just on the object's distance, but on its size. Figure 10.7 shows the electrical patterns formed on the side of an elephantnose fish in response to a metal sphere. A small sphere will cast a smaller image than a large one, if they are at the same distance. But if the small sphere is moved farther from the fish, its image will expand, potentially matching that cast by the large, close sphere. In other words, shadow size is not an unambiguous indicator of distance.

So in judging distance, the fish must be taking other properties of the shadow into account, such as the amplitude of the shadow at its center or the sharpness of the shadow's edge. In fact, both of these properties increase as an object draws closer to the fish, but sharpness of the shadow increases faster than its maximum intensity. This means that the ratio of sharpness to maximum intensity also increases gradually as the object approaches. In fact, it provides a unambiguous indicator of distance, and is therefore likely to be the cue underlying the fish's distance perception (von der Emde, 2004).

Another experiment showed that elephantnose fish can discriminate objects on the basis of shape (von der Emde, 2004). Fish were placed in an open tank containing two objects, such as a cube and a cylinder, and were free to spend time near either or both of them over a 1-hour period. Whenever they were near one of the objects (arbitrarily chosen by the experimenter), it emitted the prerecorded EODs of another fish. The fish spent most of their time near the EOD-emitting object, and continued to favor this shape during a follow-up period in which it did not emit EODs. The experiments indicate both the precision of active electrosense and the ability of these fish to perceive object distance and shape; they also show that EODs have a social function.

In fact, perceiving the geometrical properties and location of an object is just part of the information weakly electrical fish derive from EODs. They also learn about one another. For example, the waveform of an individual's discharges can signal its sex, as shown in *Brachyhypopomus gauderio*, a recently discovered knifefish native to South American waters. (This and other weakly electric fish in the Western Hemisphere are distant cousins of those in the Old World.) Waddell

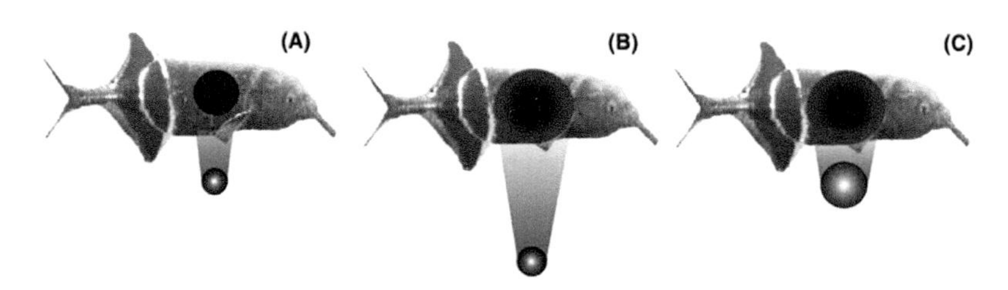

Figure 10.7 Electrical images of metal spheres formed on the surface of an elephantnose fish. The amplitude of reflected EODs is increased in the center of the image (red area) but decreased in a surrounding ring (dark area). The size of the image can be increased either by moving the object farther from the fish (B) or by substituting a larger object (C).

Source: **Reprinted with permission from von der Emde, G. (2004). Distance and shape: Perception of the 3-dimensional world by weakly electric fish. *Journal of Physiology (Paris)*, *98*, 67–80. Used with permission of Elsevier Science & Technology Journals; permission conveyed through Copyright Clearance Center, Inc.**

and Caputi (2021) tested individual fish in a Y-shaped maze, placing them in the stem of the Y and allowing them to proceed along either of the two branches. Trains of prerecorded female EODs emanated from one branch, and of male EODs from the other. Females consistently chose the male-EOD branch, and males the female-EOD branch.

Finally, weakly electric fish communicate socially by varying the timing of their electrical signals. By changing the frequency or clustering of their EODs, they convey to conspecifics situational information such as their desire to affiliate. For example, when an elephantnose fish detects a lifelike sequence of EODs played through an electrode, it typically pauses for a second or more, emits several double pulses, and then begins a regular series of EODs resembling those emanating from the electrode (Worm et al., 2018). Researchers are working to understand the meaning communicated by such responses.

One affiliative behavior observed by Worm and colleagues is for an elephantnose fish to synchronize its EODs with those of another individual. The fish initiating the interaction may follow the other fish, synchronize its EODs with those of the target, then catch up and swim alongside it. The investigators even observed some fish trying to socialize in this way with an EOD-emitting robot.

But changes in the timing of EODs do not always communicate positive messages, as Raab et al. (2021) discovered in the brown ghost knifefish, *Apteronotus leptorhynchus,* a species known for its aggressiveness. The experimenters set the stage for competition by leaving two fish alone together for 6 hours in a tank with several shelters, one of which was more desirable than the others. The fish competed for the best shelter, engaging in ritualized agonistic behavior—chasing and biting—with the larger of the two fish (regardless of sex) usually emerging the winner.

For our purposes, the most relevant finding of the study was that the losing fish did a lot of electrical signaling characterized by transient increases, called *rises,* in the frequency of EODs. These were typically followed by renewed aggression on the part of the winner, leading Raab and colleagues to conclude that the rises were expressions not of submission, but rather, of defiance, indicating that the loser did not accept defeat.

Overall, behavioral experiments with weakly electric fish demonstrate that these animals live in a perceptual world—their *Umwelt*—that is remarkably rich in both its physical and social aspects. This is probably true of many other fish as well (Balcombe, 2016); our ability to record EODs and therefore to study electrical perception may simply give us a better window into the lives of these species.

NEURAL MECHANISMS OF ELECTRORECEPTION IN WEAKLY ELECTRIC FISH

The elephantnose fish and its close relatives actually have three types of electroreceptors, which serve different functions. Like sharks, they have ampullae of Lorenzini for passive electrosense; to locate and identify objects with active electrosense they have another type called *mormyromasts;* and for social communication they have a third type called *knollenorgans.* These differ in structure, but all include a route by which current can rapidly pass from the water outside the fish to the receptor cells. Mormyromasts are smaller than ampullae of Lorenzini and are present in higher density, allowing for greater resolution of the electrical image on the surface of the fish.

The density of mormyromasts is especially high in two areas on the face. One is the region between the nostrils and the mouth, which faces forward and slightly upward as the fish swims along; the other is the Schnauzenorgan itself, which the fish swings from side to side as it moves over the ocean floor. Von der Emde (2006) has compared these "electric foveas" to the two areas of high receptor density in the retina of the pigeon (discussed in Chapter 2): one to monitor the animal's surroundings, the other to examine potential food items before ingesting them.

Axons carrying electrosensory information from all three types of electroreceptors terminate in a hindbrain structure called the *electrosensory lateral line lobe (ELL),* which is close to, and in structure resembles, the cerebellum. The different classes of receptors go to different regions of the ELL, which are involved in

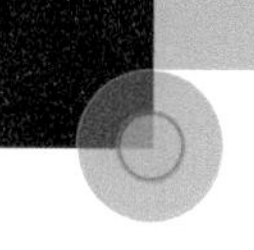

different types of behaviors (e.g., reproduction, predation) and contain separate maps of the body surface (Krahe & Maler, 2014). The ELL is a busy place with many types of neurons. One of its functions is to modify *reafferent* responses to the fish's own EODs, so that these can be more readily distinguished from external stimuli. These modifications are achieved by means of *corollary discharges*, special neural signals sent to the ELL whenever a motor command for an EOD has been issued (Bell, 1989; Fukutomi & Carlson, 2020).

Outputs of the ELL go to the midbrain—first to the torus semicircularis and then to the tectum, where electrosensory signals are integrated with visual ones within maps that are in register (Krahe & Maler, 2014). When both senses are active, the result is a rich neural representation of both the visible and the electrically detectable properties of the object. For example, the elephantnose fish may be able to perceive both the brightness and the electrical capacitance of an approaching object. Another possible benefit of the juxtaposition of these two tectal maps is that it may give electrosensory information access to neural circuitry that normally belongs to the visual system.

STRONGLY ELECTRIC FISH

Strongly electric fish are fish that have weaponized their electricity-generating power, using shocks to stun prey or potential predators. The best-known example is the electric eel, *Electrophorus electricus*, that is native to the rivers of northeastern South America (see Figure 10.8). It is more closely related to the weakly electric knifefish than to true eels and is able to generate low-voltage EODs as well as strong (600 v) shocks. The two types of discharges are produced by separate organs: High-voltage discharges are produced mainly in the body's midsection, while smaller EODs originate in an organ located in the tail. The electrocytes that make up these structures are derived from muscle cells.

Exactly how electric eels make coordinated use of these two types of electrical discharges has been revealed in a series of experiments by Catania (2014). He first observed the eels in

Figure 10.8 The electric eel, *Electrophorus electricus*, moving slowly through a cluttered underwater environment in search of prey. Sensory pits in its head house both electroreceptors and lateral line receptors.

***Source:* Used with permission of Shutterstock. Photo Contributor, Danny Ye.**

a naturalistic setting while recording their discharges. While moving slowly about on a riverbed, they emit low-voltage EODs, much as a weakly electric fish does. These presumably generate "shadows" of objects on their body surface, giving clues to each object's size, shape, location, and (based on its electrical conductivity) whether it is alive or not. If the eel finds indications of a fish, it generates two or three strong shocks, which make the fish twitch, even if it is hidden in weeds. Now apprised of its exact location, the eel produces a long volley of strong shocks, which cause contraction of all the fish's muscles, paralyzing it. This allows the eel to close in on and devour its prey.

In control experiments, Catania found that shocks did not make fish twitch if they were paralyzed with curare, which blocks neuromuscular junctions; and when fish were curarized, eels had no interest in them. Clearly it was their twitching that drew the eels' attention.

How did the eel perceive that the fish had twitched? Did it see the fish move? This seems unlikely because of the eel's poor eyesight, but to test the possibility Catania placed a sheet of transparent but vibration-blocking Plexiglas between the fish and the eel. Now when the fish twitched (after being shocked by the experimenter), the eel just ignored it. But if the Plexiglas was replaced with a thin layer of translucent agar, through which vibrations can pass, the eel promptly responded to a twitch with an attack. Apparently, the eel uses its lateral line system to detect the twitches that its shocks cause!

BEES

Electric current cannot normally flow through air, which explains why animals with the keenest electrical sense are aquatic. But electric *fields* can extend through air, and can be detected by some insects, of which the best studied is the bumblebee. The ability is useful to these pollinating insects, because many flowers have an electric charge that bees can detect at distances of up to half a meter (Clarke et al., 2013). The charge density often varies across the flower, with the tips of the petals having a stronger negative charge; this forms a distinctive electrical pattern that bees can detect.

Bees themselves generally have a positive charge, which they acquire by moving across surfaces in the environment, much as we build up static electricity in winter if we scuff our feet on the carpet. When a positively charged bee gets very close to a negatively charged flower, electrostatic forces come into play, causing pollen grains to jump between the flower and the bee (Clarke et al., 2017). So electricity makes both sensory and nonsensory contributions to pollination.

But how are electric fields actually detected by a bumblebee? Its electroreceptors are a subset of the many fine hairlike filaments that extend from the surface of its body. Deflections of these hairs were measured by Sutton et al. (2016), using *laser Doppler vibrometry*. This is a technique in which a tiny laser beam is reflected from a surface, in this case the side of a hair. If the hair is stationary at the time of the measurement, the frequency of the laser beam will be the same before and after reflection. However, if the hair is moving toward the vibrometer at the time of the measurement, the laser beam's frequency will be increased when it is reflected, because of a Doppler shift, and will be decreased if the hair is moving away. The experimenters delivered an alternating electrical charge to a metal plate 1 cm from the bee, and measured the effect it had on individual hairs, causing them to swing back and forth. The hair's displacement was small—usually less than a nm—but sufficient to cause an increase in firing in sensory neurons located at the base of the hair.

INFRARED SENSE

It may seem odd to include the infrared sense in a chapter on senses that humans don't have, since all of us can feel the warmth of the sun. But a few animals have evolved a much greater IR sensitivity than humans possess, and use the information IR provides in a qualitatively different way. It is the distinctiveness of their ability to capture and use IR information that justifies this special treatment.

It is also important to recognize the distinction between sensitivity to IR and thermal sensitivity more generally. The warmth we feel when we sip coffee is due largely to *conduction*: transfer of motion energy from the molecules of the coffee to temperature-sensitive receptor molecules in our lips and tongue. The same is true of the ability of mosquitoes to sense the warmth of a nearby person (Greppi et al., 2020). Infrared radiation, in contrast, is a form of electromagnetic radiation that can travel through empty space, just like light. It does not require transfer of kinetic energy between adjacent molecules.

We can see a lighted match some distance away, based on the visible wavelengths it radiates, but we are not IR-sensitive enough to feel the match's warmth. However, some animals are, and it is to them—all of them vertebrates, so far as is known—that we now turn.

PIT VIPERS

Infrared sensitivity is most highly developed in a group of snakes called *pit vipers*, including rattlesnakes, copperheads, and cottonmouths. All are venomous, injecting venom through their fangs, which are so long that they fold back when the snake closes its mouth. The name pit viper comes from a pit, with high infrared sensitivity, on each side of the snake's face between the nostril and the eye. The pit functions as a crude pinhole eye, but with a large opening that allows

only a rough image of the snake's surroundings to be formed. A membrane on the back of the pit, analogous to the retina, contains IR-sensitive molecules.

Sensory nerve fibers innervating this pit membrane are part of the trigeminal system, which in most vertebrates mediates sensations of touch and pain in the head and face. And the cell bodies of the nerve fibers innervating the pit membrane are located in a large brainstem structure, the trigeminal ganglion. Through genetic analysis of the trigeminal ganglia of the western diamondback rattlesnake (*Crotalus atrox*), Gracheva et al. (2010) discovered significant quantities of a gene that encodes for the receptor molecule TRPA1, one of the many members of the *transient receptor potential* family of receptor molecules that we discussed in Chapter 6. It was found in greater abundance here than in the rattlesnake's dorsal root ganglia, which mediate skin senses in other parts of the body, and in greater amounts than in the trigeminal ganglia of non-IR-sensitive snakes. These results imply that TRPA1 is specifically involved in the rattlesnake's IR sensitivity. It is produced in the trigeminal ganglion, and migrates along nerve fibers to the pit, where it is incorporated into the membranes of nerve endings as a channel protein.

TRPA1 is also found in mammals; it helps some pain-mediating neurons respond to irritating chemicals. But it plays a very different role in the rattlesnake. It will be recalled that in mammals, sensory fibers in the somatosensory system are either thick and rapidly conducting or thin and slowly conducting. In the human hand, for example, thick fibers mediate innocuous, exploratory touch, as when we try to find the right key in the dark. Thin fibers, in contrast, primarily mediate sensations of pain. Consistent with its role in detecting irritants, mammalian TRPA1 is found mainly in somatosensory neurons with thin axons. But Gracheva and her colleagues found that in the rattlesnake's trigeminal system, it is found primarily in neurons with thick fibers, implying an innocuous, information-rich role. Moreover, they showed that the rattlesnake's TRPA1 is extremely sensitive to warmth, more so than the version in a non-IR-sensitive snake.

IR-triggered activity in the rattlesnake's trigeminal fibers is conveyed into the brainstem, where the fibers terminate in the nucleus of the lateral descending trigeminal tract (LTTD), in an orderly way that creates a map of the pit membrane—and thus of the part of the environment imaged on it—on the LTTD. Sensory processing, including lateral inhibition, occurs here, and the resulting representation is passed along to another hindbrain structure, the nucleus reticularis caloris (RC). Bothe et al. (2019) have recorded from single neurons in both structures. The most notable difference between neurons in these two structures is that those in the RC are highly tuned to the direction of motion of an IR source (such as a scurrying mouse). A directionally sensitive neuron's activity is inhibited by movement in one direction but not by movement in the opposite direction. This neural algorithm is remarkably similar to that used in the visual systems of many animals, vertebrate and invertebrate.

From the hindbrain, where the LTTD and RC are located, IR-derived information is sent to the midbrain, where it converges on a visual representation constructed from retinal input (see Figure 10.9). The two sets of signals are in adjacent layers of the tectum, and the two maps of the scene observed by the snake are approximately in register. Connections between the layers result in many bimodal cells that respond to both visual and IR inputs (Newman & Hartline, 1981). These anatomical and electrophysiological findings suggest a multimodal representation that allows the snake both to see a moving leaf on the ground and to feel the warmth of a vole hiding under it. The juxtaposition of the two maps implies that the same motor activities, such as moving forward and striking, can be activated regardless of how the snake perceives its prey. This joint access is critical to the rattlesnake when it hunts after dark.

So far our description of a rattlesnake's behavior has been limited to its search for rodents and birds. But in fact snakes' lives are much more complicated. Like other animals, they mate, seek shelter, and avoid predators. And their prey do not wait passively to be devoured; they have developed countermeasures to use against rattlesnakes.

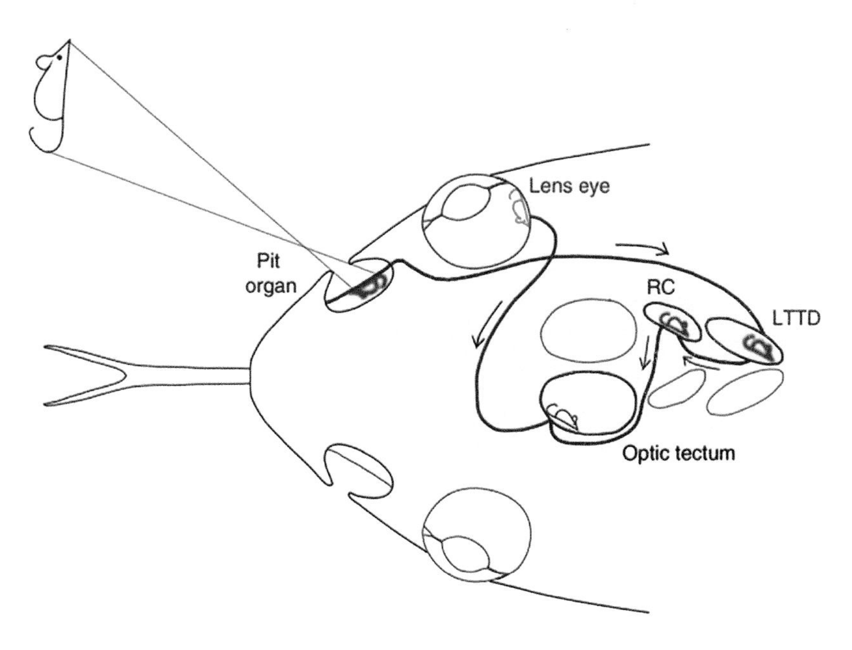

Figure 10.9 Diagram of the rattlesnake head, viewed from above, showing the convergence of visual and infrared signals at the optic tectum. The visual pathway (shown in green) leads directly from the eye to the tectum. The infrared pathway (red) goes first to the LTTD and RC, two hindbrain structures, and from there to the tectum.

Source: **Reprinted with permission from Kelber, A. (2019). Infrared imaging: A motion detection circuit for rattlesnake thermal vision. *Current Biology, 29*, R403–R405. Used with permission of Elsevier Science & Technology Journals; permission conveyed through Copyright Clearance Center, Inc.**

The complex interaction between rattlesnakes (*Crotalus oreganus*) and California ground squirrels (*Spermophilus beecheyi*) has been studied experimentally by Rundus et al. (2007). Baby squirrels in an underground burrow would make easy prey for the snake, but their mother, guarding the nest, is a ferocious antagonist in whom the snake has met its match. She needs to be on guard against not only rattlesnakes, but other snakes as well, such as gopher snakes (*Pituophis melanoleucus*). In laboratory encounters between a female squirrel and a snake—with wire mesh preventing them from coming into direct contact—the squirrel cautiously approached the snake and flagged her tail rapidly from side to side as a warning. Squirrels' responses to the two types of snakes looked similar, but infrared imaging told a more complex story. In the presence of the rattlesnake, the squirrel increased the temperature of her tail, causing it to give off IR radiation that would be observed by the rattlesnake. But the squirrel did not do this when confronting the gopher snake, which does not have IR sense. She can distinguish between the two snake species and adjust her behavior accordingly, adding a metabolically expensive IR warning only in the presence of the rattlesnake.

In a second set of experiments, Rundus and colleagues studied the responses of rattlesnakes to a robot squirrel with the ability to flag its tail and increase its temperature. A simulated squirrel burrow was added to the testing chamber, near the robot, to increase the snake's motivation. The robot intermittently flagged its tail, and did so faster and faster as the snake approached. This maneuver caused the snake to hesitate and to assume a more defensive posture. And this behavioral shift was especially marked on trials in which the robot's tail was warmed. IR signaling with the tail caused the snake to coil up and rattle, to more often assume a ready-to-strike position, and to take longer to enter the burrow. By warning off a snake from a risky confrontation, a squirrel's IR signaling benefits both animals.

VAMPIRE BATS

Vampire bats consume only blood. Native to Central and South America, these night hunters stealthily approach their prey—usually livestock—and use their IR sensitivity to find well-vascularized areas of skin. Rather than sucking blood, they make a cut in the skin with their incisors and lap up the blood that oozes out. Anticoagulants in the bat's saliva keep the blood flowing.

The vampire bat's nasal area is oddly shaped, with ridges separating flat areas containing the thermoreceptors that respond to IR (see Figure 10.10). As in pit vipers, these receptors are the endings of sensory fibers of the trigeminal nerve. They respond when warmed by the prey's infrared radiation to a temperature of 30°C or above, a sensitivity comparable to that of pit vipers. To put this value in context, human skin temperature is about 33°C.

To detect the slight temperature rise caused by IR absorption, vampire bats use a different transient receptor potential molecule from pit vipers: TRPV1 rather than TRPA1. In other mammals, TRPV1 plays an important role in signaling painful levels of heat, as described in Chapter 6, but the TRPV1 variant in bats is much more sensitive (Gracheva et al., 2011).

From an ecological point of view, a more important difference between the IR capabilities of vampire bats and rattlesnakes is that, in

Figure 10.10 The common vampire bat, *Desmodus rotundus*. Note the sharp incisors and the flattened, leaf-like nasal area, which contains IR receptors.

***Source:* Used with permission of Shutterstock. Photo Contributor, Mendesbio.**

the bats, no infrared image of the environment is formed. Pit vipers can use IR images to locate warm objects up to a meter away; in contrast, vampire bats use their IR sensitivity only to find warm areas of skin on an animal that they are already close to.

As global warming proceeds, the range of vampire bats will probably expand farther into both North and South America, so people in once-cooler areas are likely to become more familiar with these unique mammals in the coming decades.

SUMMARY

Magnetic sensitivity is a sensory modality that is widely distributed among animals; it has been demonstrated in some members of all five classes of vertebrates, and in some arthropods and molluscs. It is used to detect Earth's magnetic field and thus to assist in navigation.

Research on sea turtles shows that these migrating animals depend on magnetic information as they navigate ocean currents, finding their way along a complex route until they ultimately return to their natal beach. Migratory birds also rely on their magnetic sense during seasonal journeys.

There are two known ways in which magnetic fields are detected. One is by means of iron-based particles associated with sensory nerve fibers in, for example, the beaks of pigeons. It is believed that when the geomagnetic field exerts torque on these elongated particles, the nerve fibers are activated. The other mechanism involves cryptochrome, a molecule that is sensitive to both light and magnetic force. Located in retinal receptors, cryptochrome's response to blue light is modulated by the magnetic field. Both mechanisms have been most extensively studied in birds.

Two other senses, while not entirely lacking in people, are so highly developed in some animals as to constitute separate modalities. One is electrosense, found mostly in fish, since water is an electrical conductor. Sharks have receptors on their face with which they passively detect the electrical currents produced by other organisms. Some other fish use electricity in more active ways. Weakly electric fish generate pulses of electricity that are reflected back to the sender by objects in the environment, providing information about the size, shape, and location of those objects. Strongly electric fish, such as the electric eel, send out intense shocks that cause prey animals to twitch and ultimately to succumb. Among invertebrates, bumblebees have been found to locate flowers because of their standing electric charges.

The final modality we considered is infrared (IR) sensitivity. We can all feel the warmth of the sun, but some animals are able to detect much lower levels of IR. Most remarkable are the pit vipers, named for the pits on their faces lined with IR-sensitive molecules. A crude IR image is formed in the pit and is centrally routed to the tectum, where it is integrated with visual information. This sense helps vipers hunt for warm-blooded animals at night.

Final Thoughts

INFORMATION AND LIFE

The Information Age is said to have begun in the middle of the twentieth century. But in fact, Earth has been in an information age since the dawn of life. The behavior of the earliest living things was influenced by information about the molecules around them and the forces acting on them. One-celled organisms swim toward or away from the light, orient with respect to gravity, and shrink back or move toward an object based on its chemical composition.

Animals—especially complex ones such as vertebrates, arthropods, and molluscs—also seize upon and use more complicated forms of information, such as the patterns formed by light or sound. Animals have special body parts for capturing this information and extracting its meaning. These specialized structures, the sense organs and related parts of the brain, constitute the animals' *sensory systems.*

In this book, we have examined the sensory systems of a wide variety of animals. The most striking aspect of this survey is the remarkable diversity of sensory abilities found in the animal kingdom.

This diversity is of two kinds. First, for a given modality, some animals have greater and more wide-ranging sensitivity than others. For example, dogs are extremely sensitive to a wider variety of odors than we are, enabling them to follow the trail of another animal or locate an unconscious person in the rubble of a collapsed building. And animals can hear sounds that are higher (bats) or lower (elephants) in frequency than the human audible range.

Second, there are whole modalities of stimulation that some animals can detect while others (including humans) cannot. Pigeons, for example, can sense magnetic fields using specialized receptors in their beaks and in their eyes. And sharks can zero in on prey even in murky water thanks to their electrical sense.

Realization of this diversity is deeply humbling. We are used to thinking of our senses as the canonical ones that set the bar in sensitivity and precision, but in fact we are just one among millions of unique species, with a serviceable but not highly specialized set of sensory abilities.

THE EVOLUTION OF SENSES

The explanation for all this diversity is, of course, evolution. Adaptive radiation and natural selection have worked together, over hundreds of millions of years, to permit the survival of animals with combinations of sensory abilities that are well suited to their respective environments. The ability of bees to detect and decipher the polarization of skylight helps them navigate when the sun is not visible, thus increasing their chances of survival. The ability of bats to hear ultrasound enables them to hunt using echolocation, thus increasing their chances of survival. And because they survive, they pass on these abilities to succeeding generations.

So why don't all animals—including humans—have all of these abilities? After all,

DOI: 10.1201/9781003362319-11

an ability to perceive Earth's magnetic field would certainly have been useful to our remote ancestors foraging in the woods or navigating across an ocean. The reason these abilities are not universal is that every sensory ability has a cost, in the metabolic energy needed to maintain it, the extra weight of the neural machinery that mediates it, the space (e.g., inside the skull) that that machinery occupies, the attentional resources it monopolizes, and so on. These costs weigh against the potential benefits of a sensory system, empirically determining in real time whether accommodating that system would increase or decrease the species' chances of survival. The existence of a species, at least if it continues for many generations, proves that its combination of sensory abilities is both necessary and sufficient for its survival.

Some insects live for as little as a few hours; some vertebrates for as long as a century. In contrast, a species may continue for hundreds of thousands of years. However, even this pales in comparison with the length of time for which some genes have survived. As species die out, usually due to changes in the environment, they often are succeeded by related species which carry a large percentage of their genes forward. Richard Dawkins (2016) has argued that animals are "survival machines" that perpetuate the genes they house. Each of us is a vast repository of genes, not all of which are activated in a given individual, but which are available for use should they be needed.

We have not evolved from arthropods, nor they from us. The legs of a lobster are not homologous to our own. But they are analogous, in that they help both animals get around. Such independent occurrence of analogous structures in animals widely separated on the tree of life is called *convergent evolution*. We have encountered it frequently in this book.

But when evolution proceeds in parallel along independent tracks, it is not always starting from scratch, because very old genes common to the two branches may participate. An example of this *deep evolution* is the use of certain families of genes, such the *Hox* genes that contribute to limb formation, that have survived (albeit with modifications) for more than half a billion years, that is, from before the separation between chordates and arthropods (Shubin et al., 1997). Other ancient gene families play roles in vision, touch, and perhaps other senses. No two animals are completely unrelated.

THEORETICAL LIMITS OF SENSITIVITY

When a given sensory system is of great survival value to an animal, its sensitivity can be remarkable. This has been most clearly demonstrated in people, where sensitivity can be measured (using psychophysics) with a precision hard to achieve in animals.

Consider vision. The absolute threshold of vision has been carefully measured by asking people to report the occasional occurrence of a faint flash of light in a dark room. Under optimal conditions, a person can detect a flash even if it delivers only 100 photons to the cornea. Since considerable light is absorbed by the ocular media, only about 20 of these photons reach the receptors, and are randomly distributed across a patch of several hundred rods. There is very little likelihood that any individual rod will catch more than one of these photons; yet the flash is detected. This finding shows that a single rod can generate a sensory signal after absorbing just one photon. The sensitivity of rods is therefore at the theoretical limit, since there is no amount of light smaller than a photon.

A similar result obtains in hearing. Under optimal conditions, a person can hear a sound that moves the eardrum only 10^{-11} m, less than the diameter of a hydrogen atom. If our hearing were any more sensitive, we would hear a soft rumble from the random movement of air molecules. Likewise in touch, where we can feel a tiny bump, only a micron high, on a smooth surface. And in olfaction, where some odorants are detectable when only a few dozen molecules activate receptor cells. Only recently have artificial instruments been able to match the sensitivity of humans and some other animals. The point is that natural selection exerts strong evolutionary pressure on animals to push sensory abilities with survival value to the limit.

And sensitivity is only one aspect of sensory ability. The senses can adjust, when necessary, to different levels of environmental stimulation. This allows us, for example, to see in bright sunlight as well as by moonlight, thanks to changes in pupil size, the availability of both rods and cones, neural adaptation, and other factors. Moreover, the sensory systems (made up both of sense organs and of related parts of the nervous system) routinely make complex calculations to provide animals with the information they need to act on their environment. Examples range from an owl's brain comparing signals from the two ears to locate a scurrying mouse, to a honeybee orienting her dance to direct nestmates to flowers she has discovered. Each animal's senses have been shaped by evolution to provide exactly the information that it needs—no more and no less—to survive and reproduce.

THE ANTHROPOCENE EPOCH

As the environment changes, some species will become extinct. This has always been the case. But the rate at which species are disappearing has increased dramatically in the last century: We are now in the midst of a major extinction—one of the six largest in Earth's history. And this one is caused primarily by human activity.

At the present time, the greatest overall risk to species diversity is climate change and the resulting destruction of habitat. Climate change causes the submergence of islands and coastal areas by rising seas, the burning of vast forests by wildfires, and the death of coral reefs. The main cause of climate change is the release of large amounts of heat-trapping gases into the atmosphere by the burning of fossil fuels.

This is one of many types of *pollution*—contamination of the environment with anything harmful. Examples of pollution are the release of greenhouse gases such as methane, the use of intense outdoor lighting at night, the production of high levels of sound, and electromagnetic noise of human origin.

Pollution affects animals by interfering in a variety of ways with the functioning of their sensory systems. For example, strong illumination at night can compromise reproduction in fish (Chapter 2); it can also disorient migrating birds, causing many to collide with buildings (Chapter 5). Intense underwater sound from ships can damage the hearing of marine organisms, while military sonar can put whales to flight, sometimes resulting in their beaching and death (Chapter 7). When fish are raised in water with high levels of dissolved CO_2, some become attracted to the smell of predators (Chapter 9). And electromagnetic noise, such as radio waves, can disable birds' magnetic sense (Chapter 10). These discoveries are part of a growing research literature on the complex effects of human actions on animals.

THE VALUE OF SENSORY SCIENCE

Millions of people around the world are fighting for endangered animals, trying to reduce the severity of the mass extinction event that is currently underway. We can do this in our everyday lives, by reducing our carbon footprint, turning down outdoor lighting, and so on. We can also help by supporting some of the myriad conservation organizations, ranging from local to worldwide, that work to protect certain animals or whole ecosystems.

But essential to all of these efforts is knowledge. We need to know exactly what is harming animals (as in the examples just given), and how to remedy that harm in ways that humans will accept. Obtaining this knowledge requires research that addresses many questions, and research on the sensory systems of animals is an important part of the mix. It can be found in many journals, as the references cited in this book attest: those with a broad scientific reach (such as *Nature* and *Science*); others with a particular focus on animals (such as the *Journal of Experimental Biology* and *Animal Behaviour*) and still others devoted to environmental concerns (such as *Endangered Species Research*). As we learn more about animals' perceptual worlds and the factors that guide their increasingly constrained decision-making, we will have a better chance of saving some of them.

Let's consider an example of how research has helped an animal—the harbor porpoise—deal with an anthropogenic challenge. These echolocating cetaceans hunt for fish, often in the same areas where humans are fishing with giant gill nets. These nets, used in many fisheries, are designed to trap fish that try to squeeze through the mesh but get caught by their gills. Harbor porpoises can also get entangled in the nets, and being unable to surface for air, will drown. They are "bycatch." They can detect the nets, but ignore them when they are in hot pursuit of a fish. In psychophysical experiments, Kastelein et al. (2010) measured the audibility curve of porpoises and found that, as expected from their clicks, they can hear best in the ultrasound, their threshold being lowest at 125 kHz. Most fish, in contrast, hear best at much lower frequencies (see Chapter 7). So a high-frequency pinging device, attached to gill nets, might be able to warn away porpoises without affecting fish (Wahlberg et al., 2015).

The website *conservationevidence.com* lists more than 30 studies on this topic. For example, Larsen et al. (2013) found that pingers spaced about half a kilometer apart on a long array of gill nets in the Danish North Sea reduced harbor porpoise bycatch nearly to zero. In comparison, nets without pingers stood a better than fifty-fifty chance of entrapping a porpoise on any given haul. Fortunately for the fishing industry, if not for the fish, the ultrasonic pinger did not reduce fish catch. As a result of the study, the Danish government modified its regulations on the spacing of pingers in fisheries.

But the rewards of sensory science are not just practical ones. There is an aesthetic experience involved in discovering the elegant, surprising ways animals extract the information they need from the stimuli impinging on their sense organs. And additional modalities of perception may come to light in the future, as sensory scientists make new discoveries, and evolutionary adaptation to a changing environment continues.

WIDENING THE LENS

From its initial focus on the human senses, the field of sensory science has expanded in the past half-century to encompass the senses of animals, adding immeasurably to an understanding of how the senses have evolved and how they play critical roles in the struggle for survival. And new findings that reveal the subtlety and complexity of animals' senses have contributed to an emerging consensus that many nonhuman animals (including some invertebrates) are sentient.

Influential British ethologist C. Lloyd Morgan (1852–1936) urged caution in extrapolating from our subjective experience to the state of affairs in other species. Enunciating what came to be known as Lloyd Morgan's Canon, he wrote (1894) that we should not attribute an animal's actions to a "higher" psychological process such as perception if they can be interpreted in terms of a "lower" process such as a set of reflexes.

But to believe that of the millions of species on the planet, only the human species is capable of perception now seems—in light of what we have learned about the senses of animals—not cautious at all, but radical and implausible.

However, even if many animals are sentient, each is perceptually restricted to its own Umwelt. Humans can never really know to a certainty "what it is like to be a bat," as Nagel (1974) observed. In some ways, such as color vision, perception is clearly not comparable across species: Humans cannot perceive all the colors a bird can, and a dog cannot perceive all the colors we can.

But for some other types of perception, such as the perception of spatial relationships, there may be a degree of consistency across species, at least for animals with highly developed nervous systems. When we throw a stick and the dog jumps to catch it, the dog, like us, appears to perceive both that it is a stick and that it is traveling in a certain direction with a certain speed.

Consider the fact that some animals respond to optical illusions the same way we do. For example, pigeons are fooled by the motion aftereffect (Xiao & Güntürkün, 2008), some reptiles' perception of size is distorted by context (Santacà et al., 2019), fruit flies experience sawtooth gradients as moving (Agrochao et al., 2020), and cuttlefish are susceptible to the pictorial depth cue of texture gradient (Josef et al., 2014). Are these across-species parallels just isolated curiosities, or do they imply a broader similarity in the way diverse animals perceive spatial relationships? Many of the research findings described in this book (especially Chapter 5) suggest such a similarity.

An animal's vocalizations are another source of information about its perceptions. While no other species has language skills comparable to ours, many have vocal repertoires that hold meaning for conspecifics. For example, vervet monkeys emit alarm calls that indicate whether they have spotted a leopard, an eagle, or a snake (Seyfarth et al., 1980).

Some animals emit more elaborate patterns of sound that might give us—if only we understood them—a more nuanced understanding of their Umwelt. The click patterns emitted by sperm whales constitute a prime example. There is now an effort underway to obtain a large database of these click patterns, and to analyze them using machine language technology (Andreas et al., 2022). By relating the occurrence of specific sounds to behaviors or situations, it may be possible to learn more about a sperm whale's perceptions—how they resemble and how they differ from our own.

Consider a final example of a perceptual resonance between species. In the elaborate blue display of the male bowerbird, the trill of the songbird and the circular sand sculptures of the male pufferfish (Kawase et al., 2015), there is a quality that serves to attract mates, presumably by causing a positive affective experience. It is not surprising that these courting behaviors exist: They have evolved through sexual selection. What is surprising is that the displays elicit an affective response *in humans*. We consider them beautiful. The cross-species appeal of these displays has no obvious survival value for either species. But it does suggest that species are not entirely siloed in their own Umwelten.

In summary, the study of the senses of animals is a fascinating, challenging, and urgently needed field in which people from many disciplines are participating. And students in different academic departments, who are studying different sensory modalities in different animals, can benefit from a shared basic understanding of sensory science as a whole. It is my hope that this book has helped you to develop such a conceptual framework.

References

Abramson, A. S., & Whalen, D. H. (2017). Voice onset time (VOT) at 50: Theoretical and practical issues in measuring voicing distinctions. *Journal of Phonetics, 63*, 75–86.

Abramson, J. Z., Hernández-Lloreda, M. V., García, L., Colmenares, F., Aboitiz, F., & Call, J. (2018). Imitation of novel conspecific and human speech sounds in the killer whale (*Orcinus orca*). *Proceedings of the Royal Society B, 285*, 20172171.

Adamo, S. A. (2016). Do insects feel pain? A question at the intersection of animal behaviour philosophy and robotics. *Animal Behaviour, 118*, 75–79.

Agrochao, M., Tanaka, R., Salazar-Gatzimas, E., & Clark, D. A. (2020). Mechanism for analogous illusory motion perception in flies and humans. *PNAS, 117*, 23044–23053.

Aihara, Y., Yasuoka, A., Iwamoto, S., Yoshida, Y., Misaka, T., & Abe, K. (2008). Construction of a taste-blind medaka fish and quantitative assay of its preference-aversion behavior. *Genes, Brain and Behavior, 7*, 924–932.

Aksoy, V., & Camlitepe, Y. (2012). Behavioural analysis of chromatic and achromatic vision in the ant *Formica cunicularia* (Hymenoptera: Formicidae). *Vision Research, 67*, 28–36.

Alert, B., Michalik, A., Thiele, N., Bottesch, M., & Mouritsen, H. (2015). Re-calibration of the magnetic compass in hand-raised European robins (*Erithacus rubecula*). *Scientific Reports, 5*, 14323.

Amoore, J. E. (1963). Stereochemical theory of olfaction. *Nature, 198*, 271.

Anderson, J. B., & Vander Meer, R. K. (1993). Magnetic orientation in the fire ant, *Solenopsis invicta*. *Naturwissenschaften, 80*, 568–570.

Andouche, A., Valera, S., & Baratte, S. (2021). Exploration of chemosensory ionotropic receptors in cephalopods: The IR25 gene is expressed in the olfactory organs, suckers, and fins of *Sepia officinalis*. *Chemical Senses, 46*, 1–16.

Andreas, J., Beguš, G., Bronstein, M. M., Diamant, R., Delaney, D., Gero, S., Goldwasser, S., Gruber, D. F., de Haas, S., Malkin, P., Pavlov, N., Payne, R., Petri, G., Rus, D., Sharma, P., Tchernov, D., Tønnesen, P., Torralba, A., Vogt, D., & Wood, R. J. (2022). Toward understanding the communication in sperm whales. *iScience, 25*, 104393.

Andres, K. H., & von Düring, M. (1988). Comparative anatomy of vertebrate electroreceptors. *Progress in Brain Research, 74*, 113–131.

Arch, V. S., & Narins, P. M. (2009). Sexual hearing: The influence of sex hormones on acoustic communication in frogs. *Hearing Research, 252*, 15–20.

Arikawa, K. (2003). Spectral organization of the eye of a butterfly, *Papilio*. *Journal of Comparative Physiology A, 189*, 791–800.

Arnold, K., & Neumeyer, C. (1987). Wavelength discrimination in the turtle *Pseudemys scripta elegans*. *Vision Research, 27*, 1501–1511.

Arnott, S. R., Binns, M. A., Grady, C. L., & Alain, C. (2004). Assessing the auditory dual-pathway model in humans. *NeuroImage, 22*, 401–408.

Ashley, P. J., Ringrose, S., Edwards, K. L., Wallington, E., McCrohan, C. R., & Sneddon, L. U. (2009). Effect of noxious stimulation upon antipredator responses and dominance status in rainbow trout. *Animal Behaviour, 77*, 403–410.

Augee, M., Gooden, B., & Musser, A. (2006). *Echidna: Extraordinary egg-laying mammal*. CSIRO Publishing.

Avarguès-Weber, A., Portelli, G., Bernard, J., Dyer, A., & Giurfa, M. (2010). Configural

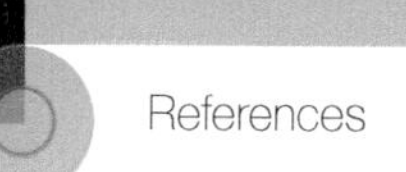

processing enables discrimination and categorization of face-like stimuli in honeybees. *Journal of Experimental Biology, 213*, 593–601.

Baatrup, E. (1985). Physiological studies on the pharyngeal terminal buds in the larval brook lamprey, *Lampetra planeri* (Bloch). *Chemical Senses, 10*, 549–558.

Baker, R. R. (1980). Goal-orientation by blindfolded humans after long-distance displacement: Possible involvement of a magnetic sense. *Science, 210*, 555–557.

Balcombe, J. (2016). *What a fish knows: The inner lives of our underwater cousins*. Scientific American/Farrar, Straus and Giroux.

Baltzley, M. J., & Nabity, M. W. (2018). Reanalysis of an oft-cited paper on honeybee magnetoreception reveals random behavior. *Journal of Experimental Biology, 221*, jeb185454.

Banks, M. S., Sprague, W. W., Schmoll, J., Parnell, J. A. Q., & Love, G. D. (2015). Why do animal eyes have pupils of different shapes? *Science Advances, 1*, e1500391.

Barbosa, A., Mäthger, L. M., Buresch, K. C., Kelly, J., Chubb, C., Chiao, C.-C., & Hanlon, R. T. (2008). Cuttlefish camouflage: The effects of substrate contrast and size in evoking uniform, mottle or disruptive body patterns. *Vision Research, 48*, 1242–1253.

Barreiro-Iglesias, A., Anadón, R., & Rodicio, M. C. (2010). The gustatory system of lampreys. *Brain, Behavior and Evolution, 75*, 241–250.

Barrios, A. W., Sánchez-Quinteiro, P., & Salazar, I. (2014). Dog and mouse: Toward a balanced view of the mammalian olfactory system. *Frontiers in Neuroanatomy, 8*, Article 106.

Bartos, M., & Minias, P. (2016). Visual cues used in directing predatory strikes by the jumping spider *Yllenus arenarius* (Araneae, Salticidae). *Animal Behavior, 120*, 51–59.

Bartoshuk, L. M., Harned, M. A., & Parks, L. H. (1971). Taste of water in the cat: Effects on sucrose preference. *Science, 171*, 699–701.

Bates, M. E., Stamper, S. A., & Simmons, J. A. (2008). Jamming avoidance response of big brown bats in target detection. *Journal of Experimental Biology, 211*, 106–113.

Beauchamp, G. K. (2009). Sensory and receptor responses to umami: An overview of pioneering work. *American Journal of Clinical Nutrition, 90*(Suppl), 723S–727S.

Beauchamp, G. K., Maller, O., & Rogers, J. G., Jr. (1977). Flavor preferences in cats (*Felis catus* and *Panthera* sp.). *Journal of Comparative and Physiological Psychology, 91*, 1118–1127.

Békésy, G. v. (1960). *Experiments in hearing* (E. G. Wever, Trans. and Ed.). McGraw-Hill.

Békésy, G. v. (1967). *Sensory inhibition*. Princeton University Press.

Bell, C. C. (1989). Sensory coding and corollary discharge effects in mormyrid electric fish. *Journal of Experimental Biology, 146*, 229–253.

Bellono, N. W., Leitch, D. B., & Julius, D. (2018). Molecular tuning of electroreception in sharks and skates. *Nature, 558*, 122–126, 126A–126O.

Bensmaïa, S. J., & Hollins, M. (2003). The vibrations of texture. *Somatosensory & Motor Research, 20*, 33–43.

Bensmaïa, S. J., & Hollins, M. (2005). Pacinian representations of fine surface texture. *Perception & Psychophysics, 67*, 842–854.

Benton, R. (2022). *Drosophila* olfaction: Past, present and future. *Proceedings of the Royal Society B, 289*, 20222054.

Blake, R., Cool, S. J., & Crawford, M. L. J. (1974). Visual resolution in the cat. *Vision Research, 14*, 1211–1217.

Blazing, R. M., & Franks, K. M. (2020). Odor coding in piriform cortex: Mechanistic insights into distributed coding. *Current Opinion in Neurobiology, 64*, 96–102.

Blest, A. D., Hardie, R. C., McIntyre, P., & Williams, D. S. (1981). The spectral sensitivities of identified receptors and the function of retina tiering in the principal eyes of a jumping spider. *Journal of Comparative Physiology, 145*, 227–239.

Blough, D. S. (1956). Dark adaptation in the pigeon. *Journal of Comparative and Physiological Psychology, 49*, 425–430.

Blough, P. M. (1971). The visual acuity of the pigeon for distant targets. *Journal of the Experimental Analysis of Behavior, 15*, 57–67.

Boal, J. G., & Golden, D. K. (1999). Distance chemoreception in the common cuttlefish, *Sepia officinalis* (Mollusca, Cephalopoda). *Journal of Experimental Marine Biology and Ecology, 235*, 307–317.

Boles, L. C., & Lohmann, K. J. (2003). True navigation and magnetic maps in spiny lobsters. *Nature, 421*, 60–63.

Bonadonna, F., & Nevitt, G. A. (2004). Partner-specific odor recognition in an Antarctic seabird. *Science, 306*, 835.

Bonadonna, F., Villafane, M., Bajzak, C., & Jouventin, P. (2004). Recognition of burrow's olfactory signature in blue petrels, *Halobaena caerulea*: An efficient discrimination mechanism in the dark. *Animal Behaviour, 67*, 893–898.

Borgia, G., & Gore, M. A. (1986). Feather stealing in the satin bowerbird (*Ptilonorhynchus violaceus*): Male competition and the quality of display. *Animal Behaviour, 34*, 727–738.

Borgia, G., & Keagy, J. (2006). An inverse relationship between decoration and food colour preferences in satin bowerbirds does not support the sensory drive hypothesis. *Animal Behaviour, 72*, 1125–1133.

Born, G., Grützner, P., & Hemminger, H. (1976). Evidenz für eine Mosaikstruktur der Netzhaut bei Konduktorinnen für Dichromasie [Evidence for a mosaic structure of the retina in carriers of dichromacy]. *Human Genetics, 32*, 189–196.

Bortolotti, L., & Costa, C. (2014). Chemical communication in the honey bee society. In C. Mucignat-Caretta (Ed.), *Neurobiology of chemical communication* (pp. 147–210). CRC Press.

Bothe, M. S., Luksch, H., Straka, H., & Kohl, T. (2019). Neuronal substrates for infrared contrast enhancement and motion detection in rattlesnakes. *Current Biology, 29*, 1827–1832.

Bowmaker, J. K., Thorpe, A., & Douglas, R. H. (1991). Ultraviolet-sensitive cones in the goldfish. *Vision Research, 31*, 349–352.

Brothers, J. R., & Lohmann, K. J. (2015). Evidence for geomagnetic imprinting and magnetic navigation in the natal homing of sea turtles. *Current Biology, 25*, 392–396.

Brown, P. K., & Wald, G. (1964). Visual pigments in single rods and cones of the human retina. *Science, 144*, 45–52.

Brownell, P., & Farley, R. D. (1979a). Detection of vibrations in sand by tarsal sense organs of the nocturnal scorpion, *Paruroctonus mesaensis*. *Journal of Comparative Physiology, 131*, 23–30.

Brownell, P., & Farley, R. D. (1979b). Prey-localizing behaviour of the nocturnal desert scorpion, *Paruroctonus mesaensis*: Orientation to substrate vibrations. *Animal Behaviour, 27*, 185–193.

Brüning, A., Kloas, W., Preuer, T., & Hölker, F. (2018). Influence of artificially induced light pollution on the hormone system of two common fish species, perch and roach, in a rural habitat. *Conservation Physiology, 6*(1), coy016.

Buchmann, S. L. (2023). *What a bee knows: Exploring the thoughts, memories, and personalities of bees*. Island Press.

Buck, L., & Axel, R. (1991). A novel multigene family may encode odorant receptors: A molecular basis for odor recognition. *Cell, 65*, 175–187.

Buck, L. B. (2005). Unraveling the sense of smell. *Angewandte Chemie (International ed. in English), 44*, 6128–6140.

Budelmann, B. U., & Bleckmann, H. (1988). A lateral line analogue in cephalopods: Water waves generate microphonic potentials in the epidermal head lines of *Sepia* and *Lolliguncula*. *Journal of Comparative Physiology A, 164*, 1–5.

Budzynski, C. A., & Bingman, V. P. (2004). Participation of the thalamofugal visual pathway in a coarse pattern discrimination task in an open arena. *Behavioural Brain Research, 153*, 543–556.

Budzynski, C. A., Gagliardo, A., Ioale, P., & Bingman, V. P. (2002). Participation of the homing pigeon thalamofugal visual pathway in sun-compass associative learning. *European Journal of Neuroscience, 15*, 197–210.

Buresch, K. C., Sklar, K., Chen, J. Y., Madden, S. R., Mongil, A. S., Wise, G. V., Boal, J. G., & Hanlon, R. T. (2022). Contact chemoreception in multi-modal sensing of prey by

Octopus. *Journal of Comparative Physiology A, 208*, 435–442.
Butler, C. G., & Gibbons, D. A. (1958). The inhibition of queen rearing by feeding queenless worker honeybees (*A. mellifera*) with an extract of "queen substance." *Journal of Insect Physiology, 2*, 61–64.
Caine, N. G., & Mundy, N. I. (2000). Demonstration of a foraging advantage for trichromatic marmosets (*Callithrix geoffroyi*) dependent on food colour. *Proceedings of the Royal Society of London B, 267*, 439–444.
Caprio, J., Brand, J. G., Teeter, J. H., Valentincic, T., Kalinoski, D. L., Kohbara, J., Kumazawa, T., & Wegert, S. (1993). The taste system of the channel catfish: From biophysics to behavior. *TINS, 16*, 192–197.
Carleton, K. L., Escobar-Camacho, D., Stieb, S. M., Cortesi, F., & Marshall, N. J. (2020). Seeing the rainbow: Mechanisms underlying spectral sensitivity in teleost fishes. *Journal of Experimental Biology, 223*, jeb193334.
Carr, C. E., & Konishi, M. (1990). A circuit for detection of interaural time differences in the brain stem of the barn owl. *Journal of Neuroscience, 10*, 3227–3246.
Carvell, G. E., & Simons, D. J. (1995). Task- and subject-related differences in sensorimotor behavior during active touch. *Somatosensory & Motor Research, 12*, 1–9.
Catania, K. (2014). The shocking predatory strike of the electric eel. *Science, 346*(6214), 1231–1234.
Catania, K. C. (2010). Born knowing: Tentacled snakes innately predict future prey behavior. *PLoS ONE, 5*, e10953.
Catania, K. C. (2011). The sense of touch in the star-nosed mole: From mechanoreceptors to the brain. *Philosophical Transactions of the Royal Society B, 366*, 3016–3025.
Catania, K. C. (2012). A nose for touch. *The Scientist, 26*(9), 29–33.
Catania, K. C., & Kaas, J. H. (1997). Somatosensory fovea in the star-nosed mole: Behavioral use of the star in relation to innervation patterns and cortical representation. *Journal of Comparative Neurology, 387*, 215–233.
Catania, K. C., Leitch, D. B., & Gauthier, D. (2010). Function of the appendages in tentacled snakes (*Erpeton tentaculatus*). *Journal of Experimental Biology, 213*, 359–367.
Catania, K. C., & Remple, F. E. (2005). Asymptotic prey profitability drives star-nosed moles to the foraging speed limit. *Nature, 433*, 519–522.
Caterina, M. J., Leffler, A., Malmberg, A. B., Martin, W. J., Trafton, J., Petersen-Zeitz, K. R., Koltzenburg, M., Basbaum, A. I., & Julius, D. (2000). Impaired nociception and pain sensation in mice lacking the capsaicin receptor. *Science, 288*, 306–313.
Caterina, M. J., Schumacher, M. A., Tominaga, M., Rosen, T. A., Levine, J. D., & Julius, D. (1997). The capsaicin receptor: A heat-activated ion channel in the pain pathway. *Nature, 389*, 816–824.
Cavoto, B. R., & Cook, R. G. (2006). The contribution of monocular depth cues to scene perception by pigeons. *Psychological Science, 17*, 628–634.
Cerveira, A. M., Nelson, X. J., & Jackson, R. R. (2021). Spatial acuity-sensitivity trade-off in the principal eyes of a jumping spider: Possible adaptations to a 'blended' lifestyle. *Journal of Comparative Physiology A, 207*, 437–448.
Chae, K.-S., Oh, I.-T., Lee, S.-H., & Kim, S.-C. (2019). Blue light-dependent human magnetoreception in geomagnetic food orientation. *PLoS ONE, 14*, e0211826.
Chandrashekar, J., Hoon, M. A., Ryba, N. J. P., & Zuker, C. S. (2006). The receptors and cells for mammalian taste. *Nature, 444*, 288–294.
Chang, R. B., Waters, H., & Liman, E. R. (2010). A proton current drives action potentials in genetically identified sour taste cells. *PNAS, 107*, 22320–22325.
Chávez, A. E., Bozinovic, F., Peichl, L., & Palacios, A. G. (2003). Retinal spectral sensitivity, fur coloration, and urine reflectance in the genus *Octodon* (Rodentia): Implications for visual ecology. *Investigative Ophthalmology & Visual Science, 44*, 2290–2296.

Childs, E., & de Wit, H. (2009). Amphetamine-induced place preference in humans. *Biological Psychiatry, 65*, 900–904.

Choi, C. Q. (2007). Can a cockroach live without its head? *Scientific American, 297*(5), 116.

Clark, D. L., & Clark, R. A. (2016). Neutral point testing of color vision in the domestic cat. *Experimental Eye Research, 153*, 23–26.

Clark, W. J., & Colombo, M. (2020). The functional architecture, receptive field characteristics, and representation of objects in the visual network of the pigeon brain. *Progress in Neurobiology, 195*, 101781.

Clarke, D., Morley, E., & Robert, D. (2017). The bee, the flower, and the electric field: Electric ecology and aerial electroreception. *Journal of Comparative Physiology A, 203*, 737–748.

Clarke, D., Whitney, H., Sutton, G., & Robert, D. (2013). Detection and learning of floral electric fields by bumblebees. *Science, 340*(6128), 66–69.

Clarke, S., Bellmann, A., Meuli, R. A., Assal, G., & Steck, A. J. (2000). Auditory agnosia and auditory spatial deficits following left hemispheric lesions: Evidence for distinct processing pathways. *Neuropsychologia, 38*, 797–807.

Cohen, K. E., Flammang, B. E., Crawford, C. H., & Hernandez, L. P. (2020). Knowing when to stick: Touch receptors found in the remora adhesive disc. *Royal Society Open Science, 7*, 190990.

Collett, T. (1977). Stereopsis in toads. *Nature, 267*, 349–351.

Corcoran, A. J., Barber, J. R., & Conner, W. E. (2009). Tiger moth jams bat sonar. *Science, 325*, 325–327.

Corcoran, A. J., Barber, J. R., Hristov, N. I., & Conner, W. E. (2011). How do tiger moths jam bat sonar? *Journal of Experimental Biology, 214*, 2416–2425.

Corcoran, A. J., & Conner, W. E. (2014). Bats jamming bats: Food competition through sonar interference. *Science, 346*, 745–747.

Corey, E. A., Bobkov, Y., Ukhanov, K., & Ache, B. W. (2013). Ionotropic crustacean olfactory receptors. *PLoS ONE, 8*, e60551.

Coste, B., Mathur, J., Schmidt, M., Earley, T. J., Ranade, S., Petrus, M. J., Dubin, A. E., & Patapoutian, A. (2010). Piezo1 and Piezo2 are essential components of distinct mechanically activated cation channels. *Science, 330*, 55–60.

Craig, A. D. (2002). How do you feel? Interoception: The sense of the physiological condition of the body. *Nature Reviews: Neuroscience, 3*, 655–666.

Cronin, T. W., Bok, M. J., Marshall, N. J., & Caldwell, R. L. (2014). Filtering and polychromatic vision in mantis shrimps: Themes in visible and ultraviolet vision. *Philosophical Transactions of the Royal Society B, 369*, 20130032.

Cronin, T. W., Marshall, N. J., & Caldwell, R. L. (2000). Spectral tuning and the visual ecology of mantis shrimps. *Philosophical Transactions of the Royal Society of London B, 355*, 1263–1267.

Cronin, T. W., Marshall, N. J., & Land, M. F. (1994). The unique visual system of the mantis shrimp. *American Scientist, 82*(4), 356–365.

Crook, R. J. (2021). Behavioral and neurophysiological evidence suggests affective pain experience in octopus. *iScience, 24*, 102229.

Crowe-Riddell, J. M., Williams, R., Chapuis, L., & Sanders, K. L. (2019). Ultrastructural evidence of a mechanosensory function of scale organs (sensilla) in sea snakes (Hydrophiinae). *Royal Society Open Science, 6*, 182022.

Cummins, S. F., Boal, J. G., Buresch, K. C., Kuanpradit, C., Sobhon, P., Holm, J. B., Degnan, B. M., Nagle, G. T., & Hanlon, R. T. (2011). Extreme aggression in male squid induced by a β-MSP-like pheromone. *Current Biology, 21*, 322–327.

Cuthill, I. C., Stevens, M., Sheppard, J., Maddocks, T., Párraga, C. A., & Troscianko, T. S. (2005). Disruptive coloration and background pattern masking. *Nature, 434*, 72–74.

Czech-Damal, N. U., Liebschner, A., Miersch, L., Klauer, G., Hanke, F. D., Marshall, C., Dehnhardt, G., & Hanke, W. (2012). Electroreception in the Guiana dolphin (*Sotalia guianensis*). *Proceedings of the Royal Society B, 279*, 663–668.

Dacke, M., Baird, E., Byrne, M., Scholtz, C. H., & Warrant, E. J. (2013). Dung beetles use the Milky Way for orientation. *Current Biology, 23,* 298–300.

Daly, K., Al-Rammahi, M., Moran, A., Marcello, M., Ninomiya, Y., & Shirazi-Beechey, S. P. (2013). Sensing of amino acids by the gut-expressed taste receptor T1R1-T1R3 stimulates CCK secretion. *American Journal of Physiology: Gastrointestinal and Liver Physiology, 304,* G271–G282.

Dalziel, D. J., Uthman, B. M., McGorray, S. P., & Reep, R. L. (2003). Seizure-alert dogs: A review and preliminary study. *Seizure, 12,* 115–120.

Damasio, A., & Carvalho, G. B. (2013). The nature of feelings: Evolutionary and neurobiological origins. *Nature Reviews: Neuroscience, 14,* 143–512.

Damasio, A., Damasio, H., & Tranel, D. (2013). Persistence of feelings and sentience after bilateral damage of the insula. *Cerebral Cortex, 23,* 833–846.

D'Amico, A., Gisiner, R. C., Ketten, D. R., Hammock, J. A., Johnson, C., Tyack, P. L., & Mead, J. (2009). Beaked whale strandings and naval exercises. *Aquatic Mammals, 35,* 452–472.

Damien, J., Adam, O., Cazau, D., White, P., Laitman, J. T., & Reidenberg, J. S. (2019). Anatomy and functional morphology of the mysticete rorqual whale larynx: Phonation positions of the U-fold. *Anatomical Record, 302,* 703–717.

Dartnall, H. J. A., Bowmaker, J. K., & Mollon, J. D. (1983). Human visual pigments: Microspectrophotometric results from the eyes of seven persons. *Proceedings of the Royal Society of London B, 220,* 115–130.

Darwin, C. (1859). *On the origin of species.* John Murray, Albemarle Street. Available in many later editions and on wikisource.org.

Darwin, C. (1872). *The expression of the emotions in man and animals.* John Murray, Albemarle Street. Reprinted in 1913 by D. Appleton and Company.

Datta, S. R., Vasconcelos, M. L., Ruta, V., Luo, S., Wong, A., Demir, E., Flores, J., Balonze, K., Dickson, B. J., & Axel, R. (2008). The *Drosophila* pheromone cVA activates a sexually dimorphic neural circuit. *Nature, 452,* nature06808.

Davis, P. (2017). *The investigation of human scent from epileptic patients for the identification of a biomarker for epileptic seizures.* Ph.D. dissertation, Department of Chemistry, Florida International University, Miami, FL.

Dawkins, R. (2016). *The selfish gene* (40th anniversary ed.). Oxford University Press.

De Silva, D. V. S. X., Ekmekcioglu, E., Fernando, W. A. C., & Worrall, S. T. (2011). Display dependent preprocessing of depth maps based on just noticeable depth difference modeling. *IEEE Journal of Selected Topics in Signal Processing, 5,* #2, 335–351.

De Valois, R. L., & De Valois, K. K. (1990). *Spatial vision.* Oxford University Press.

de Vries, H., & Stuiver, M. (1961). The absolute sensitivity of the human sense of smell. In W. A. Rosenblith (Ed.), *Sensory communication* (pp. 159–167). MIT Press.

DeAngelis, G. C., Cumming, B. G., & Newsome, W. T. (1998). Cortical area MT and the perception of stereoscopic depth. *Nature, 394,* 677–680.

Dear, S. P., Fritz, J., Haresign, T., Ferragamo, M., & Simmons, J. A. (1993). Tonotopic and functional organization in the auditory cortex of the big brown bat, *Eptesicus fuscus. Journal of Neurophysiology, 70,* 1988–2009.

Deora, T., Ahmed, M. A., Daniel, T. L., & Brunton, B. W. (2021). Tactile active sensing in an insect plant pollinator. *Journal of Experimental Biology, 224,* jeb239442.

Diebel, C. E., Proksch, R., Green, C. R., Neilson, P., & Walker, M. M. (2000). Magnetite defines a vertebrate magnetoreceptor. *Nature, 406,* 299–302.

Diego-Rasilla, F. J., & Phillips, J. B. (2021). Evidence for the use of a high-resolution magnetic map by a short-distance migrant, the Alpine newt (*Ichthyosaura alpestris*). *Journal of Experimental Biology, 224,* jeb238345.

Dittman, A. H., & Quinn, T. P. (1996). Homing in Pacific salmon: Mechanisms and ecological basis. *Journal of Experimental Biology, 199,* 83–91.

 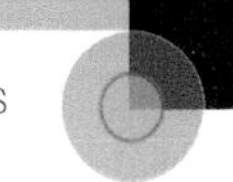

Dougherty, R. F., Koch, V. M., Brewer, A. A., Fischer, B., Modersitzki, J., & Wandell, B. A. (2003). Visual field representations and locations of visual areas V1/2/3 in human visual cortex. *Journal of Vision, 3*, 586–598.

Douglas, M. M. (1986). *The lives of butterflies*. University of Michigan Press.

Døving, K. B., & Selset, R. (1980). Behavior patterns in cod released by electrical stimulation of olfactory tract bundlets. *Science, 207*, 559–560.

Dowds, B. M., & Elwood, R. W. (1983). Shell wars: Assessment strategies and the timing of decisions in hermit crab shell fights. *Behaviour, 85*, 1–24.

Du Toit, C. J., Chinsamy, A., & Cunningham, S. J. (2020). Cretaceous origins of the vibrotactile bill-tip organ in birds. *Proceedings of the Royal Society B, 287*, 20202322.

Duarte, C. M., Chapuis, L., Collin, S. P., Costa, D. P., Devassy, R. P., Eguiluz, V. M., Erbe, C., Gordon, T. A. C., Halpern, B. S., Harding, H. R., Havlik, M. N., Meekan, M., Merchant, N. D., Miksis-Olds, J. L., Parsons, M., Predragovic, M., Radford, A. N., Radford, C. A., Simpson, S. D., . . . Juanes, F. (2021). The soundscape of the anthropocene ocean. *Science, 371*, eaba4658.

Dulac, C., & Axel, R. (1995). A novel family of genes encoding putative pheromone receptors in mammals. *Cell, 83*, 195–206.

Düring, M. V., & Seiler, W. (1974). The fine structure of lamellated receptors in the skin of *Rana esculenta*. *Zeitschrift für Anatomie und Entwicklungsgeschichte, 144*, 165–172.

D'Urso, O., & Drago, F. (2021). Pharmacological significance of extra-oral taste receptors. *European Journal of Pharmacology, 910*, 174480.

Dyer, A. G., Paulk, A. C., & Reser, D. H. (2011). Colour processing in complex environments: Insights from the visual system of bees. *Proceedings of the Royal Society B, 278*, 952–959.

Eakin, R. M. (1979). Evolutionary significance of photoreceptors: In retrospect. *American Zoologist, 19*, 647–653.

Ebeling, W., Natoli, R. C., & Hemmi, J. M. (2010). Diversity of color vision: Not all Australian marsupials are trichromatic. *PLoS ONE, 5*(12), e14231.

Ebrahim, S. A. M., Dweck, H. K. M., Stökl, J., Hofferberth, J. E., Trona, F., Weniger, K., Rybak, J., Seki, Y., Stensmyr, M. C., Sachse, S., Hansson, B. S., & Knaden, M. (2015). *Drosophila* avoids parasitoids by sensing their semiochemicals via a dedicated olfactory circuit. *PLoS Biology, 13*, e1002318.

Ehrenfeld, J. G., & Ehrenfeld, D. W. (1973). Externally secreting glands of freshwater and sea turtles. *Copeia, 1973*, 305–314.

Eisemann, C. H., Jorgensen, W. K., Merritt, D. J., Rice, M. J., Cribb, B. W., Webb, P. D., & Zalucki, M. P. (1984). Do insects feel pain? – A biological view. *Experientia, 40*, 164–167.

Ejima, A., Smith, B. P. C., Lucas, C., Van der Goes van Naters, W., Miller, C. J., Carlson, J. R., Levine, J. D., & Griffith, L. C. (2007). Generalization of courtship learning in *Drosophila* is mediated by *cis*-vaccenyl acetate. *Current Biology, 17*, 599–605.

Eklöf, J. (2020). *The darkness manifesto: On light pollution, night ecology, and the ancient rhythms that sustain life* (E. DeNoma, Trans.). Scribner.

Elliker, K. R., Sommerville, B. A., Broom, D. M., Neal, D. E., Armstrong, S., & Williams, H. C. (2014). Key considerations for the experimental training and evaluation of cancer odour detection dogs: Lessons learnt from a double-blind, controlled trial of prostate cancer detection. *BMC Urology, 14*, 22.

Elwood, R. W., & Appel, M. (2009). Pain experience in hermit crabs? *Animal Behaviour, 77*, 1243–1246.

Emery, D. G. (1975a). Ciliated sensory cells and associated neurons in the lip of *Octopus joubini* Robson. *Cell and Tissue Research, 157*, 331–340.

Emery, D. G. (1975b). The histology and fine structure of the olfactory organ of the squid *Lolliguncula brevis* Blainville. *Tissue & Cell, 7*, 357–367.

Emlen, S. T. (1975). The stellar-orientation system of a migratory bird. *Scientific American, 233*(2), 102–111.

Emlen, S. T., & Emlen, J. T. (1966). A technique for recording migratory orientation of captive birds. *The Auk, 83*, 361–367.

Engels, S., Schneider, N.-L., Lefeldt, N., Hein, C. M., Zapka, M., Michalik, A., Elbers, D., Kittel, A., Hore, P. J., & Mouritsen, H. (2014). Anthropogenic electromagnetic noise disrupts magnetic compass orientation in a migratory bird. *Nature, 509*, 353–356.

Erber, J., Masuhr, T., & Menzel, R. (1980). Localization of short-term memory in the brain of the bee, *Apis mellifera*. *Physiological Entomology, 5*, 343–358.

Erickson, R. P. (1963). Sensory neural patterns and gustation. In Y. Zotterman (Ed.), *Olfaction and taste* (pp. 205–213). Pergamon Press.

Ernst, D. A., & Lohmann, K. J. (2016). Effect of magnetic pulses on Caribbean spiny lobsters: Implications for magnetoreception. *Journal of Experimental Biology, 219*, 1827–1832.

Evangelista, C., Kraft, P., Dacke, M., Labhart, T., & Srinivasan, M. V. (2014). Honeybee navigation: Critically examining the role of the polarization compass. *Philosophical Transactions of the Royal Society B, 369*, 20130037.

Falkenberg, G., Fleissner, G., Schuchardt, K., Kuehbacher, M., Thalau, P., Mouritsen, H., Heyers, D., Wellenreuther, G., & Fleissner, G. (2010). Avian magnetoreception: Elaborate iron mineral containing dendrites in the upper beak seem to be a common feature of birds. *PLoS ONE, 5*, e9231.

Fay, R. R. (1969). Behavioral audiogram for the goldfish. *Journal of Auditory Research, 9*, 112–121.

Fay, R. R. (1970). Auditory frequency discrimination in the goldfish (*Carassius auratus*). *Journal of Comparative and Physiological Psychology, 73*, 175–180.

Fay, R. R. (1972). Perception of amplitude-modulated auditory signals by the goldfish. *Journal of the Acoustical Society of America, 52*, 660–666.

Feng, P., & Liang, S. (2018). Molecular evolution of umami/sweet taste receptor genes in reptiles. *PeerJ, 6*, e5570.

Feord, R. C., Sumner, M. E., Pusdekar, S., Kalra, L., Gonzalez-Bellido, P. T., & Wardill, T. J. (2020). Cuttlefish use stereopsis to strike at prey. *Science Advances, 6*, eaay6036.

Ferrero, D. M., Lemon, J. K., Fluegge, D., Pashkovski, S. L., Korzan, W. J., Datta, S. R., Spehr, M., Fendt, M., & Liberles, S. D. (2011). Detection and avoidance of a carnivore odor by prey. *Proceedings of the National Academy of Sciences of the United States of America, 108*, 11235–11240.

Field, T. (2014). Massage therapy research review. *Complementary Therapies in Clinical Practice, 20*, 224–229.

Field, T., Diego, M., & Hernandez-Reif, M. (2011). Potential underlying mechanisms for greater weight gain in massaged preterm infants. *Infant Behavior and Development, 34*, 383–389.

Fineran, B. A., & Nicol, J. A. C. (1978). Studies on the photoreceptors of *Anchoa mitchilli* and *A. hepsetus* (Engraulidae) with particular reference to the cones. *Philosophical Transactions of the Royal Society of London B, 283*, 25–60.

Fite, K. V. (1973). Anatomical and behavioral correlates of visual acuity in the great horned owl. *Vision Research, 13*, 219–230.

Flanigan, K. A. S., Wiegmann, D. D., Hebets, E. A., & Bingman, V. P. (2021). Multisensory integration supports configural learning of a home refuge in the whip spider *Phrynus marginemaculatus*. *Journal of Experimental Biology, 224*, jeb238444.

Fobert, E. K., Burke da Silva, K., & Swearer, S. E. (2019). Artificial light at night causes reproductive failure in clownfish. *Biology Letters, 15*, 20190272.

Foster, J. J., Tocco, C., Smolka, J., Khaldy, L., Baird, E., Byrne, M. J., Nilsson, D.-E., & Dacke, M. (2021). Light pollution forces a chane in dung beetle orientation behavior. *Current Biology, 31*, 1–8.

Fox, R., & Blake, R. R. (1971). Stereoscopic vision in the cat. *Nature, 233*, 55–56.

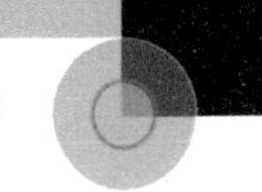

Fox, R., Lehmkuhle, S. W., & Bush, R. C. (1977). Stereopsis in the falcon. *Science, 197,* 79–81.

Frank, H. E. R., Amato, K., Trautwein, M., Maia, P., Liman, E. R., Nichols, L. M., Schwenk, K., Breslin, P. A. S., & Dunn, R. R. (2022). The evolution of sour taste. *Proceedings of the Royal Society B, 289,* 20211918.

Frank, M. (1973). An analysis of hamster afferent taste nerve response functions. *Journal of General Physiology, 61,* 588–618.

Frank, M. E., Lundy, R. F., Jr., & Contreras, R. J. (2008). Cracking taste codes by tapping into sensory neuron impulse traffic. *Progress in Neurobiology, 86,* 245–263.

Frisch, K. V. (1967). *The dance language and orientation of bees* (L. E. Chadwick, Trans.). Belknap Press of Harvard University Press. Translated from *Tanzsprache und Orientierung der Bienen,* Springer-Verlag, 1965.

Fukutomi, M., & Carlson, B. A. (2020). A history of corollary discharge: Contributions of mormyrid weakly electric fish. *Frontiers in Integrative Neuroscience, 14,* Article 42.

Furuta, T., Bush, N. E., Yang, A. E.-T., Ebara, S., Miyazaki, N., Murata, K., Hirai, D., Shibata, K., & Hartmann, M. J. Z. (2020). The cellular and mechanical basis for response characteristics of identified primary afferents in the rat vibrissal system. *Current Biology, 30,* 815–826.

Gadziola, M. A., Grimsley, J. M. S., Faure, P. A., & Wenstrup, J. J. (2012). Social vocalizations of big brown bats vary with behavioral context. *PLoS ONE, 7,* e44550.

Gaffney, M. F., & Hodos, W. (2003). The visual acuity and refractive state of the American kestrel (*Falco sparverius*). *Vision Research, 43,* 2053–2059.

Gagliardo, A. (2013). Forty years of olfactory navigation in birds. *Journal of Experimental Biology, 216,* 2165–2171.

Gagliardo, A., Odetti, F., & Ioalè, P. (2001). Relevance of visual cues for orientation at familiar sites by homing pigeons: An experiment in a circular arena. *Proceedings of the Royal Society of London B, 268,* 2065–2070.

Gagliardo, A., Pollonara, E., & Wikelski, M. (2020). Pigeons remember visual landmarks after one release and rely upon them more if they are anosmic. *Animal Behaviour, 166,* 85–94.

Garm, A., & Nilsson, D.-E. (2014). Visual navigation in starfish: First evidence for the use of vision and eyes in starfish. *Proceedings of the Royal Society B, 281,* 20133011.

Geldard, F. A. (1972). *The human senses* (2nd ed.). John Wiley & Sons.

Gentle, M. J., Tilston, V., & McKeegan, D. E. F. (2001). Mechanothermal nociceptors in the scaly skin of the chicken leg. *Neuroscience, 106,* 643–652.

Gescheider, G. A., Wright, J. H., & Verrillo, R. T. (2009). *Information-processing channels in the tactile sensory system: A psychophysical and physiological analysis.* Psychology Press.

Gibson, E. J., & Walk, R. D. (1960). The "visual cliff". *Scientific American, 202*(4), 64–71.

Gilbert, A. (2008). *What the nose knows: The science of scent in everyday life.* Crown Publishers.

Gilly, W. F., & Lucero, M. T. (1992). Behavioral responses to chemical stimulation of the olfactory organ in the squid *Loligo opalescens. Journal of Experimental Biology, 162,* 209–229.

Gleich, O., Dooling, R. J., & Manley, G. A. (2005). Audiogram, body mass, and basilar papilla length: Correlations in birds and predictions for extinct archosaurs. *Naturwissenschaften, 92,* 595–598.

Godfrey-Smith, P. (2016). *Other minds: The octopus, the sea, and the deep origins of consciousness.* Farrar, Straus and Giroux.

Godfrey-Smith, P. (2020). *Metazoa: Animal life and the birth of the mind.* Farrar, Straus and Giroux.

Goldsmith, T. H., & Butler, B. K. (2005). Color vision of the budgerigar (*Melopsittacus undulatus*): Hue matches, tetrachromacy, and intensity discrimination. *Journal of Comparative Physiology A, 191,* 933–951.

Goodman, J. M., & Bensmaia, S. J. (2018). The neural basis of haptic perception. In J. T. Wixted (Ed.), *Stevens' handbook of*

experimental psychology and cognitive neuroscience (fourth edition). Vol. 2 (Serences, J., Volume editor), Sensation, perception, & attention (pp. 201–239). Wiley.

Göpfert, M. C., & Hennig, R. M. (2016). Hearing in insects. *Annual Review of Entomology, 61,* 257–276.

Gould, J. L. (1980). The case for magnetic sensitivity in birds and bees (such as it is). *American Scientist, 68*(3), 256–267.

Gracheva, E. O., Cordero-Morales, J. F., González-Carcacía, J. A., Ingolia, N. T., Manno, C., Aranguren, C. I., Weissman, J. S., & Julius, D. (2011). Ganglion-specific splicing of TRPV1 underlies infrared sensation in vampire bats. *Nature, 476,* 88–92.

Gracheva, E. O., Ingolia, N. T., Kelly, Y. M., Cordero-Morales, J. F., Hollopeter, G., Chesler, A. T., Sánchez, E. E., Perez, J. C., Weissman, J. S., & Julius, D. (2010). Molecular basis of infrared detection by snakes. *Nature, 464,* 1006–1012.

Graham, C. H., & Hsia, Y. (1958). Color defect and color theory. *Science, 127,* 675–682.

Gray, E. G. (1970). The fine structure of the vertical lobe of the octopus brain. *Philosophical Transactions of the Royal Society of London, Series B, 258,* 379–394.

Graziadei, P. (1962). Receptors in the suckers of *Octopus. Nature, 195,* 57–59.

Graziadei, P. (1964). Electron microscopy of some primary receptors in the sucker of *Octopus vulgaris. Zeitschrift für Zellforschung, 64,* 510–522.

Greppi, C., Laursen, W. J., Budelli, G., Chang, E. C., Daniels, A. M., Giesen, L. V., Smidler, A. L., Catteruccia, F., & Garrity, P. A. (2020). Mosquito heat seeking is driven by an ancestral cooling receptor. *Science, 367,* 681–684.

Gross, C. G. (2002). Genealogy of the "grandmother cell." *Neuroscientist, 8,* 512–518.

Gross, C. G., Rocha-Miranda, C. E., & Bender, D. B. (1972). Visual properties of neurons in inferotemporal cortex of the macaque. *Journal of Neurophysiology, 5,* 96–111.

Grötzner, S. R., Rocha, F. A. de F., Corredor, V. H., Liber, A. M. P., Hamassaki, D. E., Bonci, D. M. O., & Ventura, D. F. (2020). Distribution of rods and cones in the red-eared turtle retina (*Trachemys scripta elegans*). *Journal of Comparative Neurology, 528,* 1548–1560.

Guerra, P. A., Gegear, R. J., & Reppert, S. M. (2014). A magnetic compass aids monarch butterfly migration. *Nature Communications, 5,* 4164.

Guilford, T., & Taylor, G. K. (2014). The sun compass revisited. *Animal Behavior, 97,* 135–143.

Gunderson, V. M., Yonas, A., Sargent, P. L., & Grant-Webster, K. S. (1993). Infant macaque monkeys respond to pictorial depth. *Psychological Science, 4,* 93–98.

Günther, A., Dedek, K., Haverkamp, S., Irsen, S., Briggman, K. L., & Mouritsen, H. (2021). Double cones and the diverse connectivity of photoreceptors and bipolar cells in an avian retina. *Journal of Neuroscience, 41,* 5015–5028.

Günther, A., Einwich, A., Sjulstok, E., Feederle, R., Bolte, P., Koch, K.-W., Solov'yov, I. A., & Mouritsen, H. (2018). Double-cone localization and seasonal expression pattern suggest a role in magnetoreception for European robin cryptochrome 4. *Current Biology, 28,* 211–223.

Halata, Z., Grim, M., & Bauman, K. L. (2003). Friedrich Sigmund Merkel and his "Merkel cell", morphology, development, and physiology: Review and new results. *Anatomical Record, Part A, 271A,* 225–239.

Haller, N. K., Lind, O., Steinlechner, S., & Kelber, A. (2014). Stimulus motion improves spatial contrast sensitivity in budgerigars (*Melopsittacus undulatus*). *Vision Research, 102,* 19–25.

Hamamoto, D. T., & Simone, D. A. (2003). Characterization of cutaneous primary afferent fibers excited by acetic acid in a model of nociception in frogs. *Journal of Neurophysiology, 90,* 566–577.

Hamdani, E. H., & Døving, K. B. (2007). The functional organization of the fish olfactory system. *Progress in Neurobiology, 82,* 80–86.

Hamdani, E. H., Kasumyan, A., & Døving, K. B. (2001). Is feeding behaviour in crucian carp mediated by the lateral olfactory tract? *Chemical Senses, 26,* 1133–1138.

Hamdani, E.-H., Stabell, O. B., Alexander, G., & Døving, K. B. (2000). Alarm reaction in the crucian carp is mediated by the medial bundle of the medial olfactory tract. *Chemical Senses, 25,* 103–109.

Hamdorf, K. (1979). The physiology of invertebrate visual pigments. In H. Autrum (Ed.), *Comparative physiology and evolution of vision in invertebrates. A: Invertebrate photoreceptors* (pp. 145–224). Springer-Verlag.

Hamilton, P. V. (1976). Predation on *Littorina irrorata* (Mollusca: Gastropoda) by *Callinectes sapidus* (Crustacea: Portunidae). *Bulletin of Marine Science, 26,* 403–409.

Hamilton, P. V., Ardizzoni, S. C., & Penn, J. S. (1983). Eye structure and optics in the intertidal snail, *Littorina irrorata. Journal of Comparative Physiology, 152,* 435–445.

Hamilton, P. V., & Winter, M. A. (1982). Behavioural responses to visual stimuli by the snail *Littorina irrorata. Animal Behaviour, 30,* 752–760.

Hamilton, P. V., & Winter, M. A. (1984). Behavioural responses to visual stimuli by the snails *Tectarius muricatus, Turbo castanea,* and *Helix aspersa. Animal Behaviour, 32,* 51–57.

Hanke, F. D., & Kelber, A. (2020). The eye of the common octopus (*Octopus vulgaris*). *Frontiers in Physiology, 10,* Article 1637.

Hanlon, R. T., Chiao, C.-C., Mäthger, L. M., & Marshall, N. J. (2013). A fish-eye view of cuttlefish camouflage using in situ spectrometry. *Biological Journal of the Linnean Society, 109,* 535–551.

Hanlon, R. T., & Messenger, J. B. (1988). Adaptive coloration in young cuttlefish (*Sepia officinalis* L.): The morphology and development of body patterns and their relation to behaviour. *Philosophical Transactions of the Royal Society of London B, 320,* 437–487.

Hanlon, R. T., & Messenger, J. B. (2018). *Cephalopod behavior* (2nd ed.). Cambridge University Press.

Hardy, A. R., & Hale, M. E. (2020). Sensing the structural characteristics of surfaces: Texture encoding by a bottom-dwelling fish. *Journal of Experimental Biology, 223,* jeb227280.

Hardy, A. R., Steinworth, B. M., & Hale, M. E. (2016). Touch sensation by pectoral fins of the catfish *Pimelodus pictus. Proceedings of the Royal Society B, 283,* 20152652.

Harlow, H. F. (1958). The nature of love. *American Psychologist, 13,* 673–685.

Harlow, H. F., Dodsworth, R. O., & Harlow, M. K. (1965). Total social isolation in monkeys. *Proceedings of the National Academy of Sciences, 54,* 90–97.

Hartline, H. K. (1938). The discharge of impulses in the optic nerve of *Pecten* in response to illumination of the eye. *Journal of Cellular and Comparative Physiology, 11,* 465–477.

Hartline, H. K., & Ratliff, F. (1957). Inhibitory interaction of receptor units in the eye of *Limulus. Journal of General Physiology, 40,* 357–376.

Harzsch, S., & Hansson, B. S. (2008). Brain architecture in the terrestrial hermit crab *Coenobita clypeatus* (Anomura, Coenobitidae), a crustacean with a good aerial sense of smell. *BMC Neuroscience, 9,* 58.

Hataji, Y., Kuroshima, H., & Fujita, K. (2021). Motion parallax via head movements modulates visuo-motor control in pigeons. *Journal of Experimental Biology, 224,* jeb236547 (8 pages).

Hawryshyn, C. W. (2000). Ultraviolet polarization vision in fishes: Possible mechanisms for coding e-vector. *Philosophical Transactions of the Royal Society of London B, 355,* 1187–1190.

Hawryshyn, C. W., & Beauchamp, R. (1985). Ultraviolet photosensitivity in goldfish: An independent U.V. retinal mechanism. *Vision Research, 25,* 11–20.

Haynes, K. F., Yeargan, K. V., & Gemeno, C. (2001). Detection of prey by a spider that aggressively mimics pheromone blends. *Journal of Insect Behavior, 14,* 535–544.

Hecht, S., Schlaer, S., & Pirenne, M. H. (1942). Energy, quanta, and vision. *Journal of General Physiology, 25,* 819–840.

Hedwig, B., & Poulet, J. F. A. (2004). Complex auditory behaviour emerges from simple reactive steering. *Nature, 430,* 781–785.

Heerema, J. L., Jackman, K. W., Miliano, R. C., Li, L., Zaborniak, T. S. M., Veldhoen, N., Aggelen, G. V., Parker, W. J., Pyle, G. G., & Helbing, C. C. (2018). Behavioral and molecular analyses of olfaction-mediated avoidance responses of *Rana (Lithobates) catesbeiana* tadpoles: Sensitivity to thyroid hormones, estrogen, and treated municipal wastewater effluent. *Hormones and Behavior, 101*, 85–93.

Heffner, R. S., & Heffner, H. E. (1982). Hearing in the elephant (*Elephas maximus*): Absolute sensitivity, frequency discrimination, and sound localization. *Journal of Comparative and Physiological Psychology, 96*, 926–944.

Hepper, P., & Wells, D. (2015). Olfaction in the order carnivora: Family Canidae. In R. L. Doty (Ed.), *Handbook of olfaction and gustation* (3rd ed., pp. 591–603). Wiley.

Hepper, P. G., & Wells, D. L. (2005). How many footsteps do dogs need to determine the direction of an odour trail? *Chemical Senses, 30*, 291–298.

Herrada, G., & Dulac, C. (1997). A novel family of putative pheromone receptors in mammals with a topographically organized and sexually dimorphic distribution. *Cell, 90*, 763–773.

Herz, R. S., & Schooler, J. W. (2002). A naturalistic study of autobiographical memories evoked by olfactory and visual cues: Testing the Proustian hypothesis. *American Journal of Psychology, 115*, 21–32.

Heyers, D., Manns, M., Luksch, H., Güntürkün, O., & Mouritsen, H. (2007). A visual pathway links brain striuctures active during magnetic compass orientation in migratory birds. *PLoS ONE, 2*, e937.

Heyers, D., Zapka, M., Hoffmeister, M., Wild, J. M., & Mouritsen, H. (2010). Magnetic field changes activate the trigeminal brainstem complex in a migratory bird. *PNAS, 107*, 9394–9399.

Higgs, D. M., & Radford, C. A. (2013). The contribution of the lateral line to 'hearing' in fish. *Journal of Experimental Biology, 216*, 1484–1490.

Hodos, W., Bessette, B. B., Macko, K. A., & Weiss, S. R. B. (1985). Normative data for pigeon vision. *Vision Research, 25*, 1525–1527.

Hodos, W., Miller, R. F., & Fite, K. V. (1991). Age-dependent changes in visual acuity and retinal morphology in pigeons. *Vision Research, 31*, 669–677.

Höglund, J., Mitkus, M., Olsson, P., Lind, O., Drews, A., Bloch, N. I., Kelber, A., & Strandh, M. (2019). Owls lack UV-sensitive cone opsin and red oil droplets, but see UV light at night: Retinal transcriptomes and ocular media transmittance. *Vision Research, 158*, 109–119.

Holland, R. A. (2010). Differential effects of magnetic pulses on the orientation of naturally migrating birds. *Journal of the Royal Society Interface, 7*, 1617–1625.

Holland, R. A., Borissov, I., & Siemers, B. M. (2010). A nocturnal mammal, the greater mouse-eared bat, calibrates a magnetic compass by the sun. *PNAS, 107*, 6941–6945.

Holland, R. A., Kirschvink, J. L., Doak, T. G., & Wikelski, M. (2008). Bats use magnetite to detect the Earth's magnetic field. *PLoS ONE, 3*, e1676.

Hollins, M. (1989). *Understanding blindness: An integrative approach*. Lawrence Erlbaum Associates. (Reprinted in 2022 by Routledge, an imprint of the Taylor & Francis Group)

Hollins, M. (2010). Somesthetic senses. *Annual Review of Psychology, 61*, 243–271.

Hom, K. N., Linnenschmidt, M., Simmons, J. A., & Simmons, A. M. (2016). Echolocation behavior in big brown bats is not impaired after intense broadband noise exposures. *Journal of Experimental Biology, 219*, 3253–3260.

Hore, P. J., & Mouritsen, H. (2022). The quantum nature of bird migration. *Scientific American, 326*(4), 26–31.

Howard, S. R., Prendergast, K., Symonds, M. R. E., Shrestha, M., & Dyer, A. G. (2021). Spontaneous choices for insect-pollinated flower shapes by wild non-eusocial halictid bees. *Journal of Experimental Biology, 224*, jeb242457.

Huang, A. L., Chen, X., Hoon, M. A., Chandrashekar, J., Guo, W., Tränkner, D.,

Ryba, N. J. P., & Zuker, C. S. (2006). The cells and logic for mammalian sour taste detection. *Nature, 442*, 934–938.

Huang, Y. A., Maruyama, Y., Stimac, R., & Roper, S. D. (2008). Presynaptic (Type III) cells in mouse taste buds sense sour (acid) taste. *Journal of Physiology, 586*, 2903–2912.

Hubel, D. H., & Wiesel, T. N. (1962). Receptive fields, binocular interaction and functional architecture in the cat's visual cortex. *Journal of Physiology, 160*, 106–154.

Hubel, D. H., & Wiesel, T. N. (1977). Functional architecture of macaque monkey visual cortex. *Proceedings of the Royal Society of London B, 198*, 1–59.

Hudson, S. D., Sims, C. A., Odabasi, A. Z., Colquhoun, T. A., Snyder, D. J., Stamps, J. J., Dotson, S. C., Puentes, L., & Bartoshuk, L. M. (2018). Flavor alterations associated with miracle fruit and *Gymnema sylvestre*. *Chemical Senses, 43*, 481–488.

Hulgard, K., Moss, C. F., Jakobsen, L., & Surlykke, A. (2016). Big brown bats (*Eptesicus fuscus*) emit intense search calls and fly in stereotyped flight paths as they forage in the wild. *Journal of Experimental Biology, 219*, 334–340.

Hunt, D. M., Dulai, K. S., Bowmaker, J. K., & Mollon, J. D. (1995). The chemistry of John Dalton's color blindness. *Science, 267*, 984–988.

Hunt, J. E., Bruno, J. R., & Pratt, K. G. (2020). An innate color preference displayed by *Xenopus* tadpoles is persistent and requires the tegmentum. *Frontiers in Behavioral Neuroscience, 14*, Article 71, 1–11.

Hurvich, L. M. (1981). *Color vision*. Sinauer Associates.

Hwang, R. Y., Zhong, L., Xu, Y., Johnson, T., Zhang, F., Deisseroth, K., & Tracey, W. D. (2007). Nociceptive neurons protect *Drosophila* larvae from parasitoid wasps. *Current Biology, 17*, 2105–2116.

Iggo, A., & Muir, A. R. (1969). The structure and function of a slowly adapting touch corpuscle in hairy skin. *Journal of Physiology, 200*, 763–796.

Ikeda, K. (2002). New seasonings (translated by Ogiwara, Y., & Ninomiya, Y.). *Chemical Senses, 27*, 847–849. (Original work published in the *Journal of the Chemical Society of Tokyo*, No. 30, 820–836 (1909))

Illich, P. A., & Walters, E. T. (1997). Mechanosensory neurons innervating *Aplysia* siphon encode noxious stimuli and display nociceptive sensitization. *Journal of Neuroscience, 17*, 459–469.

Inui-Yamamoto, C., Yamamoto, T., Ueda, K., Nakatsuka, M., Kumabe, S., Inui, T., & Iwai, Y. (2017). Taste preference changes throughout different life stages in male rats. *PLoS ONE, 12*, e0181650.

IUCN. (2021). *The IUCN red list of threatened species. Version 2021–3*. Retrieved April 30, 2022, from www.iucnredlist.org

Jackson, N. W., & Elwood, R. W. (1989). How animals make assessments: Information gathering by the hermit crab *Pagurus bernhardus*. *Animal Behaviour, 38*, 951–957.

Jackson, R. R., & Pollard, S. D. (1996). Predatory behavior of jumping spiders. *Annual Review of Entomology, 41*, 287–308.

Jacob, S., Zelano, B., Gungor, A., Abbott, D., Naclerio, R., & McClintock, M. K. (2000). Location and gross morphology of the nasopalatine duct in human adults. *Archives of Otolaryngology—Head & Neck Surgery, 126*, 741–748.

Jacobs, G. H. (1993). The distribution and nature of colour vision among the mammals. *Biological Review, 68*, 413–471.

Jacobs, G. H. (2009). Evolution of colour vision in mammals. *Philosophical Transactions of the Royal Society B, 364*, 2957–2967.

Jacobs, G. H., & Nathans, J. (2009). The evolution of primate color vision. *Scientific American, 300*(4), 56–63.

Jang, H.-J., Kokrashvili, Z., Theodorakis, M. J., Carlson, O. D., Kim, B.-J., Zhou, J., Kim, H. H., Xu, X., Chan, S. L., Juhaszova, M., Bernier, M., Mosinger, B., Margolskee, R. F., & Egan, J. M. (2007). Gut-expressed gustducin and taste receptors regulate secretion of glucagon-like peptide-1. *PNAS, 104*, 15069–15074.

Jen, P. H.-S., & Suga, N. (1976). Coordinated activities of middle-ear and laryngeal muscles in echolocating bats. *Science, 191*, 950–952.

Jiang, P., Josue, J., Li, X., Glaser, D., Li, W., Brand, J. G., Margolskee, R. F., Reed, D. R., & Beauchamp, G. K. (2012). Major taste loss in carnivorous mammals. *PNAS, 109,* 4956–4961.

Johansson, R. S., & Vallbo, Å. B. (1979). Tactile sensibility in the human hand: Relative and absolute densities of four types of mechanoreceptive units in glabrous skin. *Journal of Physiology, 286,* 283–300.

Johnson, K. (2002). Neural basis of haptic perception. In H. Pashler (Editor-in-Chief) & S. Yantis (Volume Ed.), *Stevens' handbook of experimental psychology, 3rd ed., volume 1: Sensation and perception* (pp. 537–583). John Wiley & Sons.

Jones, T. K., Wohlgemuth, M. J., & Conner, W. E. (2018). Active acoustic interference elicits echolocation changes in heterospecific bats. *Journal of Experimental Biology, 221,* jeb176511.

Josef, N., Mann, O., Sykes, A. V., Fiorito, G., Reis, J., Maccusker, S., & Shashar, N. (2014). Depth perception: Cuttlefish (*Sepia officinalis*) respond to visual texture density gradients. *Animal Cognition, 17,* 1393–1400.

Kalmijn, A. J. (1971). The electric sense of sharks and rays. *Journal of Experimental Biology, 55,* 371–383.

Kamermans, N., & Hawryshyn, C. (2011). Teleost polarization vision: How it might work and what it might be good for. *Philosophical Transactions of the Royal Society B, 366,* 742–756.

Kastelein, R. A., Hoek, L., de Jong, C. A. F., & Wensveen, P. J. (2010). The effect of signal duration on the underwater detection thresholds of a harbor porpoise (*Phocoena phocoena*) for single frequency-modulated tonal signals between 0.25 and 160 kHz. *Journal of the Acoustical Society of America, 128,* 3211–3222.

Katz, D. (1989). *The world of touch* (L. E. Krueger, Ed. & Trans.). Erlbaum. (Original work published 1925)

Katz, D. B., Simon, S. A., & Nicolelis, M. A. L. (2001). Dynamic and multimodal responses of gustatory cortical neurons in awake rats. *Journal of Neuroscience, 21,* 4478–4489.

Katz, I., Shomrat, T., & Nesher, N. (2021). Feel the light: Sight-independent negative phototactic response in octopus arms. *Journal of Experimental Biology, 224,* jeb237529.

Kaur, A. W., Ackels, T., Kuo, T.-H., Cichy, A., Dey, S., Hays, C., Kateri, M., Logan, D. W., Marton, T. F., Spehr, M., & Stowers, L. (2014). Murine pheromone proteins constitute a context-dependent combinatorial code governing multiple social behaviors. *Cell, 157,* 676–688.

Kavšek, M., Granrud, C. E., & Yonas, A. (2009). Infants' responsiveness to pictorial depth cues in preferential-reaching studies: A meta-analysis. *Infant Behavior and Development, 32,* 245–253.

Kawase, H., Okata, Y., & Ito, K. (2015). Role of huge geometric circular structures in the reproduction of a marine pufferfish. *Scientific Reports, 3,* 2106.

Kazial, K. A., & Masters, W. M. (2004). Female big brown bats, *Eptesicus fuscus,* recognize sex from a caller's echolocation signals. *Animal Behaviour, 67,* 855–863.

Keeling, C. I., Slessor, K. N., Higo, H. A., & Winston, M. L. (2003). New components of the honey bee (*Apis mellifera* L.) queen retinue pheromone. *PNAS, 100,* 4486–4491.

Kelber, A. (2019). Infrared imaging: A motion detection circuit for rattlesnake thermal vision. *Current Biology, 29,* R403–R405.

Kingston, A. C. N., Kuzirian, A. M., Hanlon, R. T., & Cronin, T. W. (2015). Visual phototransduction components in cephalopod chromatophores suggest dermal photoreception. *Journal of Experimental Biology, 218,* 1596–1602.

Kirschvink, J. L., Padmanabha, S., Boyce, C. K., & Oglesby, J. (1997). Measurement of the threshold sensitivity of honeybees to weak, extremely low-frequency magnetic fields. *Journal of Experimental Biology, 200,* 1363–1368.

Kitayama, C., Yamaguchi, Y., Kondo, S., Ogawa, R., Kawai, Y. K., Kayano, M., Tomiyasu, J., & Kondoh, D. (2020). Behavioral effects of scents from male mature Rathke glands

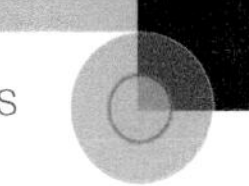

on juvenile green sea turtles (*Chelonia mydas*). *Journal of Veterinary Medical Science, 82,* 1312–1315.

Klatt, D. H., & Stefanski, R. A. (1974). How does a mynah bird imitate human speech? *Journal of the Acoustical Society of America, 55,* 822–832.

Kobylkov, D., Schwarze, S., Michalik, B., Winklhofer, M., Mouritsen, H., & Heyers, D. (2020). A newly identified trigeminal pain pathway in a night-migratory bird could be dedicated to transmitting magnetic map information. *Proceedings of the Royal Society B, 287,* 20192788.

Koch, C. (2019). *The feeling of life itself: Why consciousness is widespread but can't be computed.* MIT Press.

Koffka, K. (1935). *Principles of Gestalt psychology.* Harcourt, Brace and Company.

Konishi, M. (1973). How the owl tracks its prey. *American Scientist, 61,* 414–424. (Reprinted 2012 in *American Scientist, 100,* 494–503)

Koshitaka, H., Kinoshita, M., Vorobyev, M., & Arikawa, K. (2008). Tetrachromacy in a butterfly that has eight varieties of spectral receptors. *Proceedings of the Royal Society B, 275,* 947–954.

Kostyk, S. K., & Grobstein, P. (1987). Neuronal organization underlying visually elicited prey orienting in the frog—I. Effects of various unilateral lesions. *Neuroscience, 21,* 41–55.

Kraft, P., Evangelista, C., Dacke, M., Labhart, T., & Srinivasan, M. V. (2011). Honeybee navigation: Following routes using polarized-light cues. *Philosophical Transactions of the Royal Society of London B, 366,* 703–708.

Krahe, R., & Maler, L. (2014). Neural maps in the electrosensory system of weakly electric fish. *Current Opinion in Neurobiology, 24,* 13–21.

Kramer, G. (1952). Experiments on bird orientation. *Ibis, 94,* 265–285.

Krumm, B., Klump, G., Köppl, C., & Langemann, U. (2017). Barn owls have ageless ears. *Proceedings of the Royal Society B, 284,* 20171584.

Kurtovic, A., Widmer, A., & Dickson, B. J. (2007). A single class of olfactory neurons mediates behavioural responses to a *Drosophila* sex pheromone. *Nature, 446,* nature05672.

Land, M. F. (1965). Image formation by a concave reflector in the eye of the scallop *Pecten maximus. The Journal of Physiology, 179,* 138–153.

Land, M. F. (1969a). Structure of the retinae of the principal eyes of jumping spiders (Salticidae: Dendryphantinae) in relation to visual optics. *Journal of Experimental Biology, 51,* 443–470.

Land, M. F. (1969b). Movements of the retinae of jumping spiders (Salticidae: Dendryphantinae) in response to visual stimuli. *Journal of Experimental Biology, 51,* 471–493.

Land, M. F. (1984). Molluscs. In M. A. Ali (Ed.), *Photoreception and vision in invertebrates* (pp. 699–725). Plenum Press.

Landmann, L., & Halata, Z. (1980). Merkel cells and nerve endings in the labial epidermis of a lizard. *Cell and Tissue Research, 210,* 353–357.

Langbauer, W. R., Jr., Payne, K. B., Charif, R. A., Rapaport, L., & Osborn, F. (1991). African elephants respond to distant playbacks of low-frequency conspecific calls. *Journal of Experimental Biology, 157,* 35–46.

Larsen, F., Krog, C., & Eigaard, O. R. (2013). Determining optimal pinger spacing for harbour porpoise bycatch mitigation. *Endangered Species Research, 20,* 147–152.

Laska, M. (2017). Human and animal olfactory abilities compared. In A. Buettner (Ed.), *Springer handbook of odor* (pp. 675–689). Springer.

Lebreton, S., Borrero-Echeverry, F., Gonzalez, F., Solum, M., Wallin, E. A., Hedenström, E., Hansson, B. S., Gustavsson, A.-L., Bengtsson, M., Birgersson, G., Walker, W. B., III, Dweck, H. K. M., Becher, P. G., & Witzgall, P. (2017). A *Drosophila* female pheromone elicits species-specific long-range attraction via an olfactory channel with dual specificity for sex and food. *BMC Biology, 15,* 88.

Leclaire, S., Bourret, V., & Bonadonna, F. (2017). Blue petrels recognize the odor of their egg. *Journal of Experimental Biology, 220,* 3022–3025.

Lederman, S. J., & Klatzky, R. L. (1987). Hand movements: A window into haptic object recognition. *Cognitive Psychology, 19,* 342–368.

Lettvin, J. V., Maturana, H. R., McCulloch, W. S., & Pitts, W. H. (1959). What the frog's eye tells the frog's brain. *Proceedings of the IRE, 47*(11), 1940–1951.

Levine, J. S., & MacNichol, E. F., Jr. (1982). Color vision in fishes. *Scientific American, 246*(2), 140–149.

Liman, E. R., & Kinnamon, S. C. (2021). Sour taste: Receptors, cells and circuits. *Current Opinion in Physiology, 20,* 8–15.

Liman, E. R., Zhang, Y. V., & Montell, C. (2014). Peripheral coding of taste. *Neuron, 81,* 984–1000.

Lohmann, K. J., Goforth, K. M., Mackiewicz, A. G., Lim, D. S., & Lohmann, C. M. F. (2022). Magnetic maps in animal navigation. *Journal of Comparative Physiology A, 208,* 41–67.

Lohmann, K. J., & Lohmann, C. M. F. (1994). Detection of magnetic inclination angle by sea turtles: A possible mechanism for determining latitude. *Journal of Experimental Biology, 194,* 23–32.

Lohmann, K. J., & Lohmann, C. M. F. (1996). Detection of magnetic field intensity by sea turtles. *Nature, 380,* 59–61.

Lohmann, K. J., Lohmann, C. M. F., Ehrhart, L. M., Bagley, D. A., & Swing, T. (2004). Geomagnetic map used in sea-turtle navigation. *Nature, 428,* 909–910.

Lohmann, K. J., Putman, N. F., & Lohmann, C. M. F. (2012). The magnetic map of hatchling loggerhead sea turtles. *Current Opinion in Neurobiology, 22,* 336–342.

Lohmann, K. J., & Willows, A. O. D. (1997). Lunar-modulated geomagnetic orientation by a marine mollusk. *Science, 235,* 331–334.

Lohmann, K. J., Willows, A. O. D., & Pinter, R. B. (1991). An identifiable molluscan neuron responds to changes in Earth-strength magnetic fields. *Journal of Experimental Biology, 161,* 1–24.

Lucero, M. T., Horrigan, F. T., & Gilly, W. F. (1992). Electrical responses to chemical stimulation of squid olfactory receptor cells. *Journal of Experimental Biology, 162,* 231–249.

Maa, E., Arnold, J., Ninedorf, K., & Olsen, H. (2021a). Canine detection of volatile organic compounds unique to human epileptic seizure. *Epilepsy & Behavior, 115,* 107690.

Maa, E. H., Arnold, J., & Bush, C. K. (2021b). Epilepsy and the smell of fear. *Epilepsy & Behavior, 121,* 108078.

Madsen, P. T., Siebert, U., & Elemans, C. P. H. (2023). Toothed whales use distinct vocal registers for echolocation and communication. *Science, 379,* 928–933.

Maguire, S. E., Schmidt, M. F., & White, D. J. (2013). Social brains in context: Lesions targeted to the song control system in female cowbirds affect their social network. *PLoS ONE, 8*(5), e63239.

Malkemper, E. P., Nimpf, S., Nordmann, G. C., & Keays, D. A. (2020). Neuronal circuits and the magnetic sense: Central questions. *Journal of Experimental Biology, 223,* jeb232371.

Manley, G. A. (2017). Comparative auditory neuroscience: Understanding the evolution and function of ears. *Journal of the Association for Research in Otolaryngology, 18,* 1–24.

Manley, G. A., & Kraus, J. E. M. (2010). Exceptional high-frequency hearing and matched vocalizations in Australian pygopod geckos. *Journal of Experimental Biology, 213,* 1876–1885.

Mann, D. A., Lu, Z., & Popper, A. N. (1997). A clupeid fish can detect ultrasound. *Nature, 389,* 341.

Manton, M., Karr, A., & Ehrenfeld, D. W. (1972). Chemoreception in the migratory sea turtle, *Chelonia mydas. Biological Bulletin, 143,* 184–195.

Marasco, P. D., & Catania, K. C. (2007). Response properties of primary afferents supplying Eimer's organ. *Journal of Experimental Biology, 210,* 765–780.

Marchetti, K. (2000). Egg rejection in a passerine bird: Size does matter. *Animal Behaviour, 59,* 877–883.

Maresh, A., Gil, D. R., Whitman, M. C., & Greer, C. A. (2008). Principles of glomerular organization in the human olfactory bulb – Implications for odor processing. *PLoS ONE, 3*(7), e2640.

Marshall, N. J., Powell, S. B., Cronin, T. W., Caldwell, R. L., Johnsen, S., Gruev, V., Chiou, T.-H. S., Roberts, N. W., & How, M. J. (2019). Polarisation signals: A new currency for communication. *Journal of Experimental Biology, 222,* jeb134213.

Martin, G. R. (1974). Color vision in the tawny owl (*Strix aluco*). *Journal of Comparative and Physiological Psychology, 86,* 133–141.

Martin, K. J., Alessi, S. C., Gaspard, J. C., Tucker, A. D., Bauer, G. B., & Mann, D. A. (2012). Underwater hearing in the loggerhead turtle (*Caretta caretta*): A comparison of behavioral and auditory evoked potential audiograms. *Journal of Experimental Biology, 215,* 3001–3009.

Martínez-García, F., Martínez-Ricós, J., Agustín-Pavón, C., Martínez-Hernández, J., Novejarque, A., & Lanuza, E. (2009). Refining the dual olfactory hypothesis: Pheromone reward and odour experience. *Behavioural Brain Research, 200,* 277–286.

Maruska, K. P., & Butler, J. M. (2021). Reproductive- and social-state plasticity of multiple sensory systems in a cichlid fish. *Integrative and Comparative Biology, 61,* 249–268.

Maruska, K. P., & Fernald, R. D. (2010). Behavioral and physiological plasticity: Rapid changes during social ascent in an African cichlid fish. *Hormones and Behavior, 58,* 230–240.

Maselli, V., Al-Soudi, A.-S., Buglione, M., Aria, M., Polese, G., & Di Cosmo, A. (2020). Sensorial hierarchy in *Octopus vulgaris's* food choice: Chemical vs. visual. *Animals, 10,* 457.

Mass, A. M., & Supin, A. Y. (2007). Adaptive features of aquatic mammals' eye. *Anatomical Record, 290,* 701–715.

Masters, W. M., Raver, K. A. S., & Kazial, K. A. (1995). Sonar signals of big brown bats, *Eptesicus fuscus,* contain information about individual identity, age and family affiliation. *Animal Behaviour, 50,* 1243–1260.

Mather, J. A., & Mather, D. L. (2004). Apparent movement in a visual display: The 'passing cloud' of *Octopus cyanea* (Mollusca: Cephalopoda). *Journal of Zoology, 263,* 89–94.

Mäthger, L. M., Barbosa, A., Miner, S., & Hanlon, R. T. (2006). Color blindness and contrast perception in cuttlefish (*Sepia officinalis*) determined by a visual sensorimotor assay. *Vision Research, 46,* 1746–1753.

Mäthger, L. M., Denton, E. J., Marshall, N. J., & Hanlon, R. T. (2009). Mechanisms and behavioural functions of structural coloration in cephalopods. *Journal of the Royal Society Interface, 6,* S149–S163.

Mäthger, L. M., Roberts, S. B., & Hanlon, R. T. (2010). Evidence for distributed light sensing in the skin of cuttlefish, *Sepia officinalis. Biology Letters, 6,* 600–603.

McBurney, D. H. (1972). Gustatory cross adaptation between sweet-tasting compounds. *Perception & Psychophysics, 11,* 225–227.

McBurney, D. H., & Gent, J. F. (1979). On the nature of taste qualities. *Psychological Bulletin, 86,* 151–167.

McBurney, D. H., Smith, D. V., & Shick, T. R. (1972). Gustatory cross adaptation: Sourness and bitterness. *Perception & Psychophysics, 11,* 228–232.

McComb, K., Moss, C., Sayialel, S., & Baker, L. (2000). Unusually extensive networks of vocal recognition in African elephants. *Animal Behaviour, 59,* 1103–1109.

McKemy, D. D., Neuhausser, W. M., & Julius, D. (2002). Identification of a cold receptor reveals a general role for TRP channels in thermosensation. *Nature, 416,* 52–58.

Mellon, D., Jr. (1963). Electrical responses from the dually innervated tactile receptors on the thorax of the crayfish. *Journal of Experimental Biology, 40,* 137–148.

Mellon, D., Jr., & Hamid, O. A. A. (2012). Identified antennular near-field receptors trigger reflex flicking in the crayfish. *Journal of Experimental Biology, 215,* 1559–1566.

Mendelson, M., & Loewenstein, W. R. (1964). Mechanisms of receptor adaptation. *Science, 144,* 554–555.

Menzel, R., & Blakers, M. (1976). Colour receptors in the bee eye – Morphology and spectral sensitivity. *Journal of Comparative Physiology A, 108,* 11–33.

Menzel, R., & Manz, G. (2005). Neural plasticity of mushroom body-extrinsic neurons in the honeybee brain. *Journal of Experimental Biology, 208,* 4317–4332.

Menzel, R., & Müller, U. (1996). Learning and memory in honeybees: From behavior to neural substrates. *Annual Review of Neuroscience, 19,* 379–404.

Mercado, E., III. (2018). The sonar model for humpback whale song revised. *Frontiers in Psychology, 9,* 1156.

Mercier, D., Tsuchimoto, Y., Ohta, K., & Kazama, H. (2018). Olfactory landmark-based communication in interacting *Drosophila. Current Biology, 28,* 2624–2631.

Meyer-Rochow, V. B. (1974). Structure and function of the larval eye of the sawfly, *Perga. Journal of Insect Physiology, 20,* 1565–1591.

Michinomae, M., Masuda, H., Seidou, M., & Kito, Y. (1994). Structural basis for wavelength discrimination in the banked retina of the firefly squid *Watasenia scintillans. Journal of Experimental Biology, 193,* 1–12.

Miller, A. K., Maritz, B., McKay, S., Glaudas, X., & Alexander, G. J. (2015). An ambusher's arsenal: Chemical crypsis in the puff adder (*Bitis arietans*). *Proceedings of the Royal Society B, 282,* 20152182.

Miller, P. J. O., Isojunno, S., Siegal, E., Lam, F.-P. A., Kvadsheim, P. H., & Curé, C. (2022). Behavioral responses to predatory sounds predict sensitivity of cetaceans to anthropogenic noise within a soundscape of fear. *PNAS, 119,* e2114932119.

Miranda, J. A., & Wilczynski, W. (2009). Sex differences and androgen influences on midbrain auditory thresholds in the green treefrog, *Hyla cinerea. Hearing Research, 252,* 79–88.

Mishkin, M., Ungerleider, L. G., & Macko, K. A. (1983). Object vision and spatial vision: Two cortical pathways. *Trends in Neurosciences, 6,* 414–417.

Mobley, A. S., Michel, W. C., & Lucero, M. T. (2008). Odorant responsiveness of squid olfactory receptor neurons. *Anatomical Record, 291,* 763–774.

Moiseff, A. (1989). Bi-coordinate sound localization by the barn owl. *Journal of Comparative Physiology A, 164,* 637–644.

Moiseff, A., Pollack, G. S., & Hoy, R. R. (1978). Steering responses of flying crickets to sound and ultrasound: Mate attraction and predator avoidance. *Proceedings of the National Academy of Sciences, 75,* 4052–4056.

Mollon, J. D., Bowmaker, J. K., & Jacobs, G. H. (1984). Variations of colour vision in a New World primate can be explained by polymorphism of retinal photopigments. *Proceedings of the Royal Society of London B, 222,* 373–399.

Monroy, J. A., Carter, M. E., Miller, K. E., & Covey, E. (2011). Development of echolocation and communication vocalizations in the big brown bat, *Eptesicus fuscus. Journal of Comparative Physiology A, 197,* 459–467.

Mooney, T. A., Hanlon, R. T., Christensen-Dalsgaard, J., Madsen, P. T., Ketten, D. R., & Nachtigall, P. E. (2010). Sound detection by the longfin squid (*Loligo pealeii*) studied with auditory evoked potentials: Sensitivity to low-frequency particle motion and not pressure. *Journal of Experimental Biology, 213,* 3748–3759.

Mora, C. V., Ross, J. D., Gorsevski, P. V., Chowdhury, B., & Bingman, V. P. (2012). Evidence for discrete landmark use by pigeons during homing. *Journal of Experimental Biology, 215,* 3379–3387.

Morgan, C. L. (1894/1977). An introduction to comparative psychology. In D. N. Robinson (Ed.), *Significant contributions to the history of psychology 1750–1920, series d, volume II.* University Publications of America. (Original work published by Walter Scott, Limited (London))

Morse, P., & Huffard, C. L. (2022). Chemotactile social recognition in the blue-ringed octopus, *Hapalochlaena maculosa. Marine Biology, 169,* 99.

Moynihan, M. (1985). Why are cephalopods deaf? *American Naturalist, 125,* 465–469.

Muheim, R., Sjöberg, S., & Pinzon-Rodriguez, A. (2016). Polarized light modulates light-dependent magnetic compass orientation in birds. *PNAS, 113,* 1654–1659.

Munday, P. L., Dixson, D. L., Donelson, J. M., Jones, G. P., Pratchett, M. S., Devitsina, G. V., & Døving, K. B. (2009). Ocean acidification impairs olfactory discrimination and homing ability of a marine fish. *PNAS, 106,* 1848–1852.

Munday, P. L., Dixson, D. L., McCormick, M. I., Meekan, M., Ferrari, M. C. O., & Chivers, D. P. (2010). Replenishment of fish populations is threatened by ocean acidification. *PNAS, 107,* 12930–12934.

Muntz, W. R. A. (1963). The development of phototaxis in the frog (*Rana temporaria*). *Journal of Experimental Biology, 40,* 371–379.

Muntz, W. R. A., & Gwyther, J. (1988). Visual acuity in *Octopus pallidus* and *Octopus australis*. *Journal of Experimental Biology, 134,* 119–129.

Murray, R. W. (1962). The response of the ampullae of Lorenzini of elasmobranchs to electrical stimulation. *Journal of Experimental Biology, 39,* 119–128.

Nagel, T. (1974). What is it like to be a bat? *Philosophical Review, 83,* 435–450.

Nahmad-Rohen, L., & Vorobyev, M. (2019). Contrast sensitivity and behavioural evidence for lateral inhibition in octopus. *Biology Letters, 15,* 20190134.

Naisbett-Jones, L. C., & Lohmann, K. J. (2022). Magnetoreception and magnetic navigation in fishes: A half century of discovery. *Journal of Comparative Physiology A, 208,* 19–40.

Nakagawa, H., & Hongjian, K. (2010). Collision-sensitive neurons in the optic tectum of the bullfrog, *Rana catesbeiana*. *Journal of Neurophysiology, 104,* 2487–2499.

Nakamura, E. (2011). One hundred years since the discovery of the "umami" taste from seaweed broth by Kikunae Ikeda, who transcended his time. *Chemistry: An Asian Journal, 6,* 1659–1663.

Nathans, J., Thomas, D., & Hogness, D. S. (1986). Molecular genetics of human color vision: The genes encoding blue, green, and red pigments. *Science, 232,* 193–202.

Naumann, K., Winston, M. L., Slessor, K. N., Prestwich, G. D., & Webster, F. X. (1991). Production and transmission of honey bee queen (*Apis mellifera* L.) mandibular gland pheromone. *Behavioral Ecology and Sociobiology, 29,* 321–332.

Neitz, J., Geist, T., & Jacobs, G. H. (1989). Color vision in the dog. *Visual Neuroscience, 3,* 119–125.

Neumeyer, C. (1992). Tetrachromatic color vision in goldfish: Evidence from color mixture experiments. *Journal of Comparative Physiology A, 171,* 639–649.

Neumeyer, C., & Jäger, J. (1985). Sprectral sensitivity of the freshwater turtle Pseudemys scripta elegans: Evidence for the filter-effect of colored oil droplets. *Vision Research, 25,* 833–838.

Nevitt, G. A. (2008). Sensory ecology on the high seas: The odor world of the procellariiform seabirds. *Journal of Experimental Biology, 211,* 1706–1713.

Newman, E. A., & Hartline, P. H. (1981). Integration of visual and infrared information in bimodal neurons of the rattlesnake optic tectum. *Science, 213,* 789–791.

Ng, B. S. W., Grabska-Barwińska, A., Güntürkün, O., & Jancke, D. (2010). Dominant vertical orientation processing without clustered maps: Early visual brain dynamics imaged with voltage-sensitive dye in the pigeon visual Wulst. *Journal of Neuroscience, 30,* 6713–6725.

Nikonov, A. A., Butler, J. M., Field, K. E., Caprio, J., & Maruska, K. P. (2017). Reproductive and metabolic state differences in olfactory responses to amino acids in a mouth brooding African cichlid fish. *Journal of Experimental Biology, 220,* 2980–2992.

Nikonov, A. A., Finger, T. E., & Caprio, J. (2005). Beyond the olfactory bulb: An odotopic map in the forebrain. *PNAS, 102,* 18688–18693.

Nottebohm, F., Stokes, T. M., & Leonard, C. M. (1976). Contral control of song in the canary, *Serinus canarius*. *Journal of Comparative Neurology, 165,* 457–486.

Novales Flamarique, I. (2011). Unique photoreceptor arrangements in a fish with polarized light discrimination. *Journal of Comparative Neurology, 519,* 714–737.

Novales Flamarique, I., & Hawryshyn, C. W. (1998). Photoreceptor types and their relation to the spectral and polarization sensitivities of clupeid fishes. *Journal of Comparative Physiology A, 182,* 793–803.

Novales Flamarique, I., Hawryshyn, C. W., & Hárosi, F. I. (1998). Double-cone internal reflection as a basis for polarization detection in fish. *Journal of the Optical Society of America A, 15,* 349–358.

Nummela, S., Thewissen, J. G. M., Bajpai, S., Hussain, T., & Kumar, K. (2007). Sound transmission in archaic and modern whales: Anatomical adaptations for underwater hearing. *Anatomical Record, 290,* 716–733.

Ochoa, J., & Torebjörk, E. (1983). Sensations evoked by intraneural microstimulation of single mechanoreceptor units innervating the human hand. *Journal of Physiology, 342,* 633–654.

O'Dwyer, T. W., Ackerman, A. L., & Nevitt, G. A. (2008). Examining the development of individual recognition in a burrow-nesting procellariiform, the Leach's storm-petrel. *Journal of Experimental Biology, 211,* 337–340.

Ohms, V. R., Escudero, P., Lammers, K., & Cate, C. T. (2012). Zebra finches and Dutch adults exhibit the same cue weighting bias in vowel perception. *Animal Cognition, 15,* 155–161.

Okada, S. (2015). The taste system of small fish species. *Bioscience, Biotechnology, and Biochemistry, 79,* 1039–1043.

Okamura, S., Kawaminami, T., Matsuura, H., Fusetani, N., & Goshima, S. (2017). Behavioral assay and chemical characters of female sex pheromones in the hermit crab *Pagurus filholi. Journal of Ethology, 35,* 169–176.

Osmanski, B. F., Martin, C., Montaldo, G., Lanièce, P., Pain, F., Tanter, M., & Gurden, H. (2014). Functional ultrasound imaging reveals different odor-evoked patterns of vascular activity in the main olfactory bulb and the anterior piriform cortex. *NeuroImage, 96,* 176–184.

Packard, A., Karlsen, H. E., & Sand, O. (1990). Low frequency hearing in cephalopods. *Journal of Comparative Physiology A, 166,* 501–505.

Packmor, F., Kishkinev, D., Bittermann, F., Kofler, B., Machowetz, C., Zechmeister, T., Zawadzki, L. C., Guilford, T., & Holland, R. A. (2021). A magnet attached to the forehead disrupts magnetic compass orientation in a migratory songbird. *Journal of Experimental Biology, 224,* jeb243337.

Pados, B. F., & McGlothen-Bell, K. (2019). Benefits of infant massage for infants and parents in the NICU. *Nursing for Women's Health, 23,* 265–271.

Papi, F., Fiore, L., Fiaschi, V., & Benvenuti, S. (1973). An experiment for testing the hypothesis of olfactory navigation of homing pigeons. *Journal of Comparative Physiology, 83,* 93–102.

Papi, F., Ioalé, P., Fiaschi, V., Benvenuti, S., & Baldaccini, N. E. (1974). Olfactory navigation of pigeons: The effect of treatment with odorous air currents. *Journal of Comparative Physiology, 94,* 187–193.

Parker, R. O., McCarragher, B., Crouch, R., & Darden, A. G. (2010). Photoreceptor development in prematamorphic and metamorphic *Xenopus laevis. Anatomical Record, 293,* 383–387.

Paulk, A. C., Dacks, A. M., & Gronenberg, W. (2009a). Color processing in the medulla of the bumblebee (Apidae: *Bombus impatiens*). *Journal of Comparative Neurology, 513,* 441–456.

Paulk, A. C., Dacks, A. M., Phillips-Portillo, J., Fellous, J.-M., & Gronenberg, W. (2009b). Visual processing in the central bee brain. *Journal of Neuroscience, 29,* 9987–9999.

Paulk, A. C., Phillips-Portillo, J., Dacks, A. M., Fellous, J.-M., & Gronenberg, W. (2008). The processing of color, motion, and stimulus timing are anatomically segregated in the bumblebee brain. *Journal of Neuroscience, 28,* 6319–6332.

Pernal, S. F. (2021). The social life of honey bees. *Veterinary Clinics of North America. Food Animal Practice, 37,* 387–400.

Perrett, D. I., Rolls, E. T., & Caan, W. (1982). Visual neurones responsive to faces in the monkey temporal cortex. *Experimental Brain Research, 47,* 329–342.

Poggio, G. F. (1995). Mechanisms of stereopsis in monkey visual cortex. *Cerebral Cortex, 3,* 193–204.

Pollack, G. S. (1988). Selective attention in an insect auditory neuron. *Journal of Neuroscience, 8,* 2635–2639.

Pollack, G. S. (2015). Neurobiology of acoustically mediated predator detection. *Journal of Comparative Physiology A, 201,* 99–109.

Porteus, C. S., Hubbard, P. C., Webster, T. M. U., van Aerle, R., Canário, A. V. M., Santos, E. M., & Wilson, R. W. (2018). Near-future CO_2 levels impair the olfactory system of a marine fish. *Nature Climate Change, 8,* 737–743.

Porteus, C. S., Roggatz, C. C., Velez, Z., Hardege, J. D., & Hubbard, P. C. (2021). Acidification can directly affect olfaction in marine organisms. *Journal of Experimental Biology, 224,* jeb237941.

Puri, S., & Faulkes, Z. (2015). Can crayfish take the heat? *Procambarus clarkii* show nociceptive behaviour to high temperature stimuli, but not low temperature or chemical stimuli. *Biology Open, 4,* 441–448.

Putman, N. F., Endres, C. S., Lohmann, C. M. F., & Lohmann, K. J. (2011). Longitude perception and bicoordinate magnetic maps in sea turtles. *Current Biology, 21,* 463–466.

Putman, N. F., Scanlan, M. M., Billman, E. J., O'Neil, J. P., Couture, R. B., Quinn, T. P., Lohmann, K. J., & Noakes, D. L. G. (2014). An inherited magnetic map guides ocean navigation in juvenile Pacific salmon. *Current Biology, 24,* 446–450.

Putterill, J. F., & Soley, J. T. (2003). General morphology of the oral cavity of the Nile crocodile, *Crocodylus niloticus* (Laurenti, 1768). I. Palate and gingivae. *Onderstepoort Journal of Veterinary Research, 70,* 281–297.

Querubin, A., Lee, H. R., Provis, J. M., & O'Brien, K. M. B. (2009). Photoreceptor and ganglion cell topographies correlate with information convergence and high acuity regions in the adult pigeon (*Columba livia*) retina. *Journal of Comparative Neurology, 517,* 711–722.

Quignon, P., Giraud, M., Rimbault, M., Lavigne, P., Tacher, S., Morin, E., Retout, E., Valin, A.-S., Lindblad-Toh, K., Nicolas, J., & Galibert, F. (2005). The dog and rat olfactory receptor repertoires. *Genome Biology, 6,* R83.

Quindlen-Hotek, J. C., Bloom, E. T., Johnston, O. K., & Barocas, V. H. (2020). An interspecies computational analysis of vibrotactile sensitivity in Pacinian and Herbst corpuscles. *Royal Society Open Science, 7,* 191439 (10 pages).

Quinn, T. P. (1980). Evidence for celestial and magnetic compass orientation in lake migrating sockeye salmon fry. *Journal of Comparative Physiology, 137,* 243–248.

Raab, T., Bayezit, S., Erdle, S., & Benda, J. (2021). Electrocommunication signals indicate motivation to compete during dyadic interactions of an electric fish. *Journal of Experimental Biology, 224,* jeb242905.

Ramirez, M. D., & Oakley, T. H. (2015). Eye-independent, light-activated chromatophore expansion (LACE) and expression of phototransduction genes in the skin of *Octopus bimaculoides. Journal of Experimental Biology, 218,* 1513–1520.

Ranade, S. S., Woo, S.-H., Dubin, A. E., Moshourab, R. A., Wetzel, C., Petrus, M., Mathur, J., Bégay, V., Coste, B., Mainquist, J., Wilson, A. J., Francisco, A. G., Reddy, K., Qiu, Z., Wood, J. N., Lewin, G. R., & Patapoutian, A. (2014). Piezo2 is the major transducer of mechanical forces for touch sensation in mice. *Nature, 516,* 121–125.

Ratliff, F. (1965). *Mach bands: Quantitative studies on neural networks in the retina.* Holden-Day.

Reidenberg, J. S., & Laitman, J. T. (2007). Discovery of a low frequency sound source in mysticeti (baleen whales): Anatomical establishment of a vocal fold homolog. *Anatomical Record, 290,* 745–759.

Reinhard, J., Srinivasan, M. V., & Zhang, S. (2004). Scent-triggered navigation in honeybees. *Nature, 427,* 411.

Reiten, I., Uslu, F. E., Fore, S., Pelgrims, R., Ringers, C., Verdugo, C. D., Hoffman, M.,

Lal, P., Kawakami, K., Pekkan, K., Yaksi, E., & Jurisch-Yaksi, N. (2017). Motile-cilia-mediated flow improves sensitivity and temporal resolution of olfactory computations. *Current Biology, 27,* 166–174.

Reppert, S. M., Gegear, R. J., & Merlin, C. (2010). Navigational mechanisms of migrating monarch butterflies. *Trends in Neurosciences, 33,* 399–406.

Ressler, K. J., Sullivan, S. L., & Buck, L. B. (1993). A zonal organization of odorant receptor gene expression in the olfactory epithelium. *Cell, 73,* 597–609.

Rittschof, D., & Sutherland, J. P. (1986). Field studies on chemically mediated behavior in land hermit crabs: Volatile and nonvolatile odors. *Journal of Chemical Ecology, 12,* 1273–1284.

Roberts, S. A., Simpson, D. M., Armstrong, S. D., Davidson, A. J., Robertson, D. H., McLean, L., Beynon, R. J., & Hurst, J. L. (2010). Darcin: A male pheromone that stimulates female memory and sexual attraction to an individual male's odour. *BMC Biology, 8,* 75.

Robinson, P. P. (1988). The characteristics and regional distribution of afferent fibres in the chorda tympani of the cat. *Journal of Physiology, 406,* 345–357.

Rolen, S. H., & Caprio, J. (2008). Bile salts are effective taste stimuli in channel catfish. *Journal of Experimental Biology, 211,* 2786–2791.

Root-Gutteridge, H., Ratcliffe, V. F., Korzeniowska, A. T., & Reby, D. (2019). Dogs perceive and spontaneously normalize formant-related speaker and vowel differences in human speech sounds. *Biology Letters, 15,* 20190555.

Roper, S. D., & Chaudhari, N. (2017). Taste buds: Cells, signals and synapses. *Nature Reviews: Neuroscience, 18,* 485–497.

Rose, J. D., Arlinghaus, R., Cooke, S. J., Diggles, B. K., Sawynok, W., Stevens, E. D., & Wynne, C. D. L. (2014). Can fish really feel pain? *Fish and Fisheries, 15,* 97–133.

Rosner, R., Hadeln, J. V., Tarawneh, G., & Read, J. C. A. (2019). A neuronal correlate of insect stereopsis. *Nature Communications, 10,* 2845.

Rossel, S. (1983). Binocular stereopsis in an insect. *Nature, 302,* 821–822.

Rotjan, R. D., Chabot, J. R., & Lewis, S. M. (2010). Social context of shell acquisition in *Coenobita clypeatus* hermit crabs. *Behavioral Ecology, 21,* 639–646.

Roundsley, K. J., & McFadden, S. A. (2005). Limits of visual acuity in the frontal field of the rock pigeon (*Columba livia*). *Perception, 34,* 983–993.

Rundus, A. S., Owings, D. H., Joshi, S. S., Chinn, E., & Giannini, N. (2007). Ground squirrels use an infrared signal to deter rattlesnake predation. *Proceedings of the National Academy of Sciences, 104*(36), 14372–14376.

Sabbah, S., Troje, N. F., Gray, S. M., & Hawryshyn, C. W. (2013). High complexity of aquatic irradiance may have driven the evolution of four-dimensional colour vision in shallow-water fish. *Journal of Experimental Biology, 216,* 1670–1682.

Santacà, M., Petrazzini, M. E. M., Agrillo, C., & Wilkinson, A. (2019). Can reptiles perceive visual illusions? Delboeuf illusion in red-footed tortoise (*Chelonoidis carbonaria*) and bearded dragon (*Pogona vitticeps*). *Journal of Comparative Psychology, 133,* 419–427.

Saotome, K., Teng, B., Tsui, C. C. (A.), Lee, W.-H., Tu, Y.-H., Kaplan, J. P., Sansom, M. S. P., Liman, E. R., & Ward, A. B. (2019). Structures of the otopetrin proton channels Otop1 and Otop3. *Nature Structural and Molecular Biology, 26,* 518–525.

Sarmiento, R. F. (1975). The stereoacuity of macaque monkey. *Vision Research, 15,* 493–498.

Savard, J.-F., Keagy, J., & Borgia, G. (2011). Blue, not UV, plumage color is important in satin bowerbird *Ptilonorhychus violaceus* display. *Journal of Avian Biology, 42,* 80–84.

Schmidt, M. (2007). The olfactory pathway of decapod crustaceans—An invertebrate model for life-long neurogenesis. *Chemical Senses, 32,* 365–384.

Schneider, G. E. (1969). Two visual systems. *Science, 163,* 895–902.

Schöneich, S., & Hedwig, B. (2010). Hyperacute directional hearing and phonotactic steering in the cricket (*Gryllus bimaculatus* deGeer). *PLoS ONE, 5,* e15141.

Schöneich, S., Kostarakos, K., & Hedwig, B. (2015). An auditory feature detection circuit for sound pattern recognition. *Science Advances, 1*(8), e1500325.

Schulz-Mirbach, T., Ladich, F., Mittone, A., Olbinado, M., Bravin, A., Maiditsch, I. P., Melzer, R. R., Krysl, P., & Heß, M. (2020). Auditory chain reaction: Effects of sound pressure and particle motion on auditory structures in fishes. *PLoS ONE, 15*(3), e0230578.

Scott, K. (2004). The sweet and the bitter of mammalian taste. *Current Opinion in Neurobiology, 14,* 423–427.

Selverston, A. I., Kleindienst, H.-U., & Huber, F. (1985). Synaptic connectivity between cricket auditory interneurons as studied by selective photoinactivation. *Journal of Neuroscience, 5,* 1283–1292.

Semm, P., & Beason, R. C. (1990). Responses to small magnetic variations by the trigeminal system of the bobolink. *Brain Research Bulletin, 25,* 735–740.

Serrano-Rojas, S. J., & Pašukonis, A. (2021). Tadpole-transporting frogs use stagnant water odor to find pools in the rainforest. *Journal of Experimental Biology, 224,* jeb243122.

Seyfarth, R. M., Cheney, D. L., & Marler, P. (1980). Monkey responses to three different alarm calls: Evidence of predator classification and semantic communication. *Science, 210,* 801–803.

Shamble, P. S., Menda, G., Golden, J. R., Nitzany, E. I., Walden, K., Beatus, T., Elias, D. O., Cohen, I., Miles, R. N., & Hoy, R. R. (2016). Airborne acoustic perception by a jumping spider. *Current Biology, 26,* 2913–2920.

Shapiro, A. M. (1983). Testing visual species recognition in *Precis* (Lepidoptera: Nymphalidae) using a cold-shock phenocopy. *Psyche, 90,* 59–65.

Shomrat, T., Feinstein, N., Klein, M., & Hochner, B. (2010). Serotonin is a facilitatory neuromodulator of synaptic transmission and "reinforces" long-term potentiation induction in the vertical lobe of *Octopus vulgaris*. *Neuroscience, 169,* 52–64.

Shomrat, T., & Levin, M. (2013). An automated training paradigm reveals long-term memory in planarians and its persistence through head regeneration. *Journal of Experimental Biology, 216,* 3799–3810.

Shomrat, T., Turchetti-Maia, A. L., Stern-Mentch, N., Basil, J. A., & Hochner, B. (2015). The vertical lobe of cephalopods: An attractive brain structure for understanding the evolution of advanced learning and memory systems. *Journal of Comparative Physiology A, 201,* 947–956.

Shomrat, T., Zarrella, I., Fiorito, G., & Hochner, B. (2008). The octopus vertical lobe modulates short-term learning rate and uses LTP to acquire long-term memory. *Current Biology, 18,* 337–342.

Shubin, N., Tabin, C., & Carroll, S. (1997). Fossils, genes and the evolution of animal limbs. *Nature, 388,* 639–648.

Siebeck, U. E., Parker, A. N., Sprenger, D., Mäthger, L. M., & Wallis, G. (2010). A species of reef fish that uses ultraviolet patterns for covert face recognition. *Current Biology, 20,* 407–410.

Silberglied, R. E., & Taylor, O. R., Jr. (1978). Ultraviolet reflection and its behavioral role in the courtship of the sulfur butterflies, *Colias eurytheme* and *C. philodice* (Lepidoptera, Pieridae). *Behavioral Ecology and Sociobiology, 3,* 203–243.

Silva, L., & Antunes, A. (2017). Vomeronasal receptors in vertebrates and the evolution of pheromone detection. *Annual Review of Animal Biosciences, 5,* 535–570.

Simmons, A. M., Hom, K. N., Warnecke, M., & Simmons, J. A. (2016). Broadband noise exposure does not affect hearing sensitivity in big brown bats (*Eptesicus fuscus*). *Journal of Experimental Biology, 219,* 1031–1040.

Siniscalchi, M., Sasso, R., Pepe, A. M., Dimatteo, S., Vallortigara, G., & Quaranta, A. (2011). Sniffing with the right nostril: Lateralization of response to odour stimuli by dogs. *Animal Behaviour, 82,* 399–404.

Slessor, K. N., Kaminski, L.-A., King, G. G. S., Borden, J. H., & Winston, M. L. (1988).

Semiochemical basis of the retinue response to queen honey bees. *Nature, 332*, 354–356.

Smith, D. V., & McBurney, D. H. (1969). Gustatory cross-adaptation: Does a single mechanism code the salty taste? *Journal of Experimental Psychology, 80*, 101–105.

Smith, E. S., & Lewin, G. R. (2009). Nociceptors: A phylogenetic view. *Journal of Comparative Physiology A, 195*, 1089–1106.

Smith, J. N., Goldizen, A. W., Dunlop, R. A., & Noad, M. J. (2008). Songs of male humpback whales, *Megaptera novaeangliae*, are involved in intersexual interactions. *Animal Behaviour, 76*, 467–477.

Sneddon, L. U. (2015). Pain in aquatic animals. *Journal of Experimental Biology, 218*, 967–976.

Sneddon, L. U. (2019). Evolution of nociception and pain: Evidence from fish models. *Philosophical Transactions of the Royal Society B, 374*, 20190290.

Sneddon, L. U., Braithwaite, V. A., & Gentle, M. J. (2003). Do fishes have nociceptors? Evidence for the evolution of a vertebrate sensory system. *Proceedings of the Royal Society B, 270*, 1115–1121.

Soto-Yéber, L., Soto-Ortiz, J., Godoy, P., & Godoy-Herrera, R. (2018). The behavior of adult *Drosophila* in the wild. *PLoS ONE, 13*, e0209917.

Sovrano, V. A., & Bisazza, A. (2008). Recognition of partly occluded objects by fish. *Animal Cognition, 11*, 161–166.

Speedie, N., & Gerlai, R. (2008). Alarm substance induced behavioral responses in zebrafish (*Danio rerio*). *Behavioural Brain Research, 188*, 168–177.

Srinivasan, M. V. (1992). Distance perception in insects. *Current Directions in Psychological Science, 1*(1), 22–26.

Srinivasan, M. V., & Lehrer, M. (1988). Spatial acuity of honeybee vision and its spectral properties. *Journal of Comparative Physiology A, 162*, 159–172.

Srinivasan, M. V., Lehrer, M., Zhang, S. W., & Horridge, G. A. (1989). How honeybees measure their distance from objects of unknown size. *Journal of Comparative Physiology A, 165*, 605–613.

Stacho, M., Herold, C., Rook, N., Wagner, H., Axer, M., Amunts, K., & Güntürkün, O. (2020). A cortex-like canonical circuit in the avian forebrain. *Science, 369*, eabc5534 (12 pages).

Stafstrom, J. A., Menda, G., Nitzany, E. I., Hebets, E. A., & Hoy, R. R. (2020). Ogre-faced, net-casting spiders use auditory cues to detect airborne prey. *Current Biology, 30*, 5033–5039.

Stapput, K., Thalau, P., Wiltschko, R., & Wiltschko, W. (2008). Orientation of birds in total darkness. *Current Biology, 18*, 602–606.

Staszko, S. M., Boughter, J. D., Jr., & Fletcher, M. L. (2020). Taste coding strategies in insular cortex. *Experimental Biology and Medicine, 245*, 448–455.

Stern, K., & McClintock, M. K. (1998). Regulation of ovulation by human pheromones. *Nature, 392*, 177–179.

Stoeger, A. S., & Baotic, A. (2017). Male African elephants discriminate and prefer vocalizations of unfamiliar females. *Scientific Reports, 7*, 46414.

Stoeger, A. S., Mietchen, D., Oh, S., Silva, S. D., Herbst, C. T., Kwon, S., & Fitch, W. T. (2012). An Asian elephant imitates human speech. *Current Biology, 22*, 2144–2148.

Stowers, L., & Kuo, T.-H. (2015). Mammalian pheromones: Emerging properties and mechanisms of detection. *Current Opinion in Neurobiology, 34*, 103–109.

Surlykke, A., Ghose, K., & Moss, C. F. (2009). Acoustic scanning of natural scenes by echolocation in the big brown bat, *Eptesicus fuscus*. *Journal of Experimental Biology, 212*, 1011–1020.

Suryanarayana, S. M., Robertson, B., Wallén, P., & Grillner, S. (2017). The lamprey pallium provides a blueprint of the mammalian layered cortex. *Current Biology, 27*, 3264–3277.

Sutherland, N. S. (1957). Visual discrimination of orientation and shape by the octopus. *Nature, 179*, 11–13.

Sutton, G. P., Clarke, D., Morley, E. L., & Robert, D. (2016). Mechanosensory hairs in bumblebees (*Bombus terrestris*) detect weak electric fields. *PNAS, 113*, 7261–7265.

Svensson, A. M., & Rydell, J. (1998). Mercury vapour lamps interfere with the bat defence of tympanate moths (*Operophtera* spp.; Geometridae). *Animal Behaviour, 55,* 223–226.

Syed, A. S., Sansone, A., Hassenklöver, T., Manzini, I., & Korsching, S. I. (2017). Coordinated shift of olfactory amino acid responses and V2R expression to an amphibian water nose during metamorphosis. *Cellular and Molecular Life Sciences, 74,* 1711–1719.

Takahashi, T., Moiseff, A., & Konishi, M. (1984). Time and intensity cues are processed independently in the auditory system of the owl. *Journal of Neuroscience, 4,* 1781–1786.

Takeuchi, H.-A., Masuda, T., & Nagai, T. (1994). Electrophysiological and behavioral studies of taste discrimination in the axolotl (*Ambystoma mexicanum*). *Physiology and Behavior, 56,* 121–127.

Tansley, K. (1965). *Vision in vertebrates.* Science Paperbacks and Chapman and Hall Ltd.

Teng, B., Wilson, C. E., Tu, Y.-H., Joshi, N. R., Kinnamon, S. C., & Liman, E. R. (2019). Cellular and neural responses to sour stimuli require the proton channel Otop1. *Current Biology, 29,* 3647–3656.

Thoen, H. H., How, M. J., Chiou, T.-H., & Marshall, J. (2014). A different form of color vision in mantis shrimp. *Science, 343,* 411–413.

Tibbetts, E. A. (2002). Visual signals of individual identity in the wasp *Polistes fuscatus. Proceedings of the Royal Society of London B, 269,* 1423–1428.

Toda, Y., Ko, M.-C., Liang, Q., Miller, E. T., Rico-Guevara, A., Nakagita, T., Sakakibara, A., Uemura, K., Sackton, T., Hayakawa, T., Sin, S. Y. W., Ishimaru, Y., Misaka, T., Oteiza, P., Crall, J., Edwards, S. V., Buttemer, W., Matsumura, S., & Baldwin, M. W. (2021). Early origin of sweet perception in the songbird radiation. *Science, 373,* 226–231.

Tønnesen, P., Oliveira, C., Johnson, M., & Madsen, P. T. (2020). The long-range echo scene of the sperm whale biosonar. *Biology Letters, 16,* 20200134.

Treisman, A. (1999). Solutions to the binding problem: Progress through controversy and convergence. *Neuron, 24,* 105–110.

Treisman, A., & Schmidt, H. (1982). Illusory conjunctions in the perception of objects. *Cognitive Psychology, 14,* 107–141.

Treisman, A. M., & Gelade, G. (1980). A feature-integration theory of attention. *Cognitive Psychology, 12,* 97–136.

Tricarico, E., Borrelli, L., Gherardi, F., & Fiorito, G. (2011). I know my neighbour: Individual recognition in *Octopus vulgaris. PLoS ONE, 6,* e18710.

Troscianko, J., Wilson-Aggarwal, J., Stevens, M., & Spottiswoode, C. N. (2016). Camouflage predicts survival in ground-nesting birds. *Scientific Reports, 6,* 19966.

Tvardíková, K., & Fuchs, R. (2010). Tits use amodal completion in predator recognition: A field experiment. *Animal Cognition, 13,* 609–615.

Tyack, P. L., Zimmer, W. M. X., Moretti, D., Southall, B. L., Claridge, D. E., Durban, J. W., Clark, C. W., D'Amico, A., DiMarzio, N., Jarvis, S., McCarthy, E., Morrissey, R., Ward, J., & Boyd, I. L. (2011). Beaked whales respond to simulated and actual navy sonar. *PLoS ONE, 6,* e17009.

Uexküll, J. von. (1992). A stroll through the worlds of animals and men: A picture book of invisible worlds (translated from German by Schiller, C. H.). *Semiotica, 89,* 319–391. Reprinted from Schiller, C. H. (ed.), *Instinctive Behavior,* pp. 5–80. Madison, CT: International Universities Press. (This essay was originally published in 1934)

Van Dijk, P., Mason, M. J., Schoffelen, R. L. M., Narins, P. M., & Meenderink, S. W. F. (2011). Mechanics of the frog ear. *Hearing Research, 273,* 46–58.

Van Doren, B. M., Horton, K. G., Dokter, A. M., Klinck, H., Elbin, S. B., & Farnsworth, A. (2017). High-intensity urban light installation dramatically alters nocturnal bird migration. *PNAS, 114,* 11175–11180.

van Giesen, L., Kilian, P. B., Allard, C. A. H., & Bellono, N. W. (2020). Molecular basis of chemotactile sensation in octopus. *Cell, 183,* 594–604.

Vereecken, N. J., & McNeil, J. N. (2010). Cheaters and liars: Chemical mimicry at its finest. *Canadian Journal of Zoology, 88*, 725–752.

Vicedo, M. (2009). Mothers, machines, and morals: Harry Harlow's work on primate love from lab to legend. *Journal of the History of the Behavioral Sciences, 45*, 193–218.

Viete, S., Peña, J. L., & Konishi, M. (1997). Effects of interaural intensity difference on the processing of interaural time difference in the owl's nucleus laminaris. *Journal of Neuroscience, 17*, 1815–1824.

von der Emde, G. (2004). Distance and shape: Perception of the 3-dimensional world by weakly electric fish. *Journal of Physiology (Paris), 98*, 67–80.

von der Emde, G. (2006). Non-visual environmental imaging and object detection through active electrolocation in weakly electric fish. *Journal of Comparative Physiology A, 192*, 601–612.

Waddell, J. C., & Caputi, A. A. (2021). The captivating effect of electric organ discharges: Species, sex and orientation are embedded in every single received image. *Journal of Experimental Biology, 224*, jeb243008.

Wahlberg, M., Linnenschmidt, M., Madsen, P. T., Wisniewska, D. M., & Miller, L. A. (2015). The acoustic world of harbor porpoises. *American Scientist, 103*(1), 46–53.

Waldrop, L. D., & Koehl, M. A. R. (2016). Do terrestrial hermit crabs sniff? Air flow and odorant capture by flicking antennules. *Journal of the Royal Society Interface, 13*, 20150850.

Walker, M. M., Diebel, C. E., Haugh, C. V., Pankhurst, P. M., Montgomery, J. C., & Green, C. R. (1997). Structure and function of the vertebrate magnetic sense. *Nature, 390*, 371–376.

Walls, G. L. (1967). *The vertebrate eye and its adaptive radiation*. Hafner Publishing Co. (Original work published in 1942 by The Cranbrook Institute of Science, Bloomfield Hills, MI)

Wan, G., Hayden, A. N., Iiams, S. E., & Merlin, C. (2021). Cryptochrome 1 mediates light-dependent inclination magnetosensing in monarch butterflies. *Nature Communications, 12*, 771.

Wang, J. H., Cain, S. D., & Lohmann, K. J. (2003). Identification of magnetically responsive neurons in the marine mollusc *Tritonia diomedea*. *Journal of Experimental Biology, 206*, 381–388.

Wang, J. H., Cain, S. D., & Lohmann, K. J. (2004). Identifiable neurons inhibited by Earth-strength magnetic stimuli in the mollusc *Tritonia diomedea*. *Journal of Experimental Biology, 207*, 1043–1049.

Wang, L., & Anderson, D. J. (2010). Identification of an aggression-promoting pheromone and its receptor neurons in *Drosophila*. *Nature, 463*, nature08678.

Warrant, E., & Dacke, M. (2011). Vision and visual navigation in nocturnal insects. *Annual Review of Entomology, 56*, 239–254.

Watson, J. B. (1994). Psychology as the behaviorist views it. *Psychological Review, 101*, 248–253. (Reprinted from *Psychological Review, 20*, 158–177 1913).

Wehner, R. (1976). Polarized-light navigation by insects. *Scientific American, 235*(1), 106–115.

Wehner, R. (1994). The polarization-vision project: Championing organismic biology. In K. Schildberger & N. Elsner (Eds.), *Neural basis of behavioural adaptation* (pp. 103–143). G. Fischer.

Weiss, L., Manzini, I., & Hassenklöver, T. (2021). Olfaction across the water-air interface in anuran amphibians. *Cell and Tissue Research, 383*, 301–325.

Weldon, P. J., & Gaffney, E. S. (1998). An ancient integumentary gland in turtles. *Naturwissenschaften, 85*, 556–557.

Wellington, W. G. (1974). Bumblebee ocelli and navigation at dusk. *Science, 183*, 550–551.

Wells, M. J. (1960). Proprioception and visual discrimination of orientation in *Octopus*. *Journal of Experimental Biology, 37*, 489–499.

Weltzien, F.-A., Höglund, E., Hamdani, E. H., & Døving, K. B. (2003). Does the lateral bundle of the medial olfactory tract mediate reproductive behavior in male crucian carp? *Chemical Senses, 28*, 293–300.

West, M. J., King, A. P., & Eastzer, D. H. (1981). Validating the female bioassay of cowbird song: Relating differences in song potency to mating success. *Animal Behavior, 29,* 490–501.

West, M. J., King, A. P., White, D. J., Gros-Louis, J., & Freed-Brown, G. (2006). The development of local song preferences in female cowbirds (*Molothrus ater*): Flock living stimulates learning. *Ethology, 112,* 1095–1107.

Westhoff, G., Boetig, M., Bleckmann, H., & Young, B. A. (2010). Target tracking during venom 'spitting' by cobras. *Journal of Experimental Biology, 213,* 1797–1802.

Westhoff, G., Tzschätzsch, K., & Bleckmann, H. (2005). The spitting behavior of two species of spitting cobras. *Journal of Comparative Physiology A, 191,* 873–881.

White, D. J., King, A. P., West, M. J., Gros-Louis, J., & Tuttle, E. M. (2010). Effects of singing on copulation success and egg production in brown-headed cowbirds, *Molothrus ater. Behavioral Ecology, 21,* 211–218.

Whitear, M. (1989). Merkel cells in lower vertebrates. *Archives of Histology and Cytology, 52*(Suppl.), 415–422.

Williamson, R., & Chrachri, A. (2007). A model biological neural network: The cephalopod vestibular system. *Philosophical Transactions of the Royal Society B, 362,* 473–481.

Willis, J., Phillips, J., Muheim, R., Diego-Rasilla, F. J., & Hobday, A. J. (2009). Spike dives of juvenile southern bluefin tuna (*Thunnus maccoyii*). *Behavioral Ecology and Sociobiology, 64,* 57–68.

Willows, A. O. D. (1971). Giant brain cells in mollusks. *Scientific American, 224*(2), 68–75.

Wilson, C. D., Serrano, G. O., Koulakov, A. A., & Rinberg, D. (2017). A primacy code for odor identity. *Nature Communications, 8,* 1477.

Wilson, M., Haga, J. Å. R., & Karlsen, H. E. (2018). Behavioural responses to infrasonic particle acceleration in cuttlefish. *Journal of Experimental Biology, 221,* jeb166074.

Wiltschko, R., Schiffner, I., Fuhrmann, P., & Wiltschko, W. (2010). The role of the magnetite-based receptors in the beak in pigeon homing. *Current Biology, 20,* 1534–1538.

Wiltschko, R., & Wiltschko, W. (2013). The magnetite-based receptors in the beak of birds and their role in avian navigation. *Journal of Comparative Physiology A, 199,* 89–98.

Wiltschko, R., & Wiltschko, W. (2019). Magnetoreception in birds. *Journal of the Royal Society Interface, 16,* 20190295.

Wiltschko, W., & Wiltschko, R. (1972). Magnetic compass of European robins. *Science, 176,* 62–64.

Winger, B. M., Weeks, B. C., Farnsworth, A., Jones, A. W., Hennen, M., & Willard, D. E. (2019). Nocturnal flight-calling behaviour predicts vulnerability to artificial light in migratory birds. *Proceedings of the Royal Society B, 286,* 20190364.

Witkovsky, P. (2000). Photoreceptor classes and transmission at the photoreceptor synapse in the retina of the clawed frog, *Xenopus laevis. Microscopy Research and Technique, 50,* 338–346.

Woo, T., Liang, X., Evans, D. A., Fernandez, O., Kretschmer, F., Reiter, S., & Laurent, G. (2023). The dynamics of pattern matching in camouflaging cuttlefish. *Nature, 619,* 122–128.

Worm, M., Kirschbaum, F., & von der Emde, G. (2018). Disembodying the invisible: Electrocommunication and social interactions by passive reception of a moving playback signal. *Journal of Experimental Biology, 221,* jeb172890.

Wu, L.-Q., & Dickman, J. D. (2012). Neural correlates of a magnetic sense. *Science, 336,* 1054–1057.

Wu, L.-Q., Niu, Y.-Q., Yang, J., & Wang, S.-R. (2005). Tectal neurons signal impending collision of looming objects in the pigeon. *European Journal of Neuroscience, 22,* 2325–2331.

Wyatt, T. D. (2014). Chapter 1: Introduction to chemical signaling in vertebrates and invertebrates. In C. Mucignat-Caretta

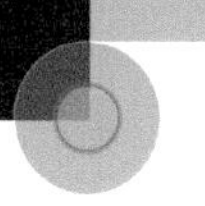

(Ed.), *Neurobiology of chemical communication*. CRC Press/Taylor & Francis.

Wyatt, T. D. (2020). Reproducible research into human chemical communication by cues and pheromones: Learning from psychology's renaissance. *Philosophical Transactions of the Royal Society B, 375*, 20190262.

Wyeth, R. C., & Willows, A. O. D. (2006). Field behavior of the nudibranch mollusc *Tritonia diomedea*. *Biological Bulletin, 210*, 81–96.

Xiao, Q., & Güntürkün, O. (2008). Do pigeons perceive the motion aftereffect? A behavioral study. *Behavioural Brain Research, 187*, 327–333.

Xu, H., Staszewski, L., Tang, H., Adler, E., Zoller, M., & Li, X. (2004). Different functional roles of T1R subunits in the heteromeric taste receptors. *PNAS, 101*, 14258–14263.

Xu, J., Jarocha, L. E., Zollitsch, T., Konowalczyk, M., Henbest, K. B., Richert, S., Golesworthy, M. J., Schmidt, J., Déjean, V., Sowood, D. J. C., Bassetto, M., Luo, J., Walton, J. R., Fleming, J., Wei, Y., Pitcher, T. L., Moise, G., Herrmann, M., Yin, H., . . . Hore, P. J. (2021). Magnetic sensitivity of cryptochrome 4 from a migratory songbird. *Nature, 594*, 535–541.

Yamaguchi, S. (1987). Fundamental properties of umami in human taste sensation. In Y. Kawamura & M. R. Kare (Eds.), *Umami: A basic taste* (pp. 41–73). Marcel Dekker.

Yamamoto, K., Nakata, M., & Nakagawa, H. (2003). Input and output characteristics of collision avoidance behavior in the frog *Rana catesbeiana*. *Brain, Behavior and Evolution, 62*, 201–211.

Yamato, M., & Pyenson, N. D. (2015). Early development and orientation of the acoustic funnel provides insight into the evolution of sound reception pathways in cetaceans. *PLoS ONE, 10*, e0118582.

York, C. A., & Bartol, I. K. (2014). Lateral line analogue aids vision in successful predator evasion for the brief squid, *Lolliguncula brevis*. *Journal of Experimental Biology, 217*, 2437–2439.

Yoshihara, Y. (2009). Molecular genetic dissection of the zebrafish olfactory system. In W. Meyerhof & S. Korsching (Eds.), *Chemosensory systems in mammals, fishes, and insects* (pp. 1–19). Springer.

Young, J. Z. (1938). The functioning of the giant nerve fibres of the squid. *Journal of Experimental Biology, 15*, 170–185.

Young, J. Z. (1971). *The anatomy of the nervous system of Octopus vulgaris*. Oxford University Press.

Yovanovich, C. A. M., Koskela, S. M., Nevala, N., Kondrashev, S. L., Kelber, A., & Donner, K. (2017). The dual rod system of amphibians supports colour discrimination at the absolute visual threshold. *Philosophical Transactions of the Royal Society B, 372*, 20160066.

Yovel, Y., Geva-Sagiv, M., & Ulanovsky, N. (2011). Click-based echolocation in bats: Not so primitive after all. *Journal of Comparative Physiology A, 197*, 515–530.

Zahl, P. A., & O'Neill, T. (1978). The four-eyed fish sees all. *National Geographic, 153*(3), 390–395.

Zapka, M., Heyers, D., Hein, C. M., Engels, S., Schneider, N.-L., Hans, J., Weiler, S., Dreyer, D., Kishkinev, D., Wild, J. M., & Mouritsen, H. (2009). Visual but not trigeminal mediation of magnetic compass information in a migratory bird. *Nature, 461*, nature08528.

Zeki, S. (1993). *A vision of the brain*. Blackwell Scientific Publications.

Zeki, S. (2015). A massively asynchronous, parallel brain. *Philosophical Transactions of the Royal Society B, 370*, 20140174.

Zhang, S. (2018). The last days of the blue-blood harvest. *The Atlantic*, May issue. https://www.theatlantic.com/science/archive/2018/05/blood-in-the-water/559229/

Zhong, H., Huang, J., Shang, S., & Yuan, B. (2021). Evolutionary insights into umami, sweet, and bitter taste receptors in amphibians. *Ecology and Evolution, 11*, 18011–18025.

Zieger, M. V., & Meyer-Rochow, V. B. (2008). Understanding the cephalic eyes of pulmonated gastropods: A review. *American Malacological Bulletin, 26*, 47–66.

Author Index

NOTE: Page numbers in *italics* denote figures on the corresponding page.

I

J

K

L

M

N

Subject Index

NOTE: Page numbers in *italics* denote figures on the corresponding page.

W

Y

Z

For Product Safety Concerns and Information please contact our EU representative GPSR@taylorandfrancis.com Taylor & Francis Verlag GmbH, Kaufingerstraße 24, 80331 München, Germany